The Minerals, Metals & Materials Series

Victoria M. Miller • Petra Maier •
J. Brian Jordon • Neale R. Neelameggham
Editors

Magnesium Technology 2021

TMS

Springer

Editors
Victoria M. Miller
University of Florida
Gainesville, FL, USA

Petra Maier
University of Applied Sciences Stralsund
Stralsund, Germany

J. Brian Jordon
The University of Alabama
Tuscaloosa, AL, USA

Neale R. Neelameggham
IND LLC
South Jordan, UT, USA

ISSN 2367-1181 ISSN 2367-1696 (electronic)
The Minerals, Metals & Materials Series
ISBN 978-3-030-65527-3 ISBN 978-3-030-65528-0 (eBook)
https://doi.org/10.1007/978-3-030-65528-0

This Springer imprint is published by the registered company Springer Nature Switzerland AG
The registered company address is: Gewerbestrasse 11, 6330 Cham, Switzerland

Preface

The Magnesium Technology Symposium held during TMS 2021 Virtual Annual Meeting & Exhibition, March 15–18, 2021, will be occurring in the midst of a global pandemic. Despite all these changes and upsets, research in the magnesium field has progressed and the global magnesium alloy market continues to grow year over year. This growth is driven by the promise of excellent mechanical properties at low density, vibration damping, recyclability, electromagnetic shielding, and low toxicity.

However, there are a wide range of research challenges that face magnesium alloys, from primary production to manufacturability to performance in service environments. Researchers around the globe and from academia, government, and industry turn to the TMS Magnesium Committee as a platform to keep abreast of the latest developments and showcase their own research within the annual Magnesium Technology symposium and this associated proceedings series.

This year, the symposium opens with two keynote speakers addressing one of the most central challenges in magnesium alloys: limited plastic deformation. Dr. Irene J. Beyerlein from the University of California Santa Barbara will open the symposium discussion on deformation twinning in Mg/Nb laminate structures. She will be followed by Dr. Warren Poole from the University of British Columbia who will present a method for measuring a crucial plasticity parameter—critical resolved shear stresses—using instrumented indentation. The sort of fundamental insight into the deformation behavior presented by the keynotes is a vital component of designing new alloy and process strategies to overcome the inherent plasticity limitations of Mg and its alloys.

Finally, the 2020–2021 Magnesium Committee would like to offer its gratitude to all the authors contributing to the symposium, to the reviewers, to the session chairs, and to judges, TMS staff members, and other volunteers. With the contributions of all of these people, Magnesium Technology has been able to produce a valuable proceedings volume and a symposium, even in the light of rapid changes!

Victoria M. Miller
Chair

Petra Maier
Vice Chair

J. Brian Jordon
Past Chair

Neale R. Neelameggham
Advisor

Contents

About the Editors

Victoria M. Miller is an Assistant Professor in the Department of Materials Science and Engineering at the University of Florida, a position she started in September 2019. She was previously an assistant professor at North Carolina State University from 2017 to 2019. Originally from Michigan, she received her B.S.E. in Materials Science and Engineering from the University of Michigan in 2011 and completed her Ph.D. in Materials at the University of California Santa Barbara in 2016.

After graduate school, she worked for a year at UES, Inc. onsite in the Materials and Manufacturing Directorate of the Air Force Research Laboratory in Dayton, Ohio. She also previously worked at Ford Motor Company, Toyota Engineering and Manufacturing, and Lockheed Martin Aeronautics.

Her primary research interest is microstructural evolution during thermomechanical processing of metals and alloys, particularly for those with low symmetry crystal structures. She has been researching Mg alloys since the age of 16.

Professionally, Dr. Miller has served on many committees within TMS and is a Key Reader for *Metallurgical and Materials Transactions A*. She was a recipient of the 2017 TMS Young Leaders Professional Development Award and the 2020 ASM Bronze Medal Award. She is the current chair of the TMS Magnesium Committee.

Petra Maier received her doctoral degree from Loughborough University, UK, in 2002 in Materials Science, in the field of grain boundary segregation in steel. After completing her Ph.D., she worked at the University of Applied Sciences Wildau, Germany, as a postdoctoral fellow under the supervision of Prof. Asta Richter with a focus on mechanical properties by nanoindentation. From 2004 to 2006 she worked as a research associate in the MagIC under the supervision of Dr. Norbert Hort at the Helmholtz-Zentrum Geesthacht, Germany, being a part of the Institute of Materials Research lead by Prof. Karl Ulrich Kainer. Her focus was on magnesium recycling and high-temperature alloy development.

From 2006 to 2008 Dr. Maier was a research associate at the Technical University Berlin, Germany, in the Institute of Material Sciences and Technologies, Department of Materials Engineering of Prof. Claudia Fleck. There, her research specialties included corrosion fatigue on magnesium.

Since 2008, Dr. Maier has been a professor of Materials and Production Engineering in the School of Mechanical Engineering at the University of Applied Sciences Stralsund in Germany. She enjoys working in the field of Mg-based biodegradable implants. Her research is focused on corrosion under stress and crack propagation, mostly on innovative magnesium-rare earth alloys. She is currently the vice chair of the TMS Magnesium Committee.

J. Brian Jordon is an Associate Professor in the Department of Mechanical Engineering at The University of Alabama. Dr. Jordon has extensive experience in fatigue and fracture of metals and, in particular, he has studied fatigue in solid-state processing for over a decade. His other interests include process-structure-property relationships of lightweight metals, process modeling of welding and joining, and fracture of cold spray depositions. Recently, Dr. Jordon has been leading efforts in Additive Friction Stir Deposition (AFSD), a solid-state additive manufacturing technique, for fabrication and repairs of structural components. In particular, he is developing the AFSD process for hard to weld metals such as magnesium alloys.

Dr. Jordon has published over 100 refereed journal articles and conference proceedings papers, and recently published a book *Fatigue in Friction Stir Welding*. His research has been supported by the Department of Energy, the Department of Defense, the State of Alabama, and various private industries. Professionally, Dr. Jordon has chaired committees and organized symposia at various international conferences including The Minerals, Metals, & Materials Society (TMS) meetings. Dr. Jordon was a recipient of the 2014 TMS Young Leaders Professional Development Award, and a finalist for the 2017 University of Alabama President's Faculty Research Award. He currently serves on the editorial board of *Materials and Manufacturing Processes*. Prior to coming to The University of Alabama, Dr. Jordon was an interim associate director and an assistant research professor at the Center for Advanced Vehicular Systems at Mississippi State University. He holds a Ph.D., M.S., and B.S. in Mechanical Engineering from Mississippi State University. Presently, Dr. Jordon serves as past chair of the TMS Magnesium Committee.

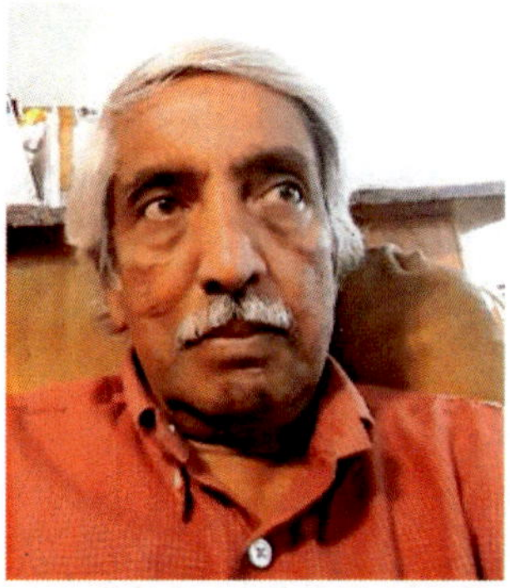

Neale R. Neelameggham is "The Guru" at IND LLC, involved in international technology and management consulting in the field of metals and associated chemicals, Thiometallurgy, energy technologies, soil biochemical reactor design, lithium ion battery design, and agricultural uses of coal.

He has more than 38 years of expertise in magnesium production and was involved in process development of its startup company NL Magnesium to the present US Magnesium LLC, UT until 2011, during which he was instrumental in process development from the solar ponds to magnesium metal foundry. His expertise includes competitive magnesium processes worldwide and related trade cases.

In 2016, Dr. Neelameggham and Brian Davis authored the ICE-JNME award winning paper "Twenty-First Century Global Anthropogenic Warming Convective Model." He is presently developing Agricoal® to greening arid soils. He authored the eBook *The Return of Manmade* CO_2 *to Earth: Ecochemistry*, published through Smashwords in November 2018.

Dr. Neelameggham holds 16 patents and patent applications and has published several technical papers. He has served in the Magnesium Committee of the TMS Light Metals Division (LMD) since its inception in 2000, chaired in 2005, and since 2007 has been a permanent co-organizer for the Magnesium Technology Symposium. He has been a member of the Reactive Metals Committee, Recycling Committee, Titanium Committee, and Program Committee for LMD and LMD council.

Dr. Neelameggham was the inaugural chair, when in 2008, LMD and the TMS Extraction and Processing Division (EPD) created the Energy Committee, and has been a co-editor of the Energy Technology Symposium through the present. He received the LMD Distinguished Service Award in 2010. As chair of the Hydrometallurgy and Electrometallurgy Committee, he initiated the Rare Metal Technology Symposium in 2014 and was a co-organizer for it through 2021. He organized the 2018 TMS Symposium on Stored Renewable Energy in Coal.

Session Chairs

Magnesium Technology

Keynote Session
Victoria M. Miller, University of Florida
Petra Maier, University of Applied Sciences Stralsund

Fundamentals of Plastic Deformation
Tracy Berman, University of Michigan
Sean Agnew, University of Virginia

Mechanical Behavior
J. Brian Jordon, The University of Alabama
Kiran Solanki, Arizona State University

Alloying & Processing/Primary Production
Vineet Joshi, Pacific Northwest National Laboratory
Joshua Caris, Terves, LLC

Reviewer Pool

Neale R. Neelameggham, IND LLC, USA
Raymond Decker, Thixomat Inc., USA
Vineet V. Joshi, Pacific Northwest National Laboratory, USA
Rajib Kalsar, Pacific Northwest National Laboratory, USA
Petra Maier, University of Applied Sciences Stralsund, Germany
Benjamin Anthony, University of Florida, USA
Victoria M. Miller, University of Florida, USA
Benjamin Begley, University of Florida, USA
Yuan Yuan, Chongqing University, China

Part I
Keynote Session

Influence of Layer Thickness on Deformation Twinning in Mg/Nb Laminates

B. Leu, M. Arul Kumar, and Irene J. Beyerlein

Abstract

Mg-based nanolayered composites can achieve remarkably high specific-strengths due to a combination of a high density of bimetal interfaces and the light-weight constituent Mg phase. However, their applicability to load-bearing applications is hindered by limited formability and the anisotropic effects of deformation twinning. Understanding the microstructural and interface properties that control twinning is crucial for the design of multilayered composites. In this study, we employ an elasto-viscoplastic fast-Fourier-transform (EVP-FFT) crystal plasticity micromechanics model to examine the effect of Mg layer thickness on the growth propensity of $\{10\bar{1}2\}$-tensile twins that span the entire Mg layer. The analysis shows that a critical Mg layer thickness exists, below which, twin growth becomes substantially harder. This critical thickness is related to the backstresses that develop along the twin boundary in the anti-twinning direction. These backstresses result from the plastic reaction of the adjacent Nb layers to the twin shear where the Mg twin lamella intersects the Mg/Nb interfaces. Below the critical layer thickness, the strong backstresses from both ends of the twin strongly interact. They increase in intensity as the layer thickness reduces. Concomitantly, increasing amounts of external loading are required to overcome the backstress and make twin growth feasible.

Keywords

Laminate material • Magnesium • Niobium • Twinning • Layer thickness

Introduction

Dual-phase nanolayered metallic composites have gained interest due to their unique ability to exhibit a multitude of outstanding properties simultaneously. For example, Cu/Nb nanolaminates have demonstrated excellent thermal stability, ductility, shock resistance, radiation resistance, and high strength [1]. These enhanced properties are attributed to the high density of bimetal interfaces that are stable, block the motion of dislocations throughout the material, and act as sources/sinks for dislocations and vacancies [2]. While interfaces are undoubtedly important, the properties of the composite will naturally depend also on its constituent phases. As such, including light-weight Mg-as one of the phases would be beneficial for lowering the overall density of the composite. The result could be a light-weight, yet strong composite material. However, one potential drawback in using Mg stems from its complex anisotropic plastic behavior, involving two inelastic mechanisms, slip and deformation twinning.

Deformation twinning is common in metals with low-symmetry hexagonal close packed (HCP) crystal structures, like Mg. The HCP structure can lead to a lack of easily activatable slip systems needed to accommodate plastic deformation along both the a- and c-axes of the crystal [3]. There are several HCP twinning modes, but the most common twinning mode in Mg and its alloys is the $\{10\bar{1}2\}$ tensile twin mode. It becomes activated when the c-axis of the crystal is subjected to tension but not compression. In this way, $\{10\bar{1}2\}$ tensile twinning helps to alleviate tensile strain in the c-axis direction, especially when it would be difficult to achieve via pyramidal $\langle c + a \rangle$-slip otherwise.

B. Leu (✉) · I. J. Beyerlein
Materials Department, University of California at Santa Barbara, Santa Barbara, CA 93106, USA
e-mail: brandonleu@ucsb.edu

M. A. Kumar
Materials Science and Technology Division, Los Alamos National Laboratory, Los Alamos, NM 87545, USA

I. J. Beyerlein
Department of Mechanical Engineering, University of California at Santa Barbara, Santa Barbara, CA 93106, USA

V. M. Miller et al. (eds.), *Magnesium Technology 2021*, The Minerals, Metals & Materials Series,
https://doi.org/10.1007/978-3-030-65528-0_1

Twinning deformation involves a lattice reorientation and localized shear in a subdomain of a grain [3]. Twins most often nucleate at the grain boundaries and/or free surfaces and propagate as fine lamellae into the grain interior [4–6]. They are often observed spanning the entire grain and, under increased loading, these twin lamellae may thicken. The formation and thickening of twins substantially affect the deformation behavior since they reorient the crystal to a plastically hard orientation and create additional interfaces [7, 8]. In addition, the characteristic shear imposed by the twin can generate localized stress concentrations ahead of the twin tip in the surrounding crystal that can lead to premature failure [9]. Thus, understanding the characteristics that influence the formation and growth of twins is crucial for understanding the design, processing, strength, ductility, and stability of Mg-based multilayered composites.

In nanolayered metallic composites, it has been observed that finer layer thicknesses yield higher strengths [10–12]. While the interactions between dislocations and the bimetal interfaces have been studied, the interactions between twins and bimetal interfaces and the constraining effects of finer layer thicknesses on twinning deformation have received less attention. Understanding layer size effects on twinning is also important for engineering Mg-based multilayered composites to be strong, ductile, and stable.

To study the effects of layer thicknesses on twinning, we employ a crystal plasticity fast-Fourier transform (EVP-FFT) model that is capable of treating multiple phases and can simulate explicit twinning. In this work, an Mg/Nb layered composite system is modeled. Using EVP-FFT, we calculate the local stresses that develop around newly formed twins in multilayers of varying Mg-layer thicknesses. These stresses are then projected onto the specific crystallographic twin system in order to evaluate whether twin growth is favorable. We found that finer layer thicknesses result in larger backstresses that impede the ability of twins to grow. Backstresses are long-range directional internal stresses that build as a result of twinning shear accommodation by the surrounding material [13]. These backstresses act as barriers to twin growth and must be overcome, for example, with external loading, in order for twin growth to continue. Conversely, larger layer thicknesses allow for easier twin growth. However, layer thicknesses beyond 60 times the twin thickness experience similar levels of backstress and thus gain no further advantages for twin growth.

Numerical Method: EVP-FFT Model

We employ a crystal-plasticity fast-Fourier-transform (EVP-FFT) model [14]. This model has been used to study the development of local stresses and effective mechanical response of heterogenous polycrystalline materials with spatial variations in orientation, and elastic and plastic properties [15–17]. This model has been adapted to study deformation twinning in a single-crystal and polycrystal with differing orientation and size [17, 18].

The model builds upon principles of equilibrium, kinematic relationships, and constitutive laws under an infinitesimal strain approximation. The model simulates deformation twins explicitly, which is briefly described here. The stress field at every material point x, or voxel, is solved for by using an implicit time discretization of the form:

$$\sigma^{t+\Delta t}(x) = C(x) : \varepsilon^{e,t+\Delta t}(x) \tag{1}$$

In the above equations, $\boldsymbol{\sigma}(\boldsymbol{x})$ is the Cauchy stress, $\mathbf{C}(\mathbf{x})$ is the elastic stiffness tensor, and $\boldsymbol{\varepsilon}^e$ is the elastic strain tensor, which is given by the total strain minus the plastic strain, $\boldsymbol{\varepsilon}^p$, and the twinning transformation strain, $\boldsymbol{\varepsilon}^{tw}$, i.e.,

$$\begin{aligned}\varepsilon^{e,t+\Delta t}(x) &= \varepsilon^{t+\Delta t}(x) - \varepsilon^{p,t}(x) - \dot{\varepsilon}^{p,t+\Delta t}(x)\Delta t \\ &\quad - \varepsilon^{tw,t}(x) - \Delta\varepsilon^{tw,t+\Delta t}(x)\end{aligned} \tag{2}$$

The plastic strain evolves due to dislocation slip on crystallographic slip systems:

$$\begin{aligned}\varepsilon^{e,t+\Delta t}(x) &= \varepsilon^{t+\Delta t}(x) - \varepsilon^{p,t}(x) - \dot{\varepsilon}^{p,t+\Delta t}(x)\Delta t \\ &\quad - \varepsilon^{tw,t}(x) - \Delta\varepsilon^{tw,t+\Delta t}(x)\end{aligned} \tag{3}$$

$$m^s = \frac{1}{2}(b^s \otimes n^s + n^s \otimes b^s) \tag{4}$$

where $\tau_c^s(\boldsymbol{x})$ is the critical resolved shear stress (CRSS) associated with the slip system s, and n is the stress exponent (inverse of the rate-sensitivity exponent). The tensor $\mathbf{m}^s$ is the symmetric part of the Schmid tensor, and $\mathbf{b}^s$ and $\mathbf{n}^s$ are the unit vectors along the slip direction and normal to the glide plane of slip system s, respectively. Twinning in the model is simulated explicitly by reorienting by following the crystallographic twinning relationship and imposing characteristic twinning shear, $\mathbf{g}^{tw}$, as an eigenstrain in a predetermined twin region. The twinning transformation strain is incremented over N^{tw} steps on a particular twinning plane in the twinning shear direction.

$$\Delta\varepsilon^{tw} = m^{tw}\Delta\gamma^{tw}(x) \tag{5}$$

$$\Delta\gamma^{tw}(x) = \frac{g^{tw}}{N^{tw}} \tag{6}$$

In simulation, $\Delta t = 10^{-4}$ s and $N^{tw} = 1000$ is kept sufficiently large to ensure convergence. Note that $\boldsymbol{\varepsilon}^{tw}$ is zero everywhere outside of the twin domain.

Results and Discussion

Figure 1a shows the model of a $\{10\bar{1}2\}$ tensile twin (red) embedded inside a single crystalline layer of Mg (blue) adjoined to a single crystalline layer of Nb (gray). With periodic boundary conditions in all directions, a composite comprised of alternative stacking of Mg and Nb layers is created. To study the effect of Mg layer thickness on twin growth, the twin thickness, t, is kept constant, while L, the Mg layer thickness, as well as the length of the twin length, is varied from 10 to 150t. Like t, the Nb layer thickness is also held constant at 12t. The simulation unit-cell size in x, y, and z ranges from $3 \times 400 \times 180$ to $3 \times 400 \times 1300$ voxels. A polycrystalline outer layer of a randomly textured, Mg/Nb composite, was used only along the top and bottom boundary (not shown in Fig. 1a) and mimics the surrounding polycrystalline composite. The polycrystalline layer is 100 voxels in thickness (in z-direction) and is sufficiently large to isolate the stress fields around the single twin from the periodic images. In the EVP-FFT model, deformation at every step of the calculations is accommodated by a combination of anisotropic elasticity and crystal plasticity. The elastic constants of Nb used in the calculations for C_{11}, C_{12}, and C_{44} are 245.6 GPa, 138.7 GPa and 29.3 GPa, respectively [19]. The critical resolved shear stress (CRSS) used for $\{110\}$ $<111>$ and $\{112\} < 111 >$ type slip in BCC Nb are both 135.0 MPa [20]. For Mg, the elastic constants C_{11}, C_{12}, C_{13}, C_{33}, and C_{44} are 58.6 GPa, 25.0 GPa, 20.8 GPa, 61.1GPa and 16.6 GPa, respectively [19]. In Mg, prismatic$\langle a \rangle$, basal $\langle a \rangle$ and pyramidal type-I $\langle c + a \rangle$ slip was allowed with CRSS values of 35.7 GPa, 3.3 GPa, and 86.2 MPa, respectively [16]. The crystallographic orientation of the Nb and Mg was (133°, 53°, 75°) and (0°, 43°, 0°) according to the Bunge convention. These orientations align the $[2\bar{1}\bar{1}0]^{Mg}$ // $[\bar{1}11]^{Nb}$ and the $(0001)^{Mg}$ // $(110)^{Nb}$ [11, 21]. The choice of Mg crystal orientation aligns the plane normal and shear direction of $(01\bar{1}2)[0\bar{1}10]$ twin variant along the z and y-directions, respectively. In these calculations, the twin was formed without external loading to analyze the stress fields that develop when a twin exists in a composite with two phases differing in elastic and plastic properties.

EVP-FFT model calculates the full stress, elastic and plastic strain tensors at every voxel comprising the unit cell. Figure 1b shows an example calculation of the twin plane resolved shear stress (TRSS) fields that develop from a twin of thickness t spanning an Mg layer with thickness 20t. The twin boundary is highlighted in dashed lines. The TRSS is the internal stress in the Mg layer resolved onto the same crystallographic twinning system as the twin being studied. It is a relevant driving force for migrating the twin boundary, resulting in twin propagation and growth. From the TRSS map in Fig. 1a, it is apparent that formation of the twin in the Mg layer leads to a strong heterogeneous stress field within and around the twin. The twin is constrained from deforming at both twin tips by the adjacent Nb layers. The Nb crystals resist the localized and directed shearing that is characteristic of the twin and generate reaction forces back on the twin and Mg layer, referred to as backstresses, as evident by the negative TRSS fields. In particular, the TRSS along the twin

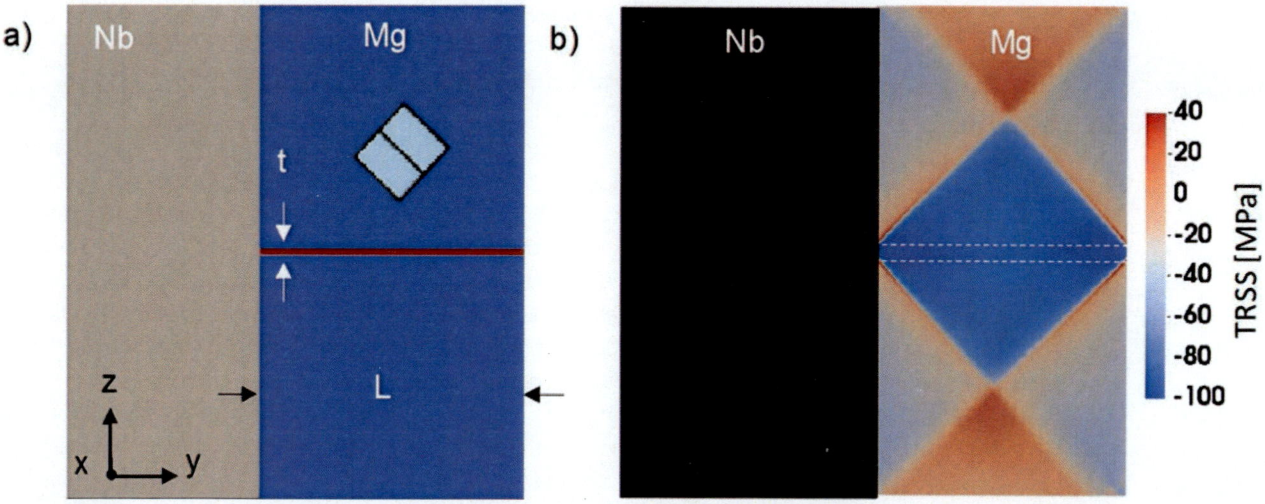

Fig. 1 **a** Schematic representation of EVPFFT unit-cell to simulate the formation of a $\{10\bar{1}2\}$ tensile twin in Mg-Nb laminate materials with different layer thicknesses. The hexagonal inset represents the crystallographic orientation of the Mg layer. The twin thickness, t, and Mg and twin length, L, are denoted. **b** An example of the TRSS distribution, in MPa, is shown for the case where L = 20t. The Nb layer is omitted and the twin boundaries are highlighted with dashed lines for clarity. (Color figure online)

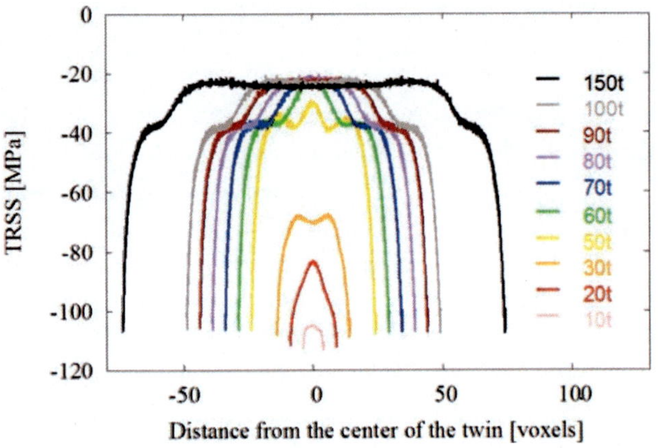

Fig. 2 TRSS profile taken along the upper twin boundary in the Mg layer for the various Mg layer thicknesses (L), ranging from 10 to 150t. Each colored line represents different values of L. (Color figure online)

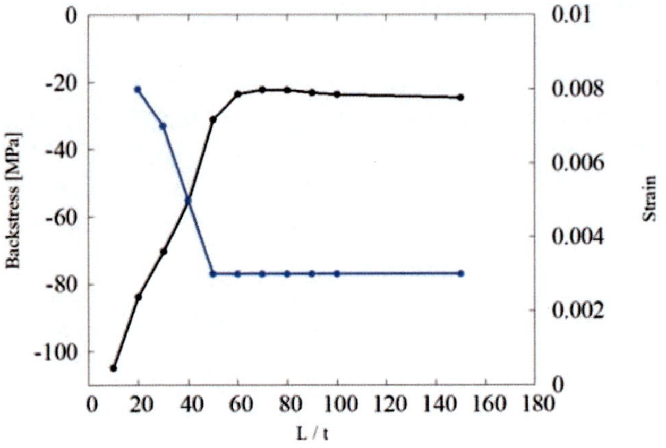

Fig. 3 Effect of Mg layer thickness on twin backstress at center of the twin (black), and additional strain required to reach positive TRSS (blue). Increase in Mg layer thickness reduces Nb layer induced stress reversal at the twin center and easy to thickens. (Color figure online)

boundaries are negative, acting in the anti-twinning sense. The backstresses act against the migration of the twin boundary. For this twin to grow, they would need to be overcome, for instance, with an externally applied load.

Figure 2 shows the TRSS profiles plotted along the twin boundary for various Mg layer thicknesses, ranging from 10 to 150t. It can be seen that in all the cases the tips of the twin, where they intersect the Mg/Nb interface experience the largest backstresses. At these areas, the TRSS levels drop below −100 MPa, going as low as −110 MPa in the smallest Mg layer thickness of 10t. Furthermore, the TRSS levels experience less backstress near the center of the twin. This implies that the twin growth is expected to be easier near the center of the twin compared to the twin tips. For layer thicknesses below 60t (green line), the backstress increases rapidly, resulting in lower levels of TRSS at the twin tip and along the center of the twin as the Mg layer thickness gets smaller. For layer thicknesses above 60t, the TRSS profiles show similar levels of backstress at the twin tip and along the center of the twin, resulting in TRSS levels approximately −110 MPa and −20 MPa, respectively.

An interesting finding from these calculations is that increasing Mg layer thicknesses above 60t does not result in further decreases in backstress. This can be seen in Fig. 3 (black), which shows the average TRSS taken from the center of the upper twin boundary where twin thickening would be easiest. The center of the twin is taken as the central 5% of the entire twin length. The calculations reveal that Mg layer thicknesses above 60t will not give further enhancement to twin growth. Meanwhile, by decreasing the Mg layer thickness below 60t, the backstresses can be enhanced, resulting in greater difficulty for existing twins to thicken.

The calculations, thus far, suggest that decreasing the Mg layer thicknesses below a critical value (i.e., below 60 times the twin thicknesses here), can suppress twin growth significantly, resulting in thinner twin domains. To further illustrate this, the Mg/Nb multilayer composite was compressed in the out of plane direction. The compression applied parallel to the $\langle a \rangle$-axis of the Mg layer in order to induce c-axis tension, which would be favorable to activate tensile twinning. Strain was applied at a rate of 1×10^{-3} s^{-1}until the average TRSS along the center of the twin was slightly greater than zero. At this point, the twin growth may be possible. In can be seen in Fig. 3 (blue) that the amount of additional external loading required for twin growth to become favorable varies depending on the Mg layer thickness. For smaller Mg layer thicknesses, more load is required to counteract the strong backstresses that have developed and thicken the twin. However, Mg layer thicknesses 50t and above require equal amounts of external strain, about 0.003, in order for twin growth to become favorable.

In prior work, a similar study on grain size effects on twin growth was conducted for polycrystalline Mg [7, 17]. As revealed here for the Mg/Nb bilayer composite, it was found that reductions in the Mg grain size increase the backstress that develops along the twin boundary. Another common feature is the size effect is severe in finer grains/layers, and almost negligible for grains/layers larger than a critical size. Consequently, the applied strain required to start the twin expansion process rises. The important difference is the value of critical grain/layer size. In the Mg/Nb laminate, the critical layer thickness is 2–3 times larger, ranging from 50–60t, than the Mg polycrystal, ranging from 17–27t. The enhanced sensitivity in the Mg/Nb stems from the fact that Nb is plastically much harder than a neighboring Mg of any crystal orientation, and therefore generates a larger reaction stress to the same twin.

The combined effect of reduced grain size and layer thickness is to reduce the ability of twins to thicken. These effects may be compounded in the case of Mg-based laminate composites. A practical way of making these composites is through severe plastic deformation techniques, such as accumulative roll bonding, which, through successive cycles, will reduce both layer thickness and grain size, making it harder for large twins to develop [22]. These effects can be seen in the work of Nie et al. on Ti/Al/Mg laminates [23]. They found that twins become progressively thinner near the Mg/Al interface, where the grain sizes are also smaller. Separating the contributions of grain size and properties of neighboring grains is a recommended area of study.

Summary

In summary, using crystal plasticity fast Fourier transform (EVP-FFT) model, we study the effect of Mg layer thickness on $\{10\bar{1}2\}$ twin growth in an Mg/Nb layered composite composites of alternating Mg and Nb crystals. The analysis reveals a critical size below which twin growth in the Mg layer would be hindered. We explain the size effect via a boost in backstresses that develop along the twin boundary in the anti-twinning direction. The backstresses are a reaction to the neighboring Nb layers to the intense shear strain imposed by the twin lamellae that impinges on the Mg/Nb interface. At a critical thickness, the backstresses from each end of the twin start to strongly interact. The backstress is influenced by the plastic properties of the neighboring layer. Compared to pure Mg, twinning is much more constrained in the Mg/Nb layer since the Nb layers provide a stiffer constraint on the abutting Mg twin than another Mg crystal. An interesting future study includes the effects of finer layer thickness on the nucleation of twins in second nearest neighbor Mg layers, such as was investigated for Zr/Nb [24].

Acknowledgements B. L. was supported by the Department of Defense (DoD) through the National Defense Science & Engineering Graduate Fellowship (NDSEG) Program. M. A. K. acknowledges the financial support from the U.S. Department of Energy, Office of Basic Energy Sciences (OBES) FWP-06SCPE401. I. J. B. acknowledges financial support from the National Science Foundation (NSF CMMI-1728224).

References

1. I. J. Beyerlein *et al.*, "Structure-property-functionality of bimetal interfaces," *JOM*, vol. 64, no. 10, pp. 1192–1207, 2012, https://doi.org/10.1007/s11837-012-0431-0.
2. J. Wang and A. Misra, "An overview of interface-dominated deformation mechanisms in metallic multilayers," *Curr. Opin. Solid State Mater. Sci.*, vol. 15, no. 1, pp. 20–28, 2011, https://doi.org/10.1016/j.cossms.2010.09.002.
3. M. H. Yoo and J. K. Lee, "Deformation twinning in h.c.p. metals and alloys," *Philos. Mag. A Phys. Condens. Matter, Struct. Defects Mech. Prop.*, vol. 63, no. 5, pp. 987–1000, 1991, https://doi.org/10.1080/01418619108213931.
4. I. J. Beyerlein, X. Zhang, and A. Misra, "Growth Twins and Deformation Twins in Metals," *Annu. Rev. Mater. Res.*, vol. 44, no. 1, pp. 329–363, 2014, https://doi.org/10.1146/annurev-matsci-070813-113304.
5. J. Jeong, M. Alfreider, R. Konetschnik, D. Kiener, and S. H. Oh, "In-situ TEM observation of {101¯2} twin-dominated deformation of Mg pillars: Twinning mechanism, size effects and rate dependency," *Acta Mater.*, vol. 158, pp. 407–421, 2018, https://doi.org/10.1016/j.actamat.2018.07.027.
6. A. Khosravani, D. T. Fullwood, B. L. Adams, T. M. Rampton, M. P. Miles, and R. K. Mishra, "Nucleation and propagation of { 1 0 1 ¯ 2 } twins in AZ31 magnesium alloy," *Acta Mater.*, vol. 100, pp. 202–214, 2015, https://doi.org/10.1016/j.actamat.2015.08.024.
7. M. Arul Kumar, I. J. Beyerlein, R. J. McCabe, and C. N. Tomé, "Grain neighbour effects on twin transmission in hexagonal close-packed materials," *Nat. Commun.*, vol. 7, no. May, p. 13826, 2016, https://doi.org/10.1038/ncomms13826.
8. L. Jiang *et al.*, "Twin formation from a twin boundary in Mg during in-situ nanomechanical testing," *Mater. Sci. Eng. A*, vol. 759, no. January, pp. 142–153, 2019, https://doi.org/10.1016/j.msea.2019.04.117.
9. M. Knezevic, M. R. Daymond, and I. J. Beyerlein, "Modeling discrete twin lamellae in a microstructural framework," *Scr. Mater.*, vol. 121, pp. 84–88, 2016, https://doi.org/10.1016/j.scriptamat.2016.04.026.
10. A. Misra, J. P. Hirth, and R. G. Hoagland, "Length-scale-dependent deformation mechanisms in incoherent metallic multilayered composites," *Acta Mater.*, vol. 53, no. 18, pp. 4817–4824, 2005, https://doi.org/10.1016/j.actamat.2005.06.025.
11. B. Ham and X. Zhang, "High strength Mg/Nb nanolayer composites," *Mater. Sci. Eng. A*, vol. 528, no. 4–5, pp. 2028–2033, 2011, https://doi.org/10.1016/j.msea.2010.10.101.
12. A. Misra and H. Kung, "{D}eformation {B}ehaviour of {N}anostructured {M}etallic {M}ultilayers," *Adv. Eng. Mater.*, vol. 3, no. 4, pp. 217–222, 2001, https://doi.org/10.1002/1527-2648(200104)3:4<217::AID-ADEM217>3.0.CO;2-5.
13. Y. Zhu and X. Wu, "Perspective on hetero deformation induced (HDI) hardening and backstress," *Mater. Res. Lett.*, vol. 7, no. 10, pp. 393–398, 2019.
14. R. A. Lebensohn, A. K. Kanjarla, and P. Eisenlohr, "An elasto-viscoplastic formulation based on fast Fourier transforms for the prediction of micromechanical fields in polycrystalline materials," *Int. J. Plast.*, vol. 32–33, pp. 59–69, 2012, https://doi.org/10.1016/j.ijplas.2011.12.005.
15. M. Arul Kumar, I. J. Beyerlein, and C. N. Tomé, "Effect of local stress fields on twin characteristics in HCP metals," *Acta Mater.*, vol. 116, pp. 143–154, 2016, https://doi.org/10.1016/j.actamat.2016.06.042.
16. M. Arul Kumar, I. J. Beyerlein, R. A. Lebensohn, and C. N. Tomé, "Modeling the effect of neighboring grains on twin growth in HCP polycrystals," *Model. Simul. Mater. Sci. Eng.*, vol. 25, no. 6, 2017, https://doi.org/10.1088/1361-651X/aa7bbb.
17. M. Arul Kumar, I. J. Beyerlein, and C. N. Tomé, "Grain size constraints on twin expansion in hexagonal close packed crystals," *J. Appl. Phys.*, vol. 120, no. 15, 2016, https://doi.org/10.1063/1.4965719.
18. M. Arul Kumar, B. Leu, P. Rottmann, and I. J. Beyerlein, "Characterization of Staggered Twin Formation in HCP

Magnesium," in *Minerals, Metals and Materials Series*, 2019, pp. 2007–213, https://doi.org/10.1007/978-3-030-05789-3_17.

19. G. Simmons, *Single Crystal Elastic Constants and Calculated Aggregate Progress*, vol. 34. 1965.
20. M. Knezevic, T. Nizolek, M. Ardeljan, I. J. Beyerlein, N. A. Mara, and T. M. Pollock, "Texture evolution in two-phase Zr/Nb lamellar composites during accumulative roll bonding," *Int. J. Plast.*, vol. 57, pp. 16–28, 2014, https://doi.org/10.1016/j.ijplas.2014.01.008.
21. Y. Chen, M. Y. Gong, S. Shao, N. A. Mara, and J. Wang, "Interface Facilitated Reorientation of Mg Nanolayers in Mg-Nb Nanolaminates," *JOM*, vol. 71, no. 4, pp. 1215–1220, 2019, https://doi.org/10.1007/s11837-019-03360-8.
22. B. Leu *et al.*, "Processing of Dilute Mg–Zn–Mn–Ca Alloy/Nb Multilayers by Accumulative Roll Bonding," *Adv. Eng. Mater.*, vol. 22, no. 1, pp. 1–9, 2020, https://doi.org/10.1002/adem.201900673.
23. H. Nie *et al.*, "Effect of twins and dynamic recrystallization on the microstructures and mechanical properties of Ti/Al/Mg laminates," *Mater. Des.*, vol. 181, p. 107948, 2019, https://doi.org/10.1016/j.matdes.2019.107948.
24. J.-W. Zhang, B. Leu, M. Arul Kumar, I. J. Beyerlein, and W.-Z. Han, "Twin hopping in nanolayered Zr-2.5Nb", *Materials Research Letters*, vol. 8, pp. 307–313, 2020, https://doi.org/10.1080/21663831.2020.1755902.

Measurement of the Critical Resolved Shear Stress for Slip in Mg Alloys Using Instrumented Indentation

Warren Poole, Shuheng Li, and Ghazal Nayyeri

Abstract

A critical challenge for the plasticity and fracture of magnesium and its alloys is the competition between the different deformation modes and how this varies with temperature. In the current study, the critically resolved shear stress for basal slip as a function of alloy composition and temperature has been measured using instrumented spherical indentation. Indentation offers the advantage that tests can be conducted on grains of known orientation in polycrystal samples, particularly of importance for alloys where producing single crystals is difficult. Here, it will be demonstrated that by doing tests with indenters of different radii, the critically resolved shear stress (CRSS) for can be extracted. Very good agreement was found between the CRSS values for basal slip by indentation and the literature values from single crystal tests. Finally, the contribution will also offer perspectives on quantifying additional deformation modes such as extension twining and 2nd order pyramidal slip.

Keywords

Magnesium alloys • Indentation • Critical resolved shear stress

The application of magnesium and its alloys in demanding applications such as aerospace or automotive requires robust models for the deformation behaviour under complex loading conditions [1]. Physically based models for polycrystal deformation require constitutive laws to describe the behaviour at the local level for both slip and deformation twinning. The effect of composition and deformation temperature are important as the slip systems with high critically resolved shear stress (CRSS) values (2nd order pyramidal and prismatic) are strongly dependent on these factors. Practically, this has important implications on the formability of magnesium sheet alloys. For example, Boba et al. have shown that the forming limits for ZEK100 are substantially improved when the forming temperature is raised from room temperature to 200 °C and is superior to more conventional AZ31B sheet alloy [2]. Further, Skszek and co-workers demonstrated that by using this rare earth containing alloy and pre-heating the sheet, it was possible to stamp a complex prototype magnesium door inner at stamping time of < 10 s even though this was not possible using AZ31B [3].

In general, it is challenging to obtain the CRSS for the different deformation modes from single crystals, especially for the most practically important alloys. This is the case due to the fact that i) there is such a significant difference in the CRSS values for basal, prismatic and 2nd order pyramidal slip [4] which makes aligning single crystals to isolate the high stress slip systems difficult and ii) it is difficult to grow single crystals for a wide range of alloys. As a result, there has been considerable interest in using indentation techniques on polycrystals within a grain of known orientation to study the activation of different deformation modes [5–12] and to attempt to extract values for the CRSS [13–16] at ambient temperature. For high temperature indentation tests, most work has focused on indentation creep tests, for example see [17], however, these tests are outside the scope of the current work.

For instrumented indentation tests where the force and depth of indentation can be measured, one must consider (i) the effect of indenter shape [5, 7, 8, 10, 12, 13], (ii) the indenter size (known as the indentation size effect) [5, 8–10, 13, 18–22] and (iii) the direction of the loading axis in relation to the crystal orientation [9, 11, 12, 23]. The current work has focused on results using spherical indenters of different radii as this has the advantage of well known elastic contact mechanics (Hertz solution [24, 25]). Figure 1a shows an example of the variation of the resolved shear

W. Poole (✉) · S. Li · G. Nayyeri
The University of British Columbia, Department of Materials Engineering, Vancouver, BC, Canada
e-mail: warren.poole@ubc.ca

V. M. Miller et al. (eds.), *Magnesium Technology 2021*, The Minerals, Metals & Materials Series,
https://doi.org/10.1007/978-3-030-65528-0_2

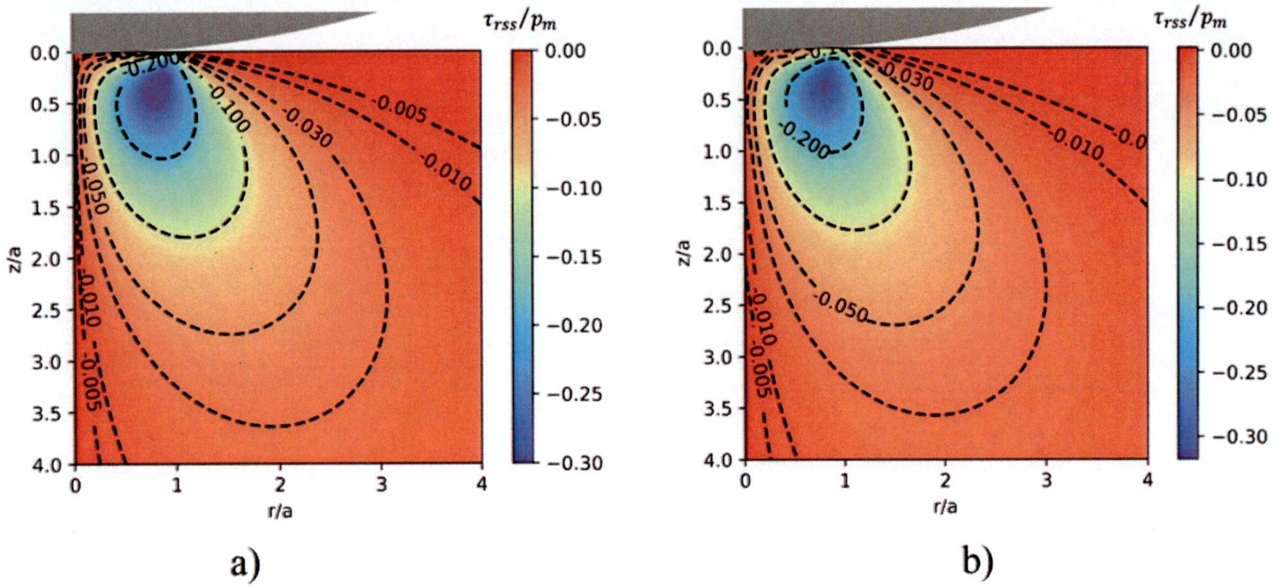

Fig. 1 Contour plots of the ratio of the resolved shear stress to the contact pressure for indentation where the indenter loading is parallel to the [0001] direction **a** analytical solution and **b** finite element method. Note, a is the contact radius between the indenter and the sample. The ratio r/a is the ratio between the horizontal distance r from the contact point and the contact radius and z/a is the ratio between the distance z from the surface of the sample radius and the contact radius. The position of r/a = 0 and z/a = 0 is the position directly under the contact point for a spherical indenter. (Color figure online)

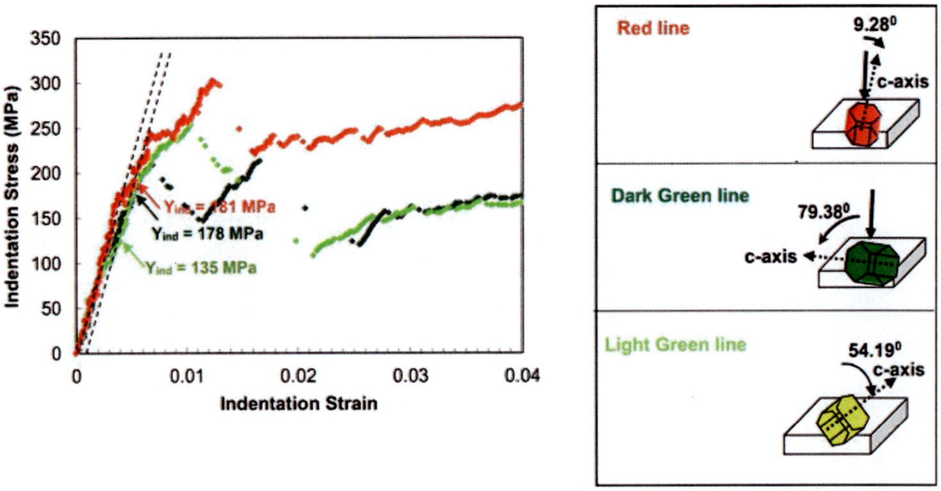

Fig. 2 Indentation stress–strain curves for high purity magnesium measured with a spherical indenter with radius of 13.5 μm. (Color figure online)

stress on the basal slip system under a spherical indenter for a loading direction of [0001] calculated from the analytical (Hertz) solution. Figure 1b shows similar results from a finite element method solution (in this case assuming isotropic elasticity). The finite element method approach has the advantage that anisotropic elastic constants can be used (noting that Mg is nearly isotropic but it is important for other materials such as iron or copper). It can be observed in Fig. 1 that the maximum resolved shear stress lies in the subsurface region below the outer edge of the contact radius. This is the region where basal slip is expected to initiate.

Figure 2 shows an example of the indentation stress–strain curve for high purity magnesium calculated from the force–displacement curves (measured with a MTS Nanoindenter XP system with a spherical indenter with radius of 13.5 μm) using the approach of Kalidindi and co-workers [26–29].

These results show that the onset of plasticity depends on the angle between the loading axis and the c-axis of the magnesium (note: in this case, the three measurements were made in different grains of a polycrystal sample where the orientation of the grain was determined by EBSD). The yield

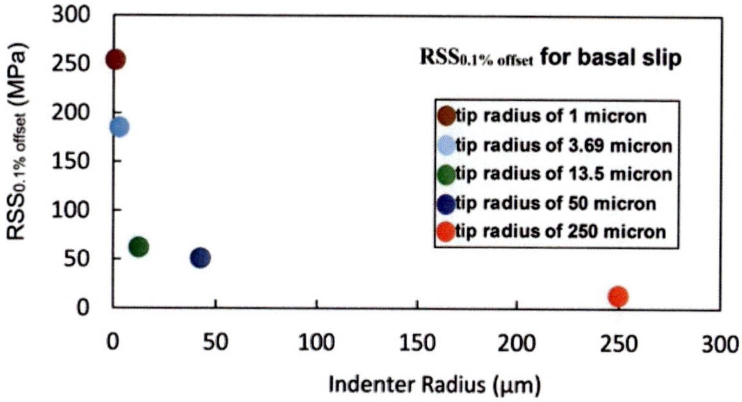

Fig. 3 The resolved shear stress (RSS) at the onset of plasticity for high purity magnesium measured with the loading axis nearly parallel to the c-axis for different radii of spherical indenters. (Color figure online)

stress for the indentation stress–strain curves was found to be similar for the case where the angle between the c-axis and the loading axis was 9.3 and 79.4° (178–181 MPa) but considerably lower for the case of 54.2°. This can be understood by calculating the resolved shear stress for these different orientations (i.e. using the approach in Fig. 1) which gives similar values of 63–68 MPa for the different orientations. This is interpreted as the resolved shear stress for basal slip for this indenter radius.

Turning to the indentation size effect, Fig. 3 summarizes experimental results for the resolved shear stress at the onset of plasticity for spherical indenters of different radii (high purity magnesium with indentation loading direction nearly parallel to the c-axis). It can be seen that there is a very strong size effect with the RSS decreasing from 250 to 21 MPa as the indenter radius increases from 1 to 250 μm. Nayyeri et al. have discussed these results in more detail [14] and shown with dislocation-based model, it is possible to measure the critically resolved shear stress by extrapolating to an infinite indenter radius. The value they obtained of 1.7 MPa for the CRSS of basal slip in high purity magnesium is in good agreement with results from single crystal experiments. Nayyeri extended this approach to study various Mg–Al and Mg-Zn and Mg-Zn-Nd alloys and again found good agreement between the indentation and single crystal results after accounting for the orientation of the magnesium crystal and the indentation size effect [15].

In summary, it has been shown that spherical indentation has the advantage of well defined contact mechanics which allows for the calculation of the resolved shear stress on the basal slip system for different loading orientations. However, there is a strong dependence of the resolved shear stress at the onset of plasticity on the indenter tip radius. In order to determine the critically resolved shear for basal slip relevant to crystal plasticity models, it is necessary to account for the size effect. Practically, this can be done by making measurements with indenters of different radii and extrapolating to an infinitely large indenter radius. Future work involves extending the work to measure the temperature dependence of the critically resolved shear stress for different alloys by doing indentation experiments at different temperatures.

Acknowledgements this work was undertaken, in part, thanks to funding from the Canada Research Chair program (Poole) and NSERC Canada.

References

1. P. Tomlinson, H. Azizi-Alizamini, W.J. Poole, C.W. Sinclair, and M.A. Gharghouri, *Metall. Mater. Trans. A.*, Biaxial Deformation of the Magnesium Alloy AZ80, 2013. **44A**: pp. 2970–2983.
2. M. Boba, C. Butcher, N. Panahi, M.J. Worswick, R.K. Mishra, and J.T. Carter, *Int J Mater Form*, Warm forming limits of rare earth-magnesium alloy ZEK100 sheet, 2017. **10**: pp. 181–191.
3. X. Niu, T. Skszek, M. Fabischek, and A. Zak, *Mat. Sci. Forum*, Low Temperature Warm Forming of Magnesium ZEK 100 Sheets for Automotive Applications, 2014. **783–786**: pp. 431–436.
4. A. Chapuis and J.H. Driver, *Acta Mater.*, Temperature dependency of slip and twinning in plane strain compressed magnesium single crystals, 2011. **59**: pp. 1986–1994.
5. T. Guo, F. Siska, and M.R. Barnett, *Scripta Mater.*, Distinguishing between slip and twinning events during nanoindentation of magnesium alloy AZ31, 2016. **110**: pp. 10–13.
6. T. Guo, F. Siska, J. Cheng, and M. Barnett, *J. Alloy. Comp.*, Initiation of basal slip and tensile twinning in magnesium alloys during nanoindentation. J. Alloy. Comp. 2018;731:620., 2018. **731**: pp. 620–630.
7. D. Catoor, Y.F. Gao, J. Geng, M.J.N.V. Prasad, E.G. Herbert, K.S. Kumar, G.M. Pharr, and E.P. George, *Acta Mater.*, Incipient plasticity and deformation mechanisms in single-crystal Mg during spherical nanoindentation, 2013. **61**: pp. 2953–2965.
8. J.H. Shin, S.H. Kim, T.K. Ha, K.H. Oh, I.S. Choi, and H.N. Han, *Scripta Mater.*, Nanoindentation study for deformation twinning of magnesium single crystal, 2013. **68**: pp. 483–486.
9. B. Selvarajou, J.H. Shin, T.K. Ha, I.S. Choi, S.P. Joshi, and H.N. Han, *Acta Mater.*, Orientation-dependent indentation response of magnesium single crystals: Modeling and experiments, 2014. **81**: pp. 358–376.
10. H. Kitahara, T. Mayama, K. Okumura, Y. Tadano, M. Tsushida, and S. Ando, *Acta Mater.*, Anisotropic deformation induced by spherical indentation of pure Mg single crystals, 2014. **78**: pp. 290–300.

11. H. Somekawa, T. Tsuru, A. Singh, S. Miura, and C.A. Schuh, *Acta Mater.*, Effect of crystal orientation on incipient plasticity during nanoindentation of magnesium, 2017. **139**: pp. 21–29.
12. H. Somekawa and C.A. Schuh, *Metall. Mater. Trans. A.*, Effect of Crystal Orientation on Nanoindentation Behavior in Magnesium, 2016. **47**: pp. 3227–3234.
13. R. Sanchez-Martin, M.T. Perez-Prado, J. Segurado, J. Bohlen, I. Gutierrez-Urrutia, J. Llorca, and J.M. Molina-Aldareguia, *Acta Mater.*, Measuring the critical resolved shear stresses in Mg alloys by instrumented nanoindentation, 2014. **71**: pp. 283–292.
14. G. Nayyeri, W.J. Poole, C.W. Sinclair, and S. Zaefferer, *Scripta Mater.*, The role of indenter radius on spherical indentation of high purity magnesium loaded nearly parallel to the c-axis, 2017. **137**: pp. 119–122.
15. G. Nayyeri, W.J. Poole, C.W. Sinclair, and S. Zaefferer, *Scripta Mater.*, Measurement of the Critical Resolved Shear Stress for Basal Slip in Magnesium Alloys Using Instrumented Indentation, 2018. **156**: pp. 37–41.
16. R. Sanchez-Martin, M.T. Perez-Prado, J. Seguradia, and J.M. Molina-Aldareguia, *Acta Mater.*, Effect of indentation size on the nucleation and propagation of tensile twinning in pure magnesium, 2015. **93**: pp. 114.
17. B. Kondori and R. Mahmudi, *Mat. Sci. Eng. A.*, Effect of Ca additions on the microstructure and creep properties of a cast Mg–Al–Mn magnesium alloy, 2017. **700**: pp. 438–447.
18. F. Siska, T. Guo, L. Stratil, J. Cizek, and M.R. Barnett, *Comp. Mater. Sci.*, Numerical study of stress distribution and size effect during AZ31 nanoindentation, 2017. **126**: pp. 393–399.
19. W.J. Poole, M.F. Ashby, and N.A. Fleck, *Scripta Materialia*, Micro-hardness of annealed and work-hardenedcopper polycrystals, 1996. **34**: pp. 559-564.
20. H. Gao, Y. Huang, W.D. Nix, and J.W. Hutchinson, *J. Mech. Phys. Sol.*, Mechanism-based strain gradient plasticity– I. Theory, 1999: pp. 1239–1263.
21. K.W. McElhaney, J.J. Vlassak, and W.D. Nix, *J. Mater. Res.*, Determination of indenter tip geometry and indentation contact area for depth-sensing indentation experiments, 1998: pp. 1300–1306.
22. W.D. Nix and H. Gao, *J. Mech. Phys. Sol.*, Indentation size effects in crystalline materials: A law for strain gradient plasticity, 1998. **46**: pp. 411–425.
23. C. Zambaldi, C. Zehnder, and D. Raabe, *Acta Mater.*, Orientation dependent deformation by slip and twinning in magnesium during single crystal indentation, 2015. **91**: pp. 267–288.
24. K.L. Johnson, *Contact Mechanics*. 1987: Cambridge University Press.
25. A.C. Fischer-Cripps, *Introduction to Contact Mechanics*. 2007: Springer US.
26. S.R. Kalidindi and S. Pathak, *Acta Mater.*, Determination of the effective zero-point and the extraction of spherical nanoindentation stress-strain curves, 2008: pp. 3523–3532.
27. D.K. Patel and S.R. Kalidindi, Correlation of spherical nanoindentation stress-strain curves to simple compression stress-strain curves for elastic-plastic isotropic materials using finite element models, 2016. **112**: pp. 295-302.
28. S. Pathak and S.R. Kalidindi, *Mat. Sci. Eng. R*, Spherical nanoindentation stress–strain curves, 2015. **91**: pp. 1–36.
29. S. Pathak, D. Stojakovic, and S.R. Kalidindi, *Acta Mater.*, Measurement of the local mechanical properties in polycrystalline samples using spherical nanoindentation and orientation imaging microscopy, 2009. **57**: pp. 3020–3028.

Development of a Low-Cost and Room-Temperature Formable Mg Alloy Sheet with In-Plane Isotropic Tensile Properties

Taiki Nakata, Chao Xu, Hideaki Ohashi, Yu Yoshida, Katsuhito Yoshida, and Shigeharu Kamado

Abstract

Twin-roll-cast Mg-3.07Al-0.25Mn (mass%) alloy was continuously rolled at a warm temperature, and the room-temperature stretch formability, tensile properties, microstructures, and texture of the annealed sheets were investigated. The sheet rolled at 220 °C exhibits large Index Erichsen values over 8 mm due to the formation of a ring-like texture feature where the basal poles are oriented ~25° from normal direction to all directions. Fine grain structure with an average grain size of ~7 μm could be obtained, so that the sheet shows moderate 0.2% proof stress of ~160 MPa and large elongation to failure over 25% in both rolling and transverse directions. The fabrication process of the alloy sheet, which consists of homogenization, warm-temperature rolling, and annealing, is viable in industrial production and accordingly, the alloy sheet developed in this work will broaden the commercial applications of wrought Mg alloys in automotive industries.

Keywords

Magnesium alloy • Rolling • Formability • Texture weakening • Recrystallization

T. Nakata (✉) · H. Ohashi · S. Kamado
Nagaoka University of Technology, 1603-1, Kamitomioka, Nagaoka, 940-2188, Japan
e-mail: nakata@mech.nagaokaut.ac.jp

C. Xu
School of Materials Science and Engineering, Harbin Institute of Technology, Harbin, 150001, China

Y. Yoshida · K. Yoshida
Sumitomo Electric Industries, Ltd, 1-1-1 Koyakita, Itami, 664-0016, Japan

Introduction

Lightweight magnesium alloys have significant potential to reduce the weight of a car. Recent efforts have successfully developed new magnesium alloys with improved extrudability and mechanical properties comparable to those of Al–Mg-Si based 6xxx aluminum alloys used in many automotive parts [1, 2]. The development of magnesium sheet alloy is also required for numerous weight-sensitive applications; however, several fundamental issues exist, and they hamper the use of magnesium alloy sheets in automobiles. For example, during rolling process, commercially available magnesium alloy sheets, such as Mg-3Al-1Zn and Mg-2Zn-1Mn (mass%, AZ31 and ZM21) alloy sheets, easily form a typical rolling texture with a strong alignment of (0001) poles to the normal direction of the sheets (ND) [3, 4]. In magnesium alloy sheets with such texture, activation of (0001) $[11\bar{2}1][11\bar{2}0]$ slips (basal slips) seldom accommodates thickness strain during forming process. Although the deformation mainly proceeds by twins, the matrix/twin interface becomes crack initiation and propagation sites. Then, the strongly textured magnesium alloy sheets exhibit poor room-temperature formability with a typical Index Erichsen (I.E.) value of 3–5 mm [5]. To randomize the texture, a high temperature rolling has been proposed by Huang et al., and they successfully increased the I.E. value of the AZ31 alloy sheet from 4.5 mm to 8.6 mm [6]. Although this is a fundamental study which helps to understand the texture weakening behavior in magnesium alloy sheets, the corresponding process requires extra costs, resulting in the increase of the cost of final products [7]. Alloying of zinc and calcium is a low-cost method to control the texture, and recent studies lead to the development of an Mg-Zn-Ca alloy sheet with a high I.E. value of 8.2 mm [8]. However, the Mg-Zn-Ca alloy sheet shows a large planar anisotropy due to the splitting of basal poles to the transverse direction (TD) of the sheet, resulting in poor 0.2% proof stress along the TD [8]. Furthermore, it has been reported

V. M. Miller et al. (eds.), *Magnesium Technology 2021*, The Minerals, Metals & Materials Series,
https://doi.org/10.1007/978-3-030-65528-0_3

that the splitting of basal poles to one direction is not an ideal texture for further improvement of the room-temperature formability [9].

We have recently reported that good balance of a room-temperature formability and tensile properties can be realized in an Mg-3Al-0.4Mn alloy sheet via industrially viable rolling process [10]. The alloy sheet exhibits a large I. E. value of 8.2 mm, this needs to be further improved to compete with Al–Mg-Si alloy sheets having an I.E. value over 10 mm [11]. In this study, we have tried to optimize rolling temperature for the further improvement of the property, and the effect of rolling temperature on tensile properties, room-temperature stretch formability, microstructures, and texture of an Mg–Al-Mn alloy sheet has been investigated.

Experimental Procedure

An Mg-3.07Al-0.25Mn (mass%, AM30) alloy sheet, 270 mm width and 4 mm thickness, was prepared by a twin-roll casting. The sheet was cut into plates with ~130 mm in width and ~ 100 mm in length, and homogenized at 415 °C for 2 h and 500 °C for 12 h in a muffle furnace under an Ar atmosphere, followed by water quenching. The homogenized plates were rolled at four different rolling temperatures, namely 220, 240, 260, and 280 °C with ~20% thickness reduction per pass. The rolling was repeated six times without reheating to obtain rolled sheets with ~1 mm thickness. The rolling speed was fixed at 5 m/min. The as-rolled samples were annealed at 250 °C for 4 h in a muffle furnace, followed by water quenching.

To evaluate the room-temperature stretch formability of the annealed samples, an Erichsen cupping test was performed on rectangular samples with 60 × 60 mm^2 using a sheet metal testing machine (ERICHSEN, Model 100) having a hemispherical punch with 20 mm in diameter. The punch speed and blank-holder force were around 6 mm/min and 10 kN, respectively. Mechanical properties of the annealed samples were evaluated by a tensile test using an Autograph AG-50kNI (Shimadzu) at an initial strain rate of 10^{-3} s^{-1} and room-temperature. The loading directions were parallel to the rolling and transverse directions of the sheets (RD and TD), and the tensile specimens have 20 mm in gauge length and 4 mm in width. Both Erichsen cupping test and tensile test were repeated three times for reproducibility.

Microstructures and textures of the as-rolled and annealed samples were evaluated by an electron backscattered diffraction (EBSD) method on a scanning electron microscope (JEOL JSM-7000F) with TSL EBSD apparatus and a TSL OIM software. Distribution of second phase particles was also observed using a scanning electron microscope (JEOL IT-500) equipped with an energy dispersive X-ray spectroscopy (EDS) detector. The EBSD- and SEM-samples were metallographically prepared by SiC papers, 0.3 μm Al_2O_3 suspension, and 0.04 μm non-drying colloidal silica suspension in an automatic polisher (IKEGAMI SEIKI co., ISPP-1000). All observations were performed on the RD-ND planes, and the EBSD data were rotated to analyze them from the RD-TD planes.

Results and Discussion

Figure 1a and b show snapshots of fractured samples after the Erichsen cupping test of the annealed AM30 samples rolled at 220 and 280 °C. The I.E. values of the annealed samples are also summarized in Fig. 1c. The rolling at 220 °C gives the highest I.E. value of $8.3^{+0.3}_{-0.3}$ mm. The stretch formability gradually deteriorates with increasing the rolling temperature, and the I.E. values of the samples rolled at 240, 260, and 280 °C are $7.6^{+0.4}_{-0.7}$ mm, $7.1^{+0.4}_{-0.2}$ mm, and $4.2^{+0.4}_{-0.3}$ mm, respectively.

Figure 2a, b show nominal tensile stress strain curves of the annealed AM30 samples. Table 1 also summarizes their ultimate tensile strength (UTS), 0.2% proof stress (PS), and

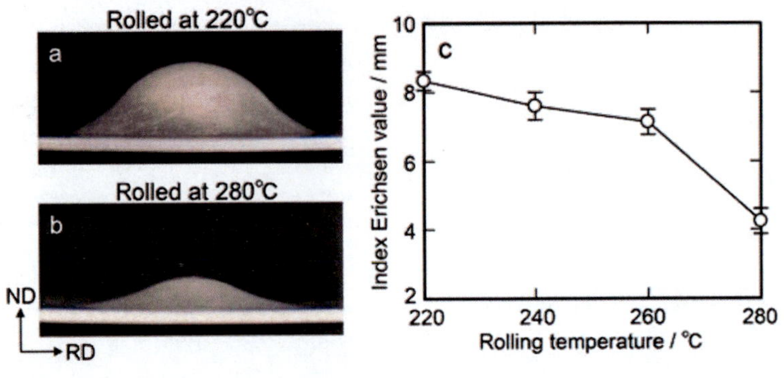

Fig. 1 Snapshots of fractured samples after the Erichsen cupping test of the annealed AM30 samples rolled at **a** 220 °C and **b** 280 °C. **c** Summary of the I.E. values of the annealed AM30 samples. (Color figure online)

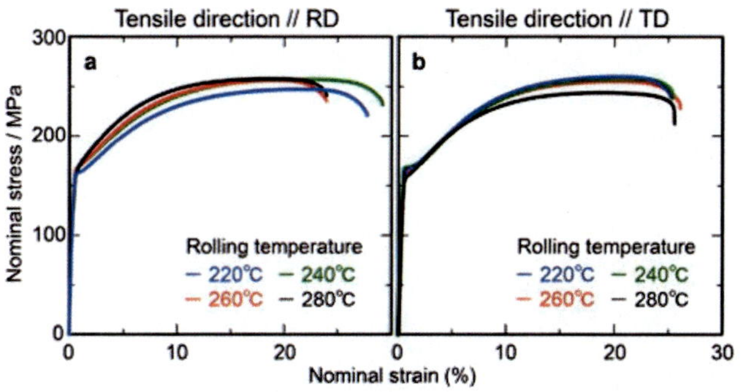

Fig. 2 Nominal tensile stress strain curves of the annealed AM30 samples stretched along the **a** RD and **b** TD. (Color figure online)

elongation to failure (EF). All samples show similar in-plane isotropic strengths. The UTS is about 250–260 MPa, and the PS is ~160 MPa. The EF in the samples rolled at 220 and 240 °C are over 25% for both RD and TD. Although the rolling at higher temperature slightly decreases the ductility, the samples still show good EF over 23% along the RD, and that along the TD is ~25%.

Figure 3 shows backscattered electron images of the annealed AM30 samples. Regardless of the rolling temperature, all samples contain secondary phase particles. The distribution of the particles is almost the same in all samples, and these particles are mostly submicron-sized and spherical-shaped. A small amount of rod-like particles also exist. They have 1–2 μm in length, and the diameter is less than 1 μm. These particles may be Al_8Mn_5 phases (rhombohedral, a = 1.2667 nm, c = 0.7942 nm [12]) as reported in [13].

Figure 4a, b, c, d show inverse pole figure maps of the annealed AM30 samples. Insets in these maps are corresponding (0001) pole figures with the maximum intensity of the (0001) poles. Intensity profiles of the (0001) poles measured along the RD are also displayed in Fig. 4e, and Table 2 summarizes average grain sizes and Schmid factors for (0001) $[11\bar{2}1]$ slips (basal slips). The grain size becomes finer by the rolling at low temperature. The sample rolled at 220 °C shows the finest grain size of 7.6 μm, while the rolling at 280 °C increases the grain size to 10.2 μm. The samples rolled at 220, 240, and 260 °C show similar RD-split texture feature with the maximum intensity of ~4 MRD. The tilting angles of the (0001) poles are ~ 25° in these samples, and they also show broadened angular distribution of the (0001) poles to the TD, leading to the relatively large Schmid factors (Table 2). Although the samples rolled at 220, 240, and 260 °C show similar texture feature, it should be mentioned that the fraction of strong basal components (grains with (0001) poles tilting 0°−10° from the ND) increases with increasing the rolling temperature. The sample rolled at 280 °C still exhibits RD-split texture feature; however, the tilting angle of the (0001) poles is only ~10°, and the sample shows the high maximum intensity of 6.6 MRD, resulting in lower Schmid factors than other samples.

To understand the microstructural changes as shown in Fig. 4, the EBSD was also done on the as-rolled condition.

Table 1 Summary of the UTS, PS, and EF of the annealed AM30 samples

Rolling temperature (°C)	Tensile direction	UTS [MPa]	PS [MPa]	EF (%)
220	RD	254^{+3}_{-7}	159^{+1}_{-1}	$26.1^{+2.0}_{-3.2}$
	TD	257^{+4}_{-6}	157^{+2}_{-1}	$26.1^{+4.8}_{-3.5}$
240	RD	255^{+2}_{-1}	158^{+3}_{-2}	$28.1^{+0.6}_{-1.1}$
	TD	258^{+1}_{-1}	164^{+1}_{-2}	$25.3^{+0.8}_{-0.5}$
260	RD	257^{+2}_{-1}	161^{+4}_{-3}	$23.3^{+0.6}_{-0.8}$
	TD	257^{+2}_{-2}	161^{+1}_{-1}	$26.1^{+1.2}_{-0.8}$
280	RD	253^{+5}_{-9}	159^{+3}_{-4}	$23.5^{+2.9}_{-2.7}$
	TD	251^{+10}_{-7}	157^{+4}_{-2}	$24.8^{+2.3}_{-2.6}$

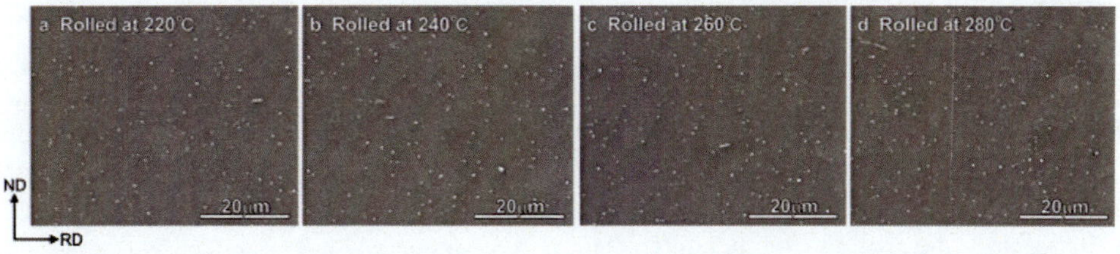

Fig. 3 Backscattered electron images of the annealed AM30 samples rolled at **a** 220 °C, **b** 240 °C, **c** 260 °C, and **d** 280 °C

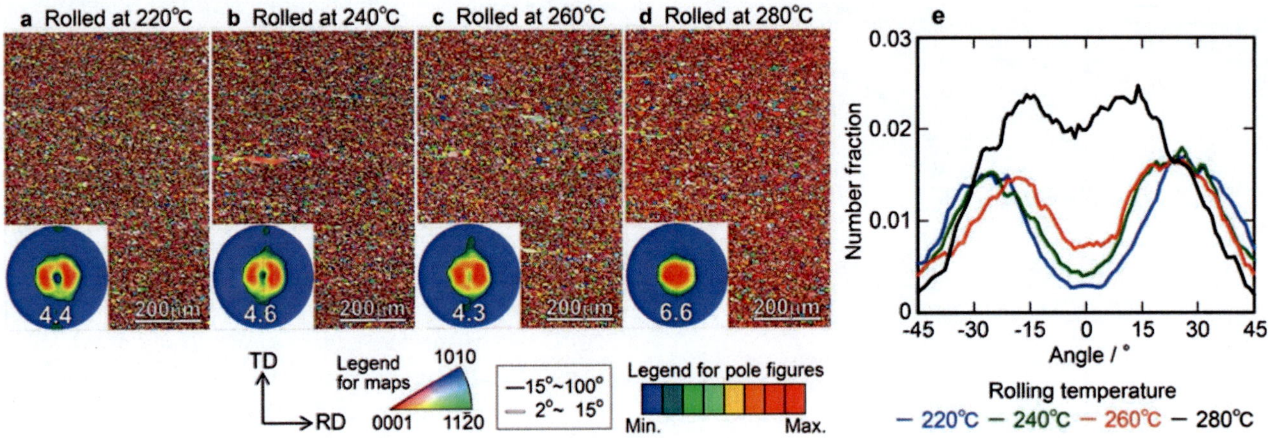

Fig. 4 Inverse pole figure maps and (0001) pole figures of the annealed AM30 samples rolled at **a** 220 °C, **b** 240 °C, **c** 260 °C, and **d** 280 °C. **e** Intensity profiles of the (0001) poles measured along the RD. (Color figure online)

Table 2 Average grain sizes and Schmid factors for basal slips in the annealed AM30 samples

Rolling temperature (°C)	Average grain size (μm)	Schmid factors for basal slips	
		Tensile direction//RD	Tensile direction//TD
220	7.6	0.29	0.28
240	7.8	0.28	0.28
260	8.9	0.27	0.27
280	10.2	0.24	0.24

Figure 5 shows inverse pole figure maps and (0001) pole figures of the as-rolled AM30 samples obtained from the whole analyzed region and the recrystallized grains (grains with grain orientation spread less than 0.5° are considered to be the recrystallized grains). All samples consist of large unrecrystallized grains and recrystallized grains, and Table 3 summarizes average sizes of the recrystallized grains, fractions of the recrystallized grains, and kernel average misorientation (KAM) values. The average recrystallized grain size and the areal fraction of recrystallized grains increase with increasing the rolling temperature, and the rolling at high temperature decreases the KAM value. They suggest that dynamic recrystallization is promoted by the rolling at high temperature. The (0001) poles of the as-rolled samples split to the RD with an angle of $\sim 10°$. The recrystallized grains show similar RD-split texture feature as the as-rolled samples, and the recrystallized grains also have strong basal texture component.

In this work, we have investigated the effect of the rolling temperature on the room-temperature stretch formability, tensile properties, and microstructures of the rolled and subsequently annealed AM30 alloy sheet. The large Index Erichsen value of $8.3^{+0.3}_{-0.3}$ mm has been obtained after the rolling at 220 °C; however, the stretch formability gradually decreases with increasing the rolling temperature. The AM30 alloy sheet rolled at 220 °C also exhibits in-plane isotropic tensile properties with moderate strengths and ductility; the UTS, PS, and EF are 254^{+3}_{-7} MPa, 159^{+1}_{-1} MPa, and $26.1^{+2.0}_{-3.2}$ % along the RD, and those for the TD are 257^{+4}_{-6} MPa, 157^{+2}_{-1} MPa, and $26.1^{+4.8}_{-3.5}$ %, respectively.

The excellent room-temperature stretch formability in the AM30 alloy sheet rolled at 220 °C is attributed to the weakly aligned (0001) poles to the RD, which promotes the basal slips during the stretch forming [14]. It is noted that the sheets rolled at 240 and 260 °C also show similar weak texture feature; however, the fraction of the strong basal texture component increases with increasing the rolling temperature. This may result in the deteriorated formability [10]. Although the splitting of the (0001) towards the RD still exists after the rolling at 280 °C, the tilting angle is $\sim 10°$, which limits the activation of basal slips. Therefore, the sheet rolled at 280 °C exhibits the limited room-temperature formability.

All alloy sheets show the in-plane isotropic properties. This is mainly attributed to the similarity of the activity of basal slips in both RD and TD. In addition, the fine grain structure with the average grain size less than 8 μm in the sheets rolled at 220 and 240 °C results in the moderate 0.2% proof stress of ~ 160 MPa and large elongation to failure over 25% [10, 14, 15]. The rolling at higher temperature

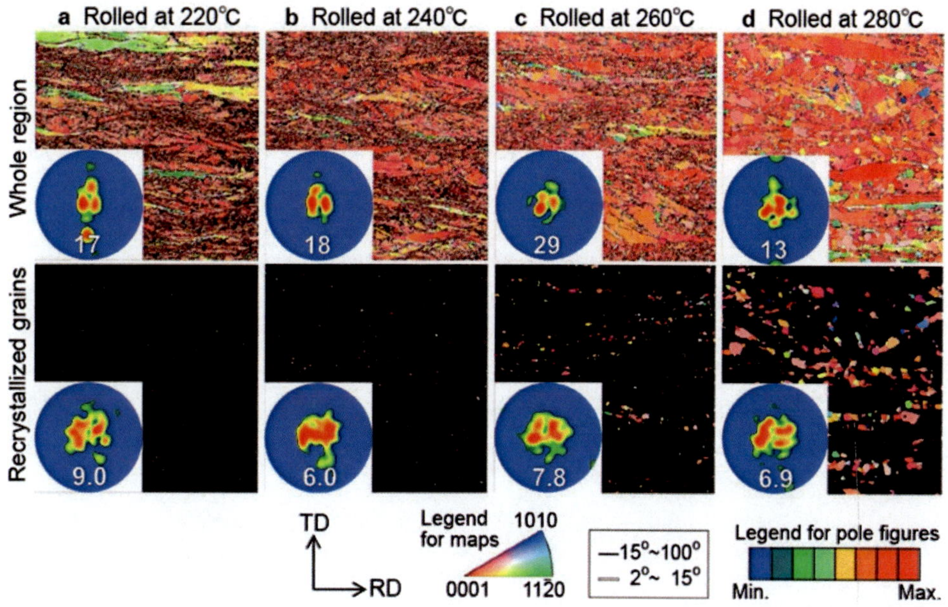

Fig. 5 Inverse pole figure maps and (0001) pole figures of the as-rolled AM30 samples obtained from the whole analyzed region and the recrystallized grains. (Color figure online)

Table 3 Average sizes of the recrystallized grains, fractions of the recrystallized grains, and kernel average misorientation (KAM) values of the as-rolled AM30 samples

Rolling temperature (°C)	Average recrystallized grain size (μm)	Areal fraction of recrystallized grains (%)	KAM value
220	0.9	0.3	1.2°
240	1.0	0.9	1.1°
260	1.9	3.6	1.0°
280	4.5	16.2	0.7°

increases the grain size; however, the sheets rolled at the temperature over 260 °C still show the moderate strengths and good ductility. This is probably due to the high fraction of strong basal texture component and activation of {10$\bar{1}$0}[11$\bar{2}$1] prismatic slips during the tensile test [15].

Formation of the strong basal texture component is strongly suppressed by the rolling at low temperature. The main reason may be the suppression of dynamic recrystallization during the rolling process. Dynamic recrystallization takes place during rolling process. The dynamically recrystallized grains generally form strong basal texture feature, and during the subsequent annealing, these grains grow larger by eating surrounding grains [10, 16], which suggest that the dynamic recrystallization and the growth of the grains promote the formation of the basal texture in magnesium alloy sheets. As shown in Fig. 5, the (0001) poles of the dynamically recrystallized grains tend to align parallel to the ND, and the fraction of the dynamically recrystallized grains increases with increasing the rolling temperature. For these reasons, the sheet rolled at 220 °C forms the lowest fraction of the basal texture component, and the basal texture component gradually increases with increasing the rolling temperature.

Summary

In this work, the effect of rolling temperature on a room-temperature stretch formability, tensile properties, microstructures, and textures in an annealed Mg-3.07Al-0.25Mn (mass%) alloy sheet was investigated. The rolling at 220 °C contributes to the development of weak alignment of the (0001) poles towards the rolling direction of the sheet. The low-temperature rolling also results in the low fraction of a strong basal texture component, leading to an excellent Index Erichsen value of 8.3 mm. Furthermore, the sheet forms fine grain structures with an average grain size of 7.6 μm; therefore, moderate 0.2% proof stress of ~160 MPa and high elongation to failure over 25% could be obtained for the rolling and transverse directions.

Acknowledgements This work was supported by JSPS KAKENHI Grant Numbers JP19K15321, JP18H03837, THE AMADA FOUNDATION (AF-2019037-C2), Advanced Low Carbon Technology Research and Development Program (ALCA), 12102886, and Nagaoka University of Technology (NUT) Presidential Research Grant.

References

1. Nakata T, Mezaki T, Xu C, Oh-ishi K, Shimizu K, Hanaki S, Kamado S (2015) Improving tensile properties of dilute Mg-0.27Al-0.13Ca-0.21Mn (at.%) alloy by low temperature high speed extrusion, J. Alloys Compd. 648:428-437
2. Nakata T, Xu C, Ajima R, Shimizu K, Hanaki S, Sasaki TT, Ma L, Hono K, Kamado S (2017) Strong and ductile age-hardening Mg-Al-Ca-Mn alloy that can be extruded as fast as aluminum alloys, Acta Mater. 130:261-270
3. Kaiser F, Bohlen J, Letzig D, Kainer KU, Styczynski A, Hartig C (2003) Influence of rolling conditions on the microstructure and mechanical properties of magnesium sheet AZ31, Adv. Eng. Mater. 5:891-896
4. Mackenzie LWF, Pekguleryuz M (2008) The influences of alloying additions and processing parameters on the rolling microstructures and textures of magnesium alloys, Mater. Sci. Eng.: A 480(1–2):189–197
5. Suh BC, Kim JH, Bae JH, Hwang JH, Shim MS, Kim NJ (2017) Effect of Sn addition on the microstructure and deformation behavior of Mg-3Al alloy, Acta Mater. 124:268-279
6. Huang X, Suzuki K, Chino Y, Mabuchi M (2011) Improvement of stretch formability of Mg–3Al–1Zn alloy sheet by high temperature rolling at finishing pass, J. Alloys Compd. 509 (28):7579-7584
7. Emley EF (1966) Principles of Magnesium Technology, Pergamon Press Ltd., Oxford
8. Yuasa M, Miyazawa N, Hayashi M, Mabuchi M, Chino Y (2015) Effects of group II elements on the cold stretch formability of Mg–Zn alloys, Acta Mater. 83:294-303
9. Suh BC, Shim MS, Shin KS, Kim NJ (2014) Current issues in magnesium sheet alloys: Where do we go from here?, Scr. Mater. 84-85:1-6
10. Nakata T, Xu C, Ohashi H, Yoshida Y, Yoshida K, Kamado S (2020) New Mg–Al based alloy sheet with good room-temperature stretch formability and tensile properties, Scr. Mater. 180:16-22
11. Hirth SM, Marchall GJ, Court SA, Lloyd DJ (2001) Effects of Si on the aging behaviour and formability of aluminium alloys based on AA6016, Mater. Sci. Eng.: A 319–321:452–456
12. Villars, P (1997) Pearson's Handbook Desk Edition Crystallographic Data for Intermetallic Phases Volume 1, ASM International, Materials Park
13. Luo AA, Zhang C, Sachdev AK (2012) Effect of eutectic temperature on the extrudability of magnesium–aluminum alloys, Scr. Mater. 66(7):491-494
14. Bian MZ, Sasaki TT, Nakata T, Kamado S, Hono (2018) Effects of rolling conditions on the microstructure and mechanical properties in a Mg–Al–Ca–Mn–Zn alloy sheet, Mater. Sci. Eng.: A. 730:147–154
15. Koike J, Kobayashi T, Mukai T, Watanabe H, Suzuki M, Maruyama K, Higashi K (2003) The activity of non-basal slip systems and dynamic recovery at room temperature in fine-grained AZ31B magnesium alloys, Acta Mater. 51(7):2055-2065
16. Nakata T, Xu C, Uehara Y, Sasaki TT, Kamado S (2019) Origin of texture weakening in a rolled ZEX4101 alloy sheet and its effect on room temperature formability and tensile property (2019) J. Alloys Compd. 782:304-314

Microstructure and Fracture Toughness of an Extruded Mg-Dy-Nd-Zn-Zr Alloy Influenced by Heat Treatment

Petra Maier, Benjamin Clausius, Charis Joy, Roman Menze, Benjamin Bittner, and Norbert Hort

Abstract

The influence of microstructural changes induced by heat treatment on fracture toughness is investigated for a resorbable Mg-Dy based alloy. The initial hot-extruded condition is a fine-grained Resoloy® (Mg–Dy–Nd–Zn–Zr) alloy consisting of lamellar LPSO structures within the matrix. Solution heat treatment causes grain growth and the formation of blocky LPSO phases. The amount of the lamellar LPSO structures reduces. Quasi-static C-ring tests with and without Ringer solution were used to evaluate force–displacement curves and their fracture energy. The coarser-grained alloys tend to twin under plastic deformation, which is influencing the crack propagation. Blocky LPSO phases clearly hinder crack growth. The fine-grained extruded condition shows the highest force and displacement values to induce the crack, the solution heat-treated microstructure consisting of a good balance of grain size, matrix, and blocky LPSO phases and twins show highest fracture energy. Even if there might be an absorption of hydrogen, the ductility under stress corrosion is high.

Keywords

Magnesium • RESOLOY® • Fracture toughness • Crack propagation • LPSO phases

P. Maier (✉) · B. Clausius · C. Joy
University of Applied Sciences Stralsund, Zur Schwedenschanze 15, 18435 Stralsund, Germany
e-mail: petra.maier@hochschule-stralsund.de

R. Menze · B. Bittner
MeKo Laserstrahl-Materialbearbeitungen E.K., Sarstedt, Germany

N. Hort
Helmholtz-Zentrum Geesthacht, Geesthacht, Germany

Introduction and Motivation

Magnesium alloys show significant promise for being applied as absorbable metals in implants, which are only needed for temporary fixation. Rare earth elements are used to achieve high strength with an acceptable biological performance, like biocompatibility and enhancement of bone formation [1–3]. Resoloy®, a Mg–Dy–Nd–Zn–Zr alloy high in Dy, was developed specifically for absorbable implants by MEKO Laser Materials Processing and the Helmholtz-Zentrum Geesthacht in Germany and internationally patented [4]. Resoloy is an alloy based on Mg-Dy with excellent strength and ductility. The corrosion rate depends on many factors, like grain size, thermomechanical history, and surface quality, but is relatively low [5]. The high elasticity of Resoloy provides excellent fatigue life in air [6, 7]. Dy, having a high solubility in Mg, can adjust in combination with other alloying elements, like Gd, Zr, both the mechanical and corrosion properties by heat treatment, and according to in vitro studies Mg-Dy alloys show good cytocompatibility [8, 9]. Also for Nd as an alloying element in Mg it has been reported, that it improves the corrosion resistance by demonstrating a good in vivo biocompatibility [10]. Mg–Nd–Zn–Zr alloy exhibits even much better mechanical properties and lower corrosion rate than WE43 and AZ31 alloys and indicates that the as-extruded condition meets the requirement of cell toxicity for biomaterials in a short-term study [11]. Mg–RE–Zn alloys, where RE are rare earth elements like Y, Gd, Tb, Dy, Ho, Er, Tm, form novel long-period stacking-ordered (LPSO) structures and show improved strength, fracture toughness, corrosion resistance, and fatigue strength and by deformation kinks also remarkable deformability [12]. Previous studies on cast and extruded Resoloy show increased amount of LPSO phase with increasing solution heat treatment time [13]. With optimized heat treatment lamellar LPSO structures within the matrix form and provide a uniform corrosion. The extruded Resoloy in study [14] improved stress corrosion

V. M. Miller et al. (eds.), *Magnesium Technology 2021*, The Minerals, Metals & Materials Series,
https://doi.org/10.1007/978-3-030-65528-0_4

behavior after a short-term annealing, whereas pitting corrosion appeared after solution heat treatment at 500 °C for 24 h, 48 h, and 72 h. The microstructure changed during solution heat treatment from fine grains with a high amount of lamellar and blocky LPSO structures to large grains with a lower amount of lamellae and new formed blocky LPSO as well as grain free of LPSO structures. However, during stress corrosion no crack initiated from these corrosion pits. Beside the requirement of a moderate, homogenous corrosion rate to avoid strong hydrogen evolution and pitting corrosion, biomedical applications need a certain fracture toughness and should not be susceptible to the stress corrosion cracking. ZX50, WZ21, and WE43, investigated by slow strain rate tensile testing in a simulated human body fluid, indicated a susceptibility to stress corrosion cracking [15], whereas EV31A is less susceptible due to a robust and stable film of mixed oxides of Zr, Nd, and Gd [16]. In general, the nature and stability of the oxide/hydroxide films and the intensity of the loading influences the ease of disruption of the protective film, which will increase the opportunity for hydrogen entry. Beside corrosion pits also microstructural features like precipitates and phase boundaries, increasing local stress level, or slip induced film rupture are responsible for crack initiation—accelerated by hydrogen embrittlement and anodic dissolution at the slip steps [17].

As much as some microstructural features cause crack initiation, low angle grain boundaries [18], twinned grains [19, 20], second phases [18], and LPSO structures [21, 22] can also hinder crack propagation and increase fracture toughness. Charpy-tests [23], slow rate tensile tests [24], or U-bent tests [17] are often used to determine fracture toughness and stress corrosion cracking. However, this study applies C-ring compression test to investigate stress corrosion and fracture toughness. Similar tests have been already done on the Mg–Dy–Nd alloy [25], but are in general rarely applied to magnesium alloys yet. The fracture energy during crack propagation, which is mainly influenced by twins, is used to discuss the fracture toughness. Like found in Resoloy, corrosion pits in Mg10Dy1Nd during static C-ring compression tests at a moderate deformation 20% did to lead to cracks. Furthermore, the displacement at fracture implies a rather ductile behavior—indicating a protective film by the passive layer. The role of LPSO phases, grain boundaries, and twins during crack propagation will be of interest. Dislocations in twinned regions or grains with lamellae LPSO structures will have short distances to move and therefore will pile up at these obstacles and can lead to crack initiation and reduced toughness [26]. The stress pile up will depend on the coherence of slip planes.

Experimental

The Resoloy alloy was cast at the Helmholtz-Zentrum in Geesthacht, Germany. After a short annealing at 500 °C for 15 min, tubes were indirectly hot-extruded at the Extrusion Research and Development Center TU Berlin at an overall temperature of 400 °C, a ram speed of 1.5 mm/s and an extrusion ratio of 19:1. The outer diameter of the as-extruded tubes was 35 mm, and the wall thickness was 5 mm. The chemical composition Dy and Zn were analyzed by using X-Ray micro fluorescence M4 Tornado (Bruker, Billerica, MA, USA), Nd and Zr by using a spark optical emission spectroscopy Spectrolab M12 Hybrid (Ametek-Spectro, Kleve, Germany) and the impurities Fe, Cu, Ni were analyzed by using atomic absorption spectrometry 240FS AA (Agilent, Santa Clara, CA, USA). Micrograph were obtained on a Leica DMi8 A (Leica Microsystems GmbH, Wetzlar, Germany). Short annealing was performed at 200 °C for 4 h, acting more as a stress relieve, and solution heat treatment were done at 500 °C for 24, 48, and 72 h.

C-ring samples with a width of 10 mm were machined from the extruded tubes with an outer diameter of 34 mm and a wall thickness of 2 mm. Figure 1a shows their position in the initial extruded tubes according to [25]. C-ring samples were compressed with 2 mm/min in a testing machine up to 17 mm in air (geometrical restriction) and up to fracture in Ringer solution, see Fig. 2b. Before corrosion, the samples were cleaned in an ultrasonic bath in ethanol. During C-ring compression tests, force–displacement curves were monitored. The deformation can be calculated in % according to the deformation of the diameter [27] by setting the displacement (the initial height of 34 mm minus the height of the deformed C-ring) in relation to the initial height, which provided information on the ductility. A maximum deformation of 50% was possible. When a deformation of 40% is exceeded, the material is supposed to have ductile behavior. A transition from tensile to compressive stresses from the outer to the inner surface developed in the C-rings upon loading and in case of a crack growth, it propagates through these corresponding areas. For the micrographic investigation special attention was paid to microstructural features like branched or secondary cracks and its interaction with twins and LPSO phases, either as lamellar matrix structures or as blocky phases.

The following values of the force–displacement curves were used to describe the fracture behavior: the force at which the fracture started (evaluated by the reduction in the force with increasing displacement), its displacement value, and its resulting fracture energy (the area underneath the force-displacement curve during crack propagation up to its

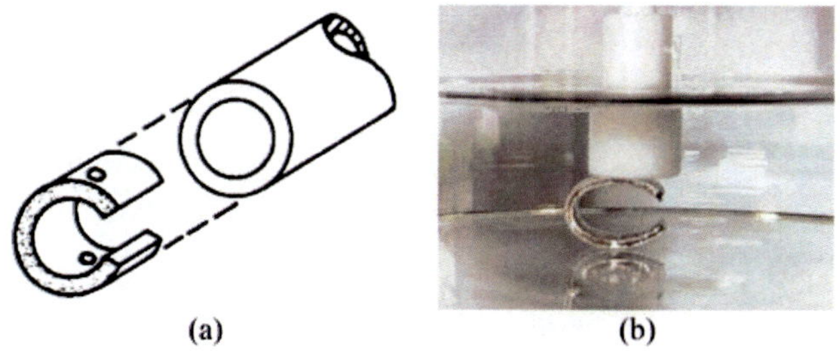

Fig. 1 C-ring sample: **a** position in the initial extruded tube and **b** in compression test in Ringer solution

turning point towards the end of the curve). Three C-ring samples were tested at each condition and the average values were reported.

Results and Discussion

The Resoloy alloy consists of 12.63 wt.% Dy, 1.05 wt.% Nd, 0.94 wt.% Zn, and 0.075 wt.% Zr. Fe, Cu, and Ni do not exceed 0.001 wt.%. Figure 2 shows the changes in microstructure by the heat treatment: the short annealing at 200 °C for 4 h does not change the microstructure significantly, see Fig. 2a, b. According to [14] the grain size increases by solution heat treatment from below 10 µm in the as-extruded condition to around 45 µm at 48 h. Due to the development of the blocky LPSO phases being considered as small grains the average grain size reduces during annealing up to 72 h, even coarser grains develop.

Figure 3a shows the force–displacement curves of representative C-ring tests of the Resoloy alloy in the different conditions in Ringer solution compared to the as-extruded Resoloy in air. Figure 3b shows each individual part of the force–displacement curves during crack propagation, from crack initiation (declining force) to turning point at the end of the curve. The force–displacement curve of one pre-deformed as-extruded Resoloy C-ring in air (compressed to 11 mm) is shown in comparison to the compression in Ringer solution in Fig. 3c. The displacement at fracture is shifted to 17 mm – so pre-deformation in air results in higher ductility before the fracture starts in Ringer solution. Figure 3c shows a small stress relaxation at 11 mm (see grey arrow at dotted line).

Without the exposure to Ringer solution at least 50% deformation (displacement of 17 mm) is possible without crack initiation. Like mentioned before, material which exceeds a C-ring compression of 40% deformation, can be classed as ductile. Testing in Ringer solution reduces the displacement at crack initiation (Fig. 3a). The passive film, which protects Resoloy at a certain deformation [14], becomes porous and allows contact between Ringer solution and Resoloy without an oxide layer. Hydrogen can penetrate

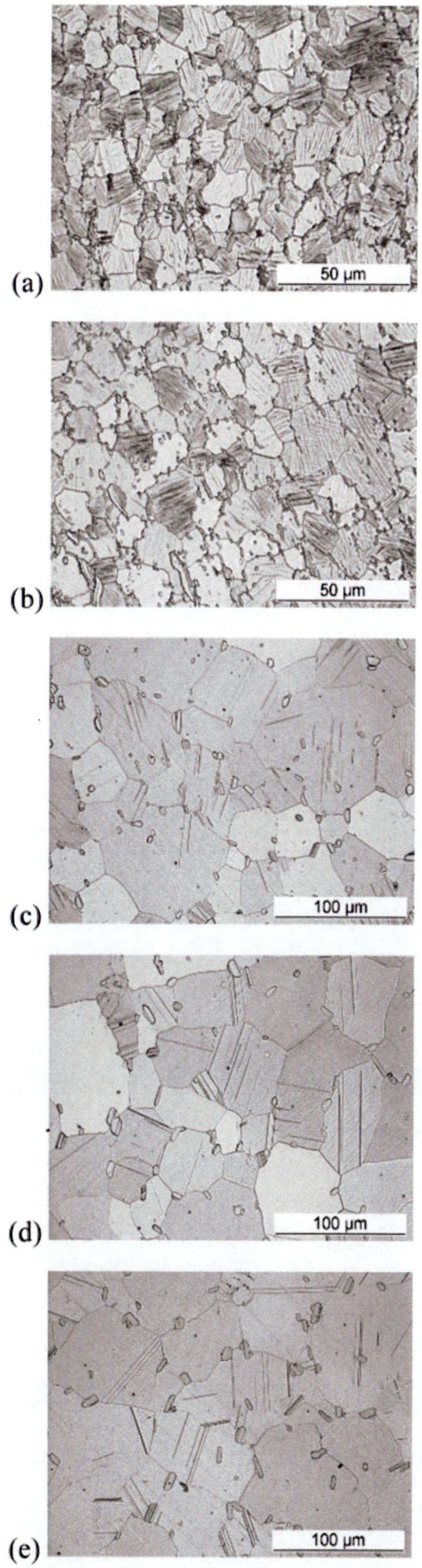

Fig. 2 Microstructure of the Resoloy alloy in the different conditions (transverse cross-section): **a** as-extruded, **b** heat-treated 200 °C 4 h, **c** solution heat-treated 500 °C 24 h, **d** solution heat-treated 500 °C 48 h and **e** solution heat-treated 500 °C 72 h

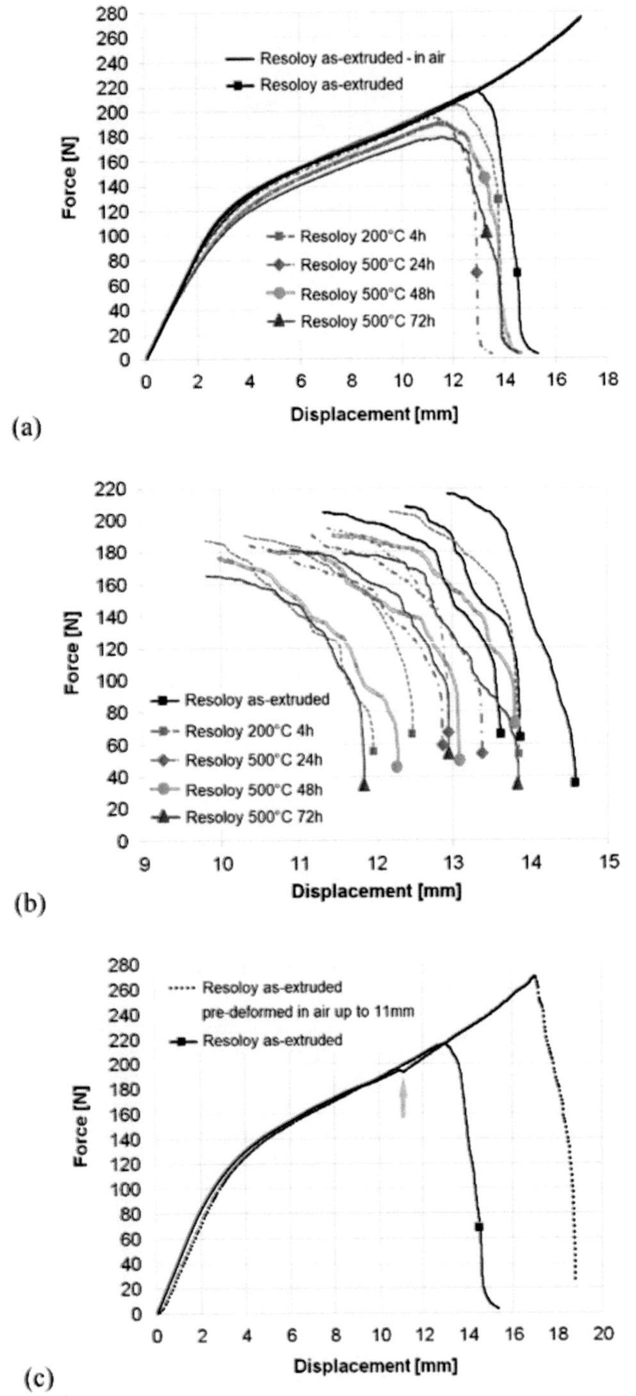

Fig. 3 Force–displacement curves of C-ring tests under compressive loading in Ringer solution for the Resoloy alloy in the different conditions: **a** compared to testing in air up to 17 mm, **b** during crack propagation from crack initiation to turning point at the end of the curve, and **c** compared to pre-deformed as-extruded Resoloy up to 11 mm and the tested up to fracture

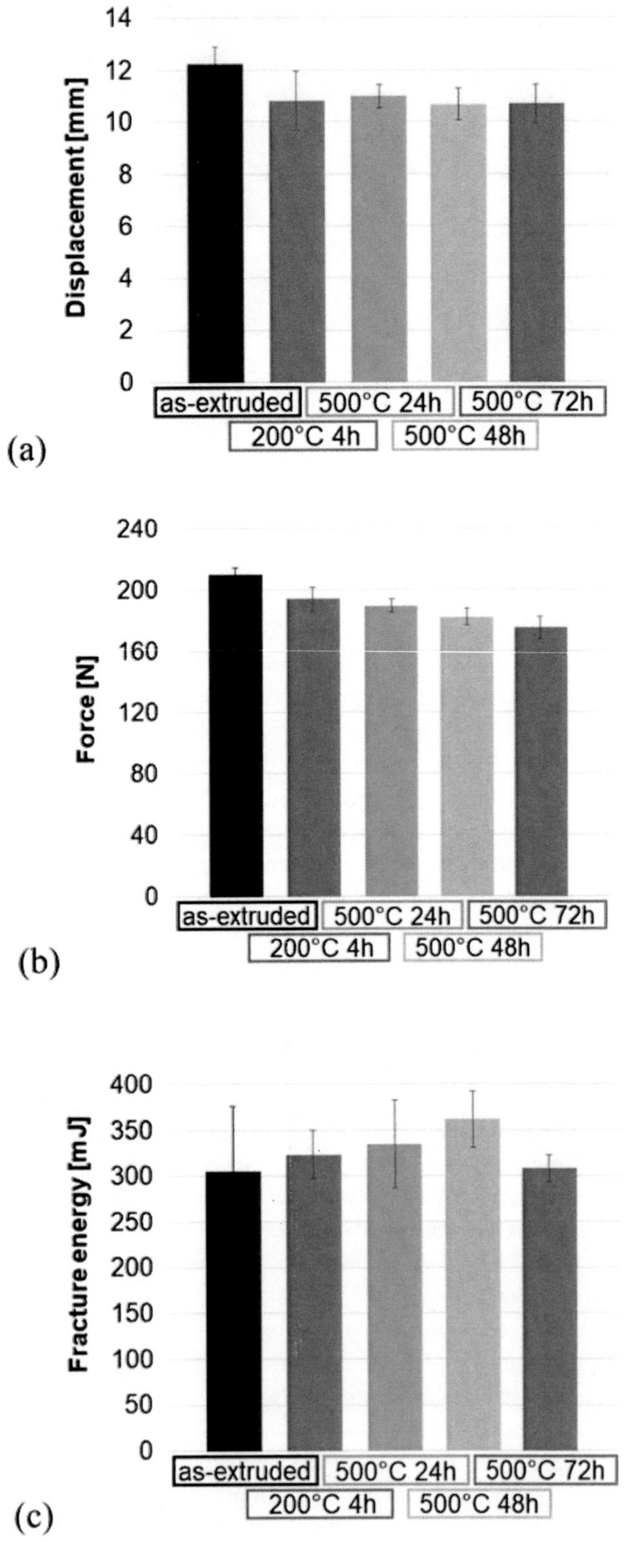

Fig. 4 Properties from force–displacement curves: **a** displacement at crack initiation, **b** force at which fracture starts, and **c** its fracture energy

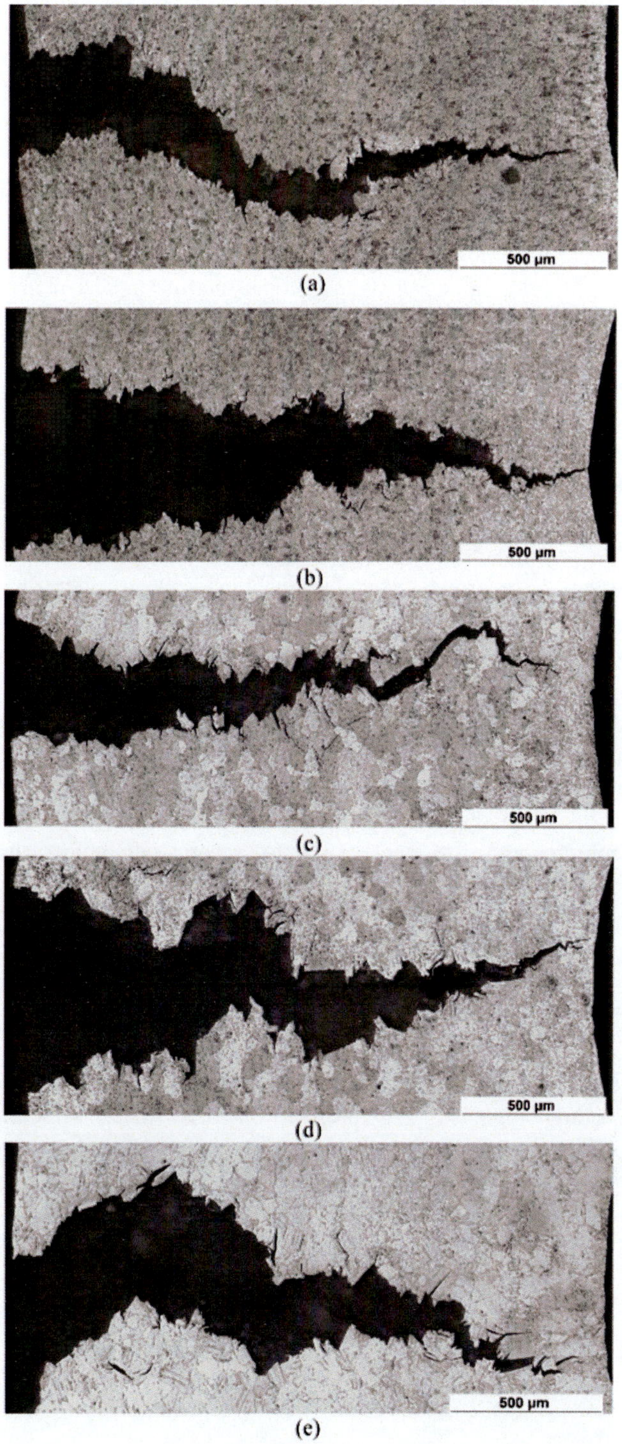

Fig. 5 Cross-sectional micrograph of the crack path in the Resoloy C-ring samples in Ringer solution in the different conditions: **a** as-extruded, **b** heat-treated 200 °C 4 h, **c** solution heat-treated 500 °C 24 h, **d** solution heat-treated 500 °C 48 h, and **e** solution heat-treated 500 °C 72 h

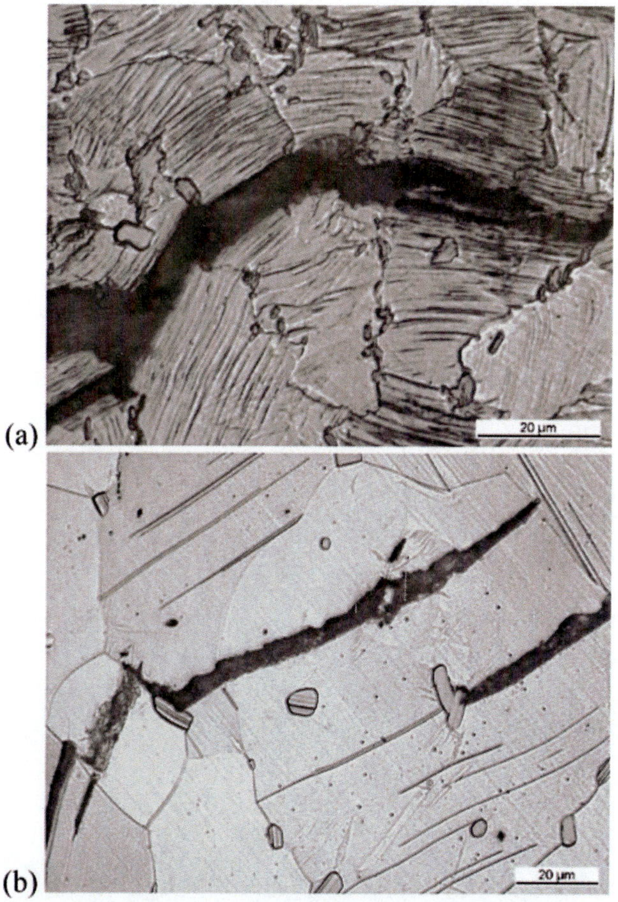

Fig. 6 Micrographs of crack path in higher magnification: **a** as-extruded and **b** solution heat-treated 500 °C 48 h

the microstructure, causing crack initiation by embrittlement and crack propagation. However, the as-extruded condition still reaches an average displacement of 12.2 mm (~36%) and the heat-treated conditions around 11 mm (~32%), see Fig. 4a. The force at which the fracture starts reduces by heat treatment (Fig. 4b) according to the hardness reduction see in previous research [14], which mainly goes back to grain growth and dissolvent of lamellar matrix LPSO structures. The fracture energy increases by solution heat treatment at 500 °C, especially at 48 h, see Fig. 4c. Comparing the fracture energy to the Mg–Dy–Nd system with similar composition (without Zn and Zr) from study [25], Resoloy reaches values twice a high. The fine-grained microstructure with lamellar matrix LPSOs in the as-extruded condition as well as the blocky LPSO phases of the heat-treated RESOLOY act positively on the fracture toughness. The reduced fracture energy at 72 h can be explained by the growth of the grains and the loss of lamellar matrix LPSOs [14].

Figure 5 shows representative cross-sectional micrographs of the crack path in the Resoloy C-rings deformed in Ringer solution. Due to stopping the compression test when the turning point at the end of the force–displacement curves was exceeded, the samples did not fail completely and the tip can be investigated.

All the cross-sectional micrographs are showing sub-cracks of the main cracks—they become very pronounced at solution heat treatment at 500 °C towards longer annealing times, see Fig. 5e. The sub-cracks branch off at grain boundaries and the crack gets redirected by blocky LPSO phases after stress pile up. When there are coarser grains free of lamellar matrix and blocky LPSO phases, the crack runs continuously across the grain and is deflected at the approaching grain boundary or rather by the other orientation of the adjacent grain. However, more often LPSO lamellae, blocky phases, twins and grain boundaries hinder crack propagation and the crack branches or get redirected [28]. In the micrographs of higher magnification it can be clearly seen, that the crack follows the direction of the LPSO lamellae of the individual grain and is sidetracked by blocky LPSO phases (Fig. 6a showing a crack path in the as-extruded condition).

Figure 6b shows the crack path in the heat-treated Resoloy at 500 °C for 48 h. The micrograph clearly shows the grain growth and that the crack follows the thinned out matrix LPSO lamellae and gets hindered by blocky LPSO phases within the grain or at grain boundaries and at grain boundaries itself.

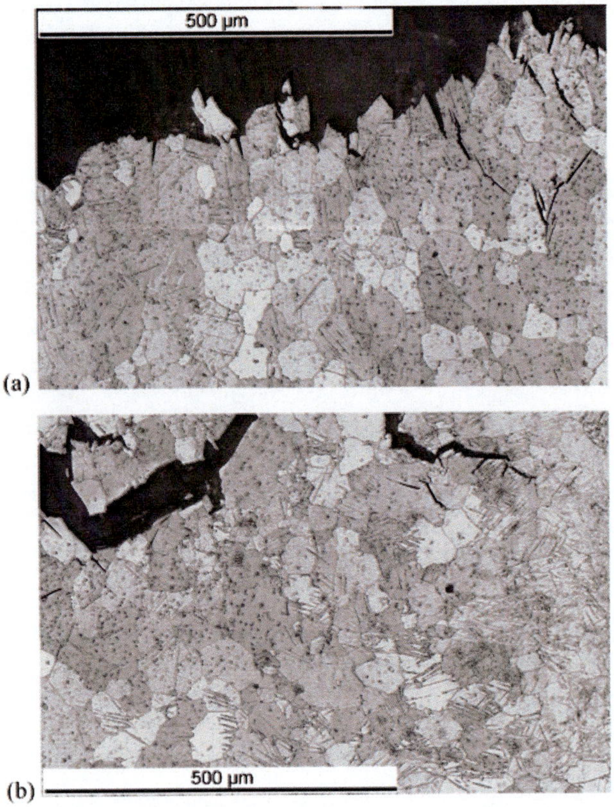

Fig. 7 Micrographs of the plastic deformed regions of the C-ring nearby the crack of solution heat-treated Resoloy at 500 °C for 24 h: **a** tensile side and **b** compression side

Some blocky LPSO phases cause crack propagation along the interface of the phase and the matrix and some guide the crack across the phase; depending on the size of the blocky phase and of the angle between the initial crack path and the orientation of the LPSO planes with the phase. According to the bar chart in Fig. 4c, the heat-treated Resoloy at 500 °C for 48 h shows the highest fracture toughness among the conditions investigated in this study. The developed microstructure, consisting of lamellar matrix and blocky LPSO phases and grains of limited size, seems to absorb the highest energy. The microstructural feature causing the highest crack deflection by crack branching, some blunting, and bridging, also seen in [28]. More detailed investigation of the role of the LPSO phases, also their mechanical properties, are in progress.

By examining the microstructure along the crack path of the solution heat-treated conditions at 500 °C, where the grains have grown, from the tensile side where the crack initiated to the compression side it can be seen that the amount of twins increase towards the last-mentioned side. RE in Mg alloys is known to reduce the tension–compression asymmetry [29] by suppression twinning under compression. During C-ring deformation crack opening occurs at the tensile side (less strain accumulation within grains by stress relief during crack growth), while the grains at the compression side are exposed to larger strain before the crack “arrives”. Therefore, these grains seem to twin more. At this point twinning will have also induced plasticity.

Representative micrographs for an annealing time of 24 h allow a good comparison: Fig. 7a shows grains without twins at the tensile side and Fig. 7b grains at the compression side, which are much stronger twinned. These twin boundaries either hinder the crack growth by redirection and leading to a zig-zag propagation or by accumulating

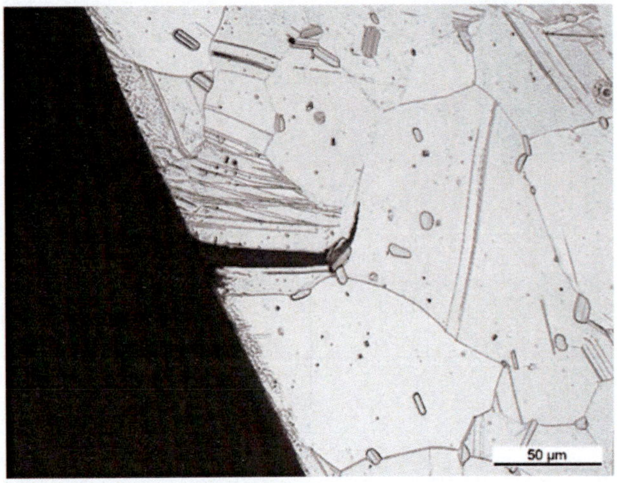

Fig. 8 Micrograph of crack initiation at lamellar matrix LPSO structures in solution heat-treated Resoloy at 500 °C for 72 h

dislocation on new slip traces, finally acting as secondary micro-crack initiation areas [30, 31], and all mechanisms are increasing the fracture energy. The drop in force within the force–displacement curve, see Fig. 3b, shifts towards higher displacements. To find out if larger grains within the solution heat-treated condition tend to twin more or if the micro-cracks also initiate at lamellar matrix LPSO structures further investigations are needed. However, the finer grained as-extruded and short-time annealed (200 °C for 4 h) microstructure does not show extensive twinning—neither at the tensile nor at the compression side. Independent of twinning, the compression side will be more strain-hardened than the tensile side where the crack initially opens. Microhardness measurement would help to establish the amount.

Like seen in Fig. 3c, pre-deformation in air up to 11 mm, which is high in the strain-hardened region, results in higher overall ductility before the fracture starts when subsequent exposure to Ringer solution (its fracture energy is with 340 mJ above the average value of 305 ± 67 mJ). The passive layer, which is ruptured during deformation to 11 mm displacement in air, immediately renews and crack initiation is delayed by preventing hydrogen embrittlement during further C-ring deformation—until the passive layer becomes porous again and bare metal surface is exposed to Ringer solution. Hydrogen diffuses into the Mg matrix and accumulates at particular sites, the so-called low surface energy planes [32]—grain boundaries and basal planes are some of them. If twins or lamellar LPSO structures in Resoloy are pathways for hydrogen, diffusion leading to micro-cracking needs to be further investigated. With increasing stress, places like these are found to act preferentially as crack initiation sites [32]. The fracture energy of the pre-deformed C-ring and then compressed in Ringer up to fracture is in the upper range compared to the initial C-ring compression. The crack is expected to propagate through a higher strain-hardened and twinned microstructure (starts at higher force and displacement).

Preliminary examination on crack initiation sites allow already the following statement: either twin boundaries or lamellar matrix LPSO structures are responsible for crack formation. They might act a slip bands and accumulate dislocations. Figure 8 shows a short crack 2 mm off the main crack, initiated at a lamellar matrix LPSO structure, whose neighboring grains are crossed by twins. The crack is, after following the linear LPSO structure for approximately 50 μm, redirected by a blocky LPSO phase and follows then a linear LPSO structure to carry on with, see its parallel lamellae ~50 μm on the right of the crack tip. The blocky LPSO phase bridged the crack from one grain into the neighboring grain due to its favorable orientation. In case of an off-orientation of the lamellae within the blocky LPSO phase, its (then weaker) interface to the Mg-matrix bridges the crack into the neighboring grain. However, the crack blunts at the obstacle before growing, which is increasing the fracture energy. With or without the LPSO structures the grain boundaries serve as an obstacle to crack growth, which implies in general to keep the grain size small; for strength and ductility.

These preliminary findings will undergo further examination, on the one hand to improve statistics and on the other hand to extent this study by fractography using scanning electron microcopy or surface topography scanner.

Summary

In this study the microstructure and its fracture toughness of RESOLOY, a magnesium resorbable alloy based on Mg–Dy alloyed with Nd, Zn, Zr, modified by annealing and solution heat treatment was investigated. C-ring compression tests were undertaken to determine the strength, ductility, and in case of a fracture the fracture energy and the interaction of the crack propagation with microstructural features were studied. The hot-extruded and short-time annealed Resoloy alloy, being fine-grained and consisting of lamellar LPSO structures within the matrix, show the highest bending strength in C-ring compression test—in general and at fracture initiation. Solution heat treatment causes grain growth and the formation of blocky LPSO phases by reduction of the amount of the lamellar LPSO structures. The ductility in air is very high and remains quite high in Ringer solution, outstanding for the as-extruded condition. Force–displacement curves offer statements on strength and ductility at crack initiation (highest for the as-extruded condition) and fracture energy during crack growth (highest for solution heat-treated condition at 500 °C for 48 h).

The coarser-grained, heat-treated conditions twin under plastic deformation, whose boundaries on top of grain boundaries and LPSO phases hinder crack growth by redirection at higher stresses. Blocky LPSO phases hinder crack growth by either bridging when phase gets crossed or by blunting followed by propagating along the interface. The microstructure consisting of a good balance of grain size being rather small, matrix, and blocky LPSO phases and twins show the highest fracture energy, in this study found in the heat-treated condition at 500 °C for 48 h.

Not to risk any inhomogeneous corrosion behavior, it is wise to apply heat treatment parameters at which the lamellar matrix LPSO structures have no completely disappeared—previous research has shown that they reduce the corrosion rate.

Acknowledgements The authors acknowledge the support of Jens Wicke (UASS) for the metallographic investigations.

References

1. M.P. Staiger, A.M. Pietak, J. Huadmai and G. Dias, Biomater, 27, 9, 1728-1734 (2006).
2. Y.F. Zheng, X.N. Gu and F. Witte, Mat. Sci. Eng. R, 77, 1 (2014).
3. D. Lui, D. Yang, X. Li and S. Hu, J Mater Res Technol, 8, 1, 1538-1549 (2019).
4. US patents: 9,566,367 B2 & 9,522,219 B2, EU patents: 2744531 & 2744532
5. Meko Laser Materials Processing, www.meko.de/RESOLOY
6. Fort Wayne Metals, www.fwmetals.com/RESOLOY-a-magnesium-alloy-for-absorbable-devices
7. P. Maier et al., Magnesium Technology, 175–181 (2019).
8. L. Yang et al. Acta Biomater. 9, 8499-8508 (2013).
9. F. Feyerabend et al., Acta Biomater. 6, 1834 (2010).
10. R. Eifler et al., J Mater Sci, 27, 25 (2016).
11. X. Zhang, G. Yuan, J. Niu, P. Fu, W. Ding, J Mech. Behavior Biomed. Mater., 9, 153-162 (2012).
12. D. Xu, E. Han and Y. Xu, Progress in Natural Science: Mater. Int., 26, 2, 117-128 (2016).
13. A Steinacker et al., Cells and Materials 32, 6, 22 (2016).
14. P. Maier, A. Steinacker, B. Clausius and N. Hort, JOM, 72, 5, 1870-1879 (2020).
15. L. Choudhary, R.K. Singh Raman, J. Hofstetter and P.J. Uggowitzer, Mater. Sci. Eng. C 42, 629–636 (2014).
16. B.S. Padekar, R.K. Singh Raman, V.S. Raja and L. Paul, Corros. Sci., 71, 1–9 (2013).
17. J. Ninlachart, N. Shrestha and K.S. Raja, Journal of Magnesium and Alloys, in press
18. X.S. Wang, in Special Issues on Magnesium alloys, W. Monteiro (Ed.), 68–108 (2011).
19. S.H. Lu et al., Mater. Design 191, 108600 (2020).
20. P. Maier et al., Mater. Sci. Forum 828-829, 311-317 (2015).
21. Z.Z. Peng et al., Mater. Sci. Eng. A 687, 211-220 (2017).
22. L. Wu, H. Li and Z. Yang, Trans. Nonferrous Met. Sci. China 27, 1026-1035 (2017).
23. Y. Chino et al. Mater. Trans. 52, 6, 1123-1126 (2011).
24. L. Choudhary and R.K. Singh Raman, Eng. Frac. Mech. 103, 94–102 (2013).
25. P. Maier, B. Clausius, J. Wicke and N. Hort, Metals, 10, 584 (2020).
26. Y. Mine et al., Mater. Sci. Eng. 570, 63-69 (2013).
27. B. Rutkowski et al., J. Eur. Ceram. Soc. 31, 493-499 (2011).
28. X. Shi et al., Materials 12, 498 (2019).
29. Y.Q. Chi et al., Mater. Design 170, 107705 (2019).
30. L. Meng, P. Yang, Q. Xie and W. Mao, Mater Transaction, 49, 4, 710-714 (2008).
31. L. Zhang et al., Mat. Sci. Eng. A, 742, 564-572 (2019).
32. S.D. Wang et al., Sci. Rep. 6, 29471 (2016).

The High-Solution Design of Magnesium Alloys

Jun Wang, Yuan Yuan, Xiongying Cheng, Tao Chen, Bin Jiang, Dajian Li, Aitao Tang, Torben Boll, and Fusheng Pan

Abstract

The broad applications of Mg alloys are still limited by poor mechanical properties and poor corrosion resistance. Our group has found strength, ductility, as well as corrosion resistance of Mg alloys can be synergic enhanced by solution elements in primary Mg phase. However, most elements only show limited solubility in α(Mg) phase. We proposed that by dissolving multi-elements in the α phase, the enhanced entropy may bring enlarged solution and cocktail effects to the final properties of Mg alloys. This solubility extension possibility was firstly investigated in this work using thermodynamic evaluation based on connected databases. Thermodynamics indicates that the solubility of Rare-earth elements (REs) in α phase can be extended by the synergetic dissolving of Li and REs in α phase. This finding is instructive for the design of high solution of Mg alloys, and provides a new strategy for improving the performance of Mg alloys by high-solution design (HSD) of primary phase in alloy.

Keywords

Mg alloys • Alloying • High-solution design • Thermodynamics

Introduction

The research and development of Mg alloys have attracted increasing attentions due to their promising application for saving energy and environment protection [1–8]. However, the broader application of Mg alloys is still limited because of its comparably poor mechanical and corrosion properties.

The most commonly and effective alloy design approach is by alloying with other elements to improve the properties. Yasi and Hector Jr. et al. considered 29 different solutes for the solution strengthening and developed a "strengthening design map" [9]. Dilute solute of second element in α(Mg) is also reported to be able to improve the formability of alloy by affecting the stacking fault energy and formation mechanism [10–12]. Our group proposed a unique strengthen synergy mechanism for Mg alloys that the dissolution of other atoms in α phase can simultaneously improve its strength and ductility [5–7, 13–16]. Moreover, it is also found solution of atoms in α(Mg) can also change the corrosion potential of α(Mg) and then reduce the possible galvanic corrosion [4].

Hence, the high-solution design (HSD) of α(Mg) is a promising approach to be able to simultaneously improve the combination properties and then advance the performance of Mg alloys. However, most of the elements have limited solubility in α(Mg) [17], and therefore any method to increase the solubility of elements in α(Mg) is interesting. There are some non-equilibrium methods to dissolve more atoms, such as fast-solidification, ion-injection, etc. But these methods are usually with high cost and inappropriate to produce large components. Thus, increasing the

J. Wang · Y. Yuan · X. Cheng · T. Chen · B. Jiang · A. Tang · F. Pan
State Key Laboratory of Mechanical Transmissions, College of Materials Science and Engineering, Chongqing University, Chongqing, 400044, China

Y. Yuan (✉) · B. Jiang · A. Tang · F. Pan
National Engineering Research Center for Magnesium Alloys, College of Materials Science and Engineering, Chongqing University, Chongqing, 400044, China
e-mail: yuan.yuan.er@gmail.com; yuanyuan17@cqu.edu.cn

D. Li
Institute for Applied Materials – Applied Materials Physics, Karlsruhe Institute of Technology, Karlsruhe, 76344, Germany

T. Boll
Karlsruhe Nano Micro Facility, Karlsruhe Institute of Technology, Karlsruhe, 76344, Germany

T. Boll
Institute for Applied Materials, Karlsruhe Institute of Technology, Karlsruhe, 76344, Germany

V. M. Miller et al. (eds.), *Magnesium Technology 2021*, The Minerals, Metals & Materials Series,
https://doi.org/10.1007/978-3-030-65528-0_5

equilibrium solubility of elements in α phase has high theoretical significance and application significance.

The multi-element dissolving in α phase can result in entropy increase and the lattice distortion. Meanwhile, the attraction and repulsion between atoms may have a complicated combination effect. Therefore, the interactions of multi-elements in the primary solution phase can be complicated and may lead to some surprising behavior, or in another saying, cocktail effects for the final properties of alloys.

As a first step, thermodynamic evaluations aiming to predict the possibility of extended solubility of some elements, specially, some low solubility Rare-earth elements (REs), in α phase, have been performed in this work.

Calculation Methods

The evaluations were based on the thermodynamic equilibrium state and corresponded to the minimization of the global Gibbs energy of the phases. The evaluations were performed by using the commercial thermodynamic software Pandat and Thermo-Calc, together with the currently available latest databases of PanMg and TCMg5. Equilibrium isothermal sections were examined for the maximum solution points of liquid in systems.

Results and Discussion

Solution of X and Li in α Phase

Based on current available Mg-based database, the evaluations of Mg-included ternary system for dozens of solution-element-pairs indicated that most elements decrease the solubility of REs in α phase. But Li was an exception that Li can strongly increase the solubility of some elements in α(Mg). Figure 1 shows the typical results, where the phase boundaries of α phase in Mg–Li–X system were shown.

Figure 1 shows the maximum solubility of X in α phase occurs for the condition of the largest concentration of Li in α phase. By extensive calculations, Table 1 shows that Li, dissolved in α(Mg), can extend the solubility range of the REs, La, Nd, Ce, Y, and the non-RE elements Sr in α(Mg), but decrease the solubility of Gd in α(Mg). It is seen, for these investigated elements with extended solubility, that the extended solubility values are around twice, or even more, of the original solubility of X in α(Mg), which is a big increment compared to the original values, as shown in Table 1.

One typical isothermal section of Mg–Li–La is shown in Fig. 2 at the defined temperature. It shows that, for Mg–Li–RE systems, the largest solubility in α(Mg) occurred at the tri-section points joining the BCC, HCP, and one Mg–RE compound regions. With the temperature decreasing, Mg–X compounds precipitated from α(Mg).

The Employed Databases and the Interaction Between Mg and Li

It is shown that Li alloyed in α(Mg) has a strong effect on the solubility of X in α(Mg). These currently employed commercial databases are encrypted, so the evaluations of these ternary systems are not possible. Probably, the evaluations in this work just represent extrapolations based on the sub-binary systems and therefore some deviations could be existed.

There are many literatures studied on the Mg–Li–X systems, especially Mg–Li–RE systems [18–22]. It is reported, with RE addition in Mg–Li-based alloys, good mechanical properties can be achieved, where some intermetallic phases, Al–RE compounds or Mg–RE compounds were presented and have effects on the grain refinement and strengthen

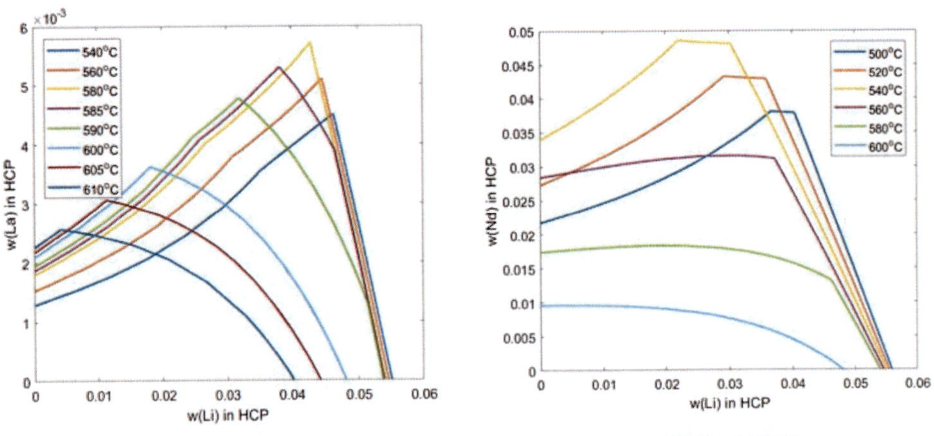

a) Mg-Li-La b) Mg-Li-Nd

Fig. 1 The weight fraction of X and Li in α(Mg) of Mg–Li–RE calculated using the Pandat software based on PanMg database. (Color figure online)

Table 1 The solubility of X in α phase calculated using PanMg database

System	Temp (°C)	w(Li) (wt%)	w(X) (wt%)	x(Li) at. %	x(X) (at. %)
Mg–Li–La	580	0	0.18	0	0.03
	580	4.30	0.57	8.82	0.07
Mg–Li–Nd	540	0	3.39	0	0.56
	540	2.21	4.85	7.6	0.81
Mg–Li–Ce	580	0	0.66	0	0.11
	580	3.02	0.98	9.9	0.16
Mg–Li–Y	560	0	13.24	0	4.00
	560	1.62	15.00	6.07	4.40
Mg–Li–Gd	550	0	22.98	0	8.13
	550	1.82	20.91	3.72	7.32
Mg–Li–Sr	500	0	0.14	0	0.04
	500	5.6	1.78	17.4	0.27

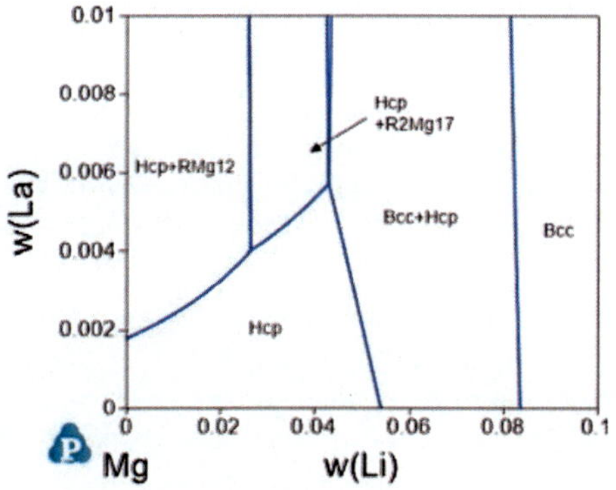

Fig. 2 The isothermal section of Mg–Li–La in Mg-rich side at 580 °C. (Color figure online)

advancement. However, because of the difficulty to identify the Li content in α(Mg) phase using normal characterization method, there is quite few studies on the composition of the α phase of the investigated alloys. It is reported in [19, 20], the selection of Y and Gd additions in Mg–Li alloy is because of the large solubility of Gd/Y in α(Mg) phase considering of the binary Mg–Gd/Y system. But no further information about the composition of α phase in ternary system where two elements are presented. Zhang and Wu et al. reported there is around 1.51 at.% Ce solid-soluted in α(Mg, Li, Ce) phase for Mg–8.5Li–xCe alloy [22], where this value is much higher than the generally accepted solubility of Ce in α(Mg, Ce) phase for the binary Mg–Ce system (around 0.1 at.%). Since the reported element content analysis is coming from EDS analysis in [22] and the Li content is also not considered in the final composition normalization, the reported content of Ce in α phase could be with low precision. Hence, we could not give more experimental evidences on the high solution of Mg–Li–X alloy here. The experimental work is in process in our group.

Considering the mutual solution of elements, most of the elements have limited solubility in α(Mg) according to the binary phase diagrams of the Mg–X system. However, the element Li is an exception, for which there is a large mutual solubility between Li and Mg, which indicates that Li and Mg atoms are similar to each other, and there is no strong attraction or repulsion between Li and Mg. Hence Li is promising to extend the dissolution of X in α(Mg) and Li did show the expected effects. Moreover, no other elements except Li show the same effect through trials on dozens of elements using current databases.

The phase diagrams of binary systems show that Y and Gd also have comparable big solubility in α(Mg). Y, Gd, and Mg can fit the empirical Hume–Rothery rules well [23]. However, the crystal structures of Li and Mg are different and the electronegativities of Li and Mg are also with a large difference. It cannot be simply concluded, that the similarity of atomic sizes of Li and Mg is the one important factor. There should be other characteristics to affect the mutual solubility of the two elements. More investigations are on the way in our group. However, these evaluations do provide promising predications. These key predictions can be tested by experiments and the exploration process is then greatly shorted.

The Implications of This Solubility Extension

The Mg–La, Mg–Nd, and Li–La, Li–Nd have little mutual solubility for each other. It is shown, even with the presence of Li addition, the solubility of La and Nd is still below 1 atom per 100 Mg atoms in α(Mg), where the interactions of nearest neighbor atom or second nearest neighbor atoms with Li (the number below 1/10 atoms) is probably ignorable. The positions of Li and X elements and the possibility of the existence of clusters or short-orders structure in crystal should be clarified and is in progress in our group.

These evaluations indicate that a large concentration of Li (> 5 at.%) in α(Mg) may increase the solubility of La and Nd in α(Mg) by some factors. Though, the higher solubility of La and Nd may have a positive effect on the mechanical properties of Mg alloys (including both strength and ductility), the Li concentration in alloys can of course change the mechanical properties accordingly. It is reported that Li dissolved in Mg alloys can lower the density of Mg alloys and enhance the ductility simultaneously [24, 25]. Hence, it can be expected that the Mg–Li–X (X = REs) alloys may have high strength and high ductility simultaneously, which are already reported by some groups [18–21, 26–29]. Of course, the strength and ductility also relate to the microstructure and deformation structure, the direct relation of mechanical properties with the concentration of RE and Li in alloys can't be predicted here. Besides, the additions of REs in Mg alloys were argued as the heavy addition of REs offset the priority of light-weight of Mg alloys. Here, with lighter element Li addition, an acceptable balance may be obtained. And with the addition of favorable REs for corrosion resistance, a combined good performance of Mg alloys is promising to be achieved with synergetic solution of Li and REs in Mg alloys.

Conclusions

Thermodynamic evaluations show the solution in α(Mg) is possible to be extended by alloying multi-element in Mg alloys. The discovery of this phenomena provides a new strategy for simultaneously improving the combined properties of Mg alloys by design a high solution of α(Mg) phase. Development of high-solution Mg alloys with a combined good performance is promising.

Acknowledgements The authors are grateful for the financial support from the National Natural Science Foundation of China (51971044 and U1910213), the Natural Science Foundation of Chongqing (cstc2019yszx-jcyjX0004), the Qinghai Provincial Science and Technology Key Program (No. 2018-GX-A1), the National Key Research and Development Program of China (2016YFB0301102), and the Fundamental Research Funds for the Central Universities (2019CDJGFCL005 and 2020CDJDPT001). Fellowship from Karlsruhe Institute of Technology for research guest stay of author Prof. Yuan Yuan during preparing this manuscript and technology support from Karlsruhe Nano Micro Facility (KNMF), KIT, is greatly appreciated.

References

1. Zeng, Z., N. Stanford, C.H.J. Davies, J.-F. Nie, and N. Birbilis, (2018) *Magnesium extrusion alloys: a review of developments and prospects.* International Materials Reviews: p. 1–36.
2. Wang, X.J., D.K. Xu, R.Z. Wu, X.B. Chen, Q.M. Peng, L. Jin, Y. C. Xin, Z.Q. Zhang, Y. Liu, X.H. Chen, G. Chen, K.K. Deng, and H.Y. Wang, (2018) *What is going on in magnesium alloys?* Journal of Materials Science & Technology, **34**(2): p. 245–247.
3. Xu, T.C., Y. Yang, X.D. Peng, J.F. Song, and F.S. Pan, (2019) *Overview of advancement and development trend on magnesium alloy.* Journal of Magnesium and Alloys, **7**(3): p. 536–544.
4. Chen, T., Y. Yuan, J. Wu, T. Liu, X. Chen, A. Tang, and F. Pan, (2019) *Alloy Design Strategies of the Native Anti-corrosion Magnesium Alloy*, in *Magnesium Technology 2019*, V. Joshi, Editor. The Minerals, Metals & Materials Series. p. 169–173.
5. Tu, T., X. Chen, C. Zhao, Y. Yuan, and F. Pan, (2020) *A simultaneous increase of elastic modulus and ductility by Al and Li additions in Mg-Gd-Zn-Zr-Ag alloy.* Materials Science and Engineering: A, **771**: p. 138576.
6. Zhao, D., X. Chen, Y. Yuan, and F. Pan, (2020) *Development of a novel Mg–Y–Zn–Al–Li alloy with high elastic modulus and damping capacity.* Materials Science and Engineering: A, **790**: p. 139744.
7. Zhao, J., B. Jiang, Y. Yuan, A. Tang, Q. Wang, T. Yang, G. Huang, D. Zhang, and F. Pan, (2020) *Influence of Ca and Zn synergistic alloying on the microstructure, tensile properties and strain hardening of Mg-1Gd alloy.* Materials Science and Engineering: A, **785**: p. 139344.
8. Chen, T., X. Xiong, Y. Yuan, A. Tang, D. Li, A. Atrens, and F. Pan, (2020) *Effect of Steels on the Purity of Molten Mg Alloys.* Advanced Engineering Materials: p. 2000338.
9. Yasi, J.A., L.G. Hector, and D.R. Trinkle, (2010) *First-principles data for solid-solution strengthening of magnesium: From geometry and chemistry to properties.* Acta Materialia, **58**(17): p. 5704–5713.
10. Wu, Z. and W.A. Curtin, (2015) *The origins of high hardening and low ductility in magnesium.* Nature, **526**(7571): p. 62–7.
11. Wu, Z., R. Ahmad, B. Yin, S. Sandlobes, and W.A. Curtin, (2018) *Mechanistic origin and prediction of enhanced ductility in magnesium alloys.* Science, **359**: p. 447–452.
12. Nie, K., Z. Zhu, P. Munroe, K. Deng, and J. Han, (2020) *The effect of Zn/Ca ratio on the microstructure, texture and mechanical properties of dilute Mg–Zn–Ca–Mn alloys that exhibit superior strength.* Journal of Materials Science, **55**(8): p. 3588–3604.
13. Pan, F., M. Yang, and X. Chen, (2016) *A Review on Casting Magnesium Alloys: Modification of Commercial Alloys and Development of New Alloys.* Journal of Materials Science & Technology, **32**(12): p. 1211–1221.
14. Wang, Q., Y. Shen, B. Jiang, A. Tang, Y. Chai, J. Song, T. Yang, G. Huang, and F. Pan, (2018) *A good balance between ductility and stretch formability of dilute Mg-Sn-Y sheet at room temperature.* Materials Science and Engineering: A, **736**: p. 404–416.
15. Wang, Q., B. Jiang, A. Tang, C. He, D. Zhang, J. Song, T. Yang, G. Huang, and F. Pan, (2019) *Formation of the elliptical texture and its effect on the mechanical properties and stretch formability of dilute Mg-Sn-Y sheet by Zn addition.* Materials Science and Engineering: A, **746**: p. 259–275.
16. Zhao, J., B. Jiang, Y. Yuan, A. Tang, H. Sheng, T. Yang, G. Huang, D. Zhang, and F. Pan, (2020) *Influence of Zn addition on the microstructure, tensile properties and work-hardening behavior of Mg-1Gd alloy.* Materials Science and Engineering: A, **772**: p. 138779.
17. Okamoto, H., (2000) *Phase diagram for binary alloys. Materials Park (OH).* ASM International.
18. Zhu, T., C. Cui, T. Zhang, R. Wu, S. Betsofen, Z. Leng, J. Zhang, and M. Zhang, (2014) *Influence of the combined addition of Y and Nd on the microstructure and mechanical properties of Mg–Li alloy.* Materials & Design, **57**: p. 245–249.

19. Zhao, J., J. Zhang, W. Liu, G. Wu, and L. Zhang, (2016) *Effect of Y content on microstructure and mechanical properties of as-cast Mg–8Li–3Al–2Zn alloy with duplex structure.* Materials Science and Engineering: A, **650**: p. 240–247.
20. Sun, Y., R. Wang, C. Peng, and Y. Feng, (2018) *Effects of Sn and Y on the microstructure, texture, and mechanical properties of as-extruded Mg-5Li-3Al-2Zn alloy.* Materials Science and Engineering: A, **733**: p. 429–439.
21. Dinesh, P., S. Manivannan, S.P. Kumaresh Babu, and S. Natarajan, (2019) *Effect of Nd on the microstructure and corrosion behaviour of Mg-9Li-3Al magnesium alloy in 3.5 wt.% NaCl solution.* Materials Today: Proceedings, **15**: p. 126–131.
22. Zhang, M.L., R.Z. Wu, and T. Wang, (2009) *Microstructure and mechanical properties of Mg–8Li–(0–3)Ce alloys.* Journal of Materials Science, **44**(5): p. 1237–1240.
23. Mizutani, U., (2016) *Hume-Rothery Rules for Structurally Complex Alloy Phase.* CRC Press.
24. Muga, C.O. and Z.W. Zhang, (2016) *Strengthening Mechanisms of Magnesium-Lithium Based Alloys and Composites.* Advances in Materials Science and Engineering, **2016**: p. 1–11.
25. Wu, R., Y. Yan, G. Wang, L.E. Murr, W. Han, Z. Zhang, and M. Zhang, (2015) *Recent progress in magnesium–lithium alloys.* International Materials Reviews, **60**(2): p. 65–100.
26. Zhong, F., H. Wu, Y. Jiao, R. Wu, J. Zhang, L. Hou, and M. Zhang, (2020) *Effect of Y and Ce on the microstructure, mechanical properties and anisotropy of as-rolled Mg-8Li-1Al alloy.* Journal of Materials Science & Technology, **39**: p. 124–134.
27. Xu, T.C., X.D. Peng, J.W. Jiang, W.D. Xie, Y.F. Chen, and G.B. Wei, (2014) *Effect of Sr content on microstructure and mechanical properties of Mg-Li-Al-Mn alloy.* Transactions of Nonferrous Metals Society of China, **24**(9): p. 2752–2760.
28. Yang, Y., X. Peng, H. Wen, B. Zheng, Y. Zhou, W. Xie, and E. J. Lavernia, (2012) *Influence of Extrusion on the Microstructure and Mechanical Behavior of Mg-9Li-3Al-xSr Alloys.* Metallurgical and Materials Transactions A, **44**(2): p. 1101–1113.
29. Tao, W., Z. Milin, N. Zhongyi, and L. Bin, (2006) *Influence of Rare Earth Elements on Microstructure and Mechanical Properties of Mg-Li Alloys.* Journal of Rare Earths, **24**(6): p. 797–800.

Part II

Fundamentals of Plastic Deformation

Accounting for the Effects of Dislocation Climb Mediated Flow in Mg Alloy ZK10 Sheet

Michael A. Ritzo and Sean R. Agnew

Abstract

Tensile samples of an Mg alloy ZK10 sheet were tested at a range of temperatures and strain rates designed to rather evenly probe a range of Zener Hollomon parameter values, from ln(Z) ≈ 15 (10^{-4} s^{-1} and 623 K) up to ln(Z) ≈ 50 (10^{-3} s^{-1} and 300 K). In contrast with more commonly examined Mg alloy AZ31B sheet material, ZK10 sheet material shows modest strain anisotropy (r-value) at low temperatures for both 45° (r_{45} ≈ 1.2) and TD (r_{TD} ≈ 1.4) sample orientations, despite showing evidence of significant prismatic slip of <a> dislocations, which often leads to high r-values at low temperatures. These low r-values become even lower (r_{45} ≈ 0.84 and r_{TD} ≈ 0.89) at high temperatures. These behaviors are hypothesized to occur due to a distinct initial texture and deformation mechanism activity, which includes a modest level of tensile twinning and <c + a> slip at both room and elevated temperature. A version of the viscoplastic self-consistent (VPSC) code, which accounts for the kinematics of dislocation climb, is used to simulate the behavior of a textured Mg alloy ZK10 sheet reveals that both the glide of pyramidal <c + a> dislocations and the climb of basal < a > dislocations are required to describe the behavior at elevated temperatures.

Keywords

Texture • Anisotropy • Dislocation • Climb • Twinning • Crystal plasticity

M. A. Ritzo · S. R. Agnew (✉)
Department of Materials Science and Engineering, University of Virginia, 22904 Charlottesville, VA, USA
e-mail: agnew@virginia.edu

M. A. Ritzo
e-mail: mar3dm@virginia.edu

Introduction

The use of polycrystal plasticity modeling has radically changed our understanding of the deformation of hexagonal close packed (hcp) Mg alloys over the past two decades. Historically, it was believed that $\{10\bar{1}2\}$ tensile twinning was a significant deformation mechanism only at low temperatures. A major reason for this was that the critical resolved shear stress (CRSS) for twinning was shown to be nearly temperature insensitive whereas the CRSSs of the hard, non-basal slip mechanisms (i.e., prism <a> and pyramidal <c + a>) tend to be highly temperature sensitive [1–4]. During room temperature deformation, $\{10\bar{1}2\}$ tensile twinning is known to be one of the few mechanisms that can accommodate strains along the <c> axis. Pyramidal slip of <c+a> dislocations, the another slip mode which can accommodate <c> axis strain, has a rather high CRSS value at room temperature, which limits its activity [5] and has led to suggestions that it may be more important at higher temperatures. Other suggestions include climb of basal <a> and non-basal [c] dislocations, grain boundary sliding, and diffusional flow. The higher stress conditions presently of interest ensure that diffusional flow will not be controlling.

The interplay between deformation twinning and dislocation slip (glide) has been studied in different contexts, but the role of dislocation climb, used to explain differences in texture evolution and r-value of AZ31 has yet to be applied broadly to the high temperature deformation of other Mg alloys, such as ZK10 (the material chosen for the current study), which is known to undergo significant twinning at room temperature [6]. One implication for the current study is that $\{10\bar{1}2\}$ twins have been shown to transmute gliding <a> matrix dislocations into <c + a> dislocations in the twin [7]. <c + a> slip has been thought to reduce the strain anisotropy in Mg alloys and promotes a characteristic texture, distinct from all other slip modes. The possible influence of twinning in the activation of <c + a> glide at high

V. M. Miller et al. (eds.), *Magnesium Technology 2021*, The Minerals, Metals & Materials Series,
https://doi.org/10.1007/978-3-030-65528-0_6

temperature is of great interest, as <a> climb has been shown to be competitive with non-basal slip modes, particularly <c + a> glide on second order pyramidal planes [8].

Dislocation climb and cross-glide are relevant to applications involving creep, load relaxation (such as bolt-load retention), and hot/warm forming operations. Indeed, the measured constitutive response of Mg and its alloys frequently exhibits power-law type constitutive behavior at temperatures and rates relevant to these applications (e.g., [1]). Rarely have crystal plasticity modelers accounted for these aspects, though the work of Staroselsky and Anand [2] notably considered the possible role of grain boundary sliding. In addition, the interplay between twinning and dislocation climb at high temperatures has not been studied previously.

Only recently has it become possible to explore the unique characteristics of Mg alloy ZK10. Several crystal plasticity models for describing twinning in polycrystalline materials have been developed, but no such models existed to predict the effect of dislocation climb on texture and strain anisotropy. A viscoplastic self-consistent (VPSC) polycrystal modeling code, which accounts for the non-conservative (climb) motion of dislocations has been developed recently, [8, 9] which will allow for the simultaneous study of twinning and climb.

We begin by providing a brief introduction to the climb and twinning models, followed by a description of experimental methods and results which highlight the effects we are hoping to explain, and then enumerate simulation results that explore the effect of tension twinning on the relative activities of slip and climb modes. The tensile deformation of ZK10 was studied as its unique rolled texture made it a model system to explore the effect of twinning. Tensile deformation parallel to and 45° away from the sheet transverse direction is considered to study the effect of sample orientation on the twin volume fraction.

Modeling Background

Crystal plasticity models generally consider dislocation glide and twinning [10, 11] as the only dissipative processes. In addition, both slip and twinning are assumed to be governed by the generalized Schmid law, which resolves the stress $\boldsymbol{\sigma}$ on the slip system as follows:

$$\tau = m : \sigma \tag{1}$$

where $\mathbf{m} = \mathrm{sym}(\hat{\mathbf{b}} \otimes \hat{\mathbf{n}})$ is the Schmid tensor, $\hat{\mathbf{n}}$ is the slip plane normal and $\hat{\mathbf{b}}$ is the slip direction [13]. The role of cross-slip is often associated with the process of dynamic recovery [14], and the role of dislocation climb treated similarly. The strain accommodated by these mechanisms is for the most part disregarded. Lebensohn et al. [9] generalized the connection between stress and dislocation motion, beyond the Schmid law of Eq. 1. by appealing to the full Peach–Koehler relationship,

$$f = (\boldsymbol{\sigma} \cdot \mathbf{b}) \times \hat{\mathbf{t}} \tag{2}$$

where $\mathbf{b}$ is the Burgers vector and $\hat{\mathbf{t}}$ is the dislocation line direction. The forces which drive dislocation glide (the Schmid stress times the magnitude of the Burgers vector) can then be parsed from those which drive climb.

$$f_g = \{(\sigma \cdot \mathbf{b}) \times \hat{\mathbf{t}}\} \cdot \hat{\chi} = \{\sigma : (\hat{\mathbf{b}} \otimes \hat{\mathbf{n}})\}|b| \tag{3}$$

where $\hat{\chi}$ is the direction of dislocation glide (within the glide plane and orthogonal to the line direction). On the other hand, the climb force is resolved along the glide plane normal direction.

$$f_{2c} = \{(\sigma \cdot \mathbf{b}) \times \hat{\mathbf{t}}\} \cdot \hat{\mathbf{n}} = -\{\sigma : (\hat{\mathbf{b}} \otimes \hat{\chi})\}|b| \tag{4}$$

Notably, the glide force only depends upon the deviatoric stress (the addition of pressure has no effect), whereas the climb force depends on the full stress tensor. For dislocation glide, the dyadic cross product $\hat{\mathbf{b}} \otimes \hat{\mathbf{n}}$ can be decomposed into symmetric (strain, $\mathbf{m}$) and antisymmetric (rotation, $\mathbf{q}$) components. Indeed, this slip-induced rotation is the basis of texture evolution during glide. For climb, the analogous tensor, $\hat{\mathbf{b}} \otimes \hat{\chi}$ can be decomposed into symmetric strain ($\mathbf{k}$) and rotation ($\mathbf{r}$) components. The distinct types of strain and rotation associated with climb and glide provide a means by which their contributions to the deformation may be parsed. For example, the strain rate within a crystal undergoing glide and climb may be expressed as a function of the applied stress as follows.

$$\dot{\boldsymbol{\varepsilon}} = \sum_s \mathbf{m}^s \left(\frac{f^s}{\tau_c^s b^s}\right)^{n_g} sign(\mathbf{m}^s : \boldsymbol{\sigma}) + \sum_s \mathbf{k}^s \left(\frac{f^c}{\sigma^c b^s}\right)^{n_c} sign(\mathbf{k}^s : \boldsymbol{\sigma}) \tag{5}$$

Relative to traditional crystal plasticity, the relations are complicated by the fact that one must track the character of dislocation populations. Indeed the "climb strain tensor," $\mathbf{k} = \mathrm{sym}(\hat{\mathbf{b}} \otimes \hat{\chi})$, can be expressed in terms of the angle between the Burgers vector and line direction.

Many approaches have been taken in order to simulate the effect of twinning; the two most important models are the Predominant Twin Reorientation (PTR) [15] and the Twinning Detwinning (TDT) [16] models. In both cases, twinning is activated per the same power-law constitutive rule and Schmid stress as described above for slip. Distinctly, a volume fraction of each activated twin variant created within each straining increment is proportional to the strain accommodated, and the proportionality constant is the characteristic twinning shear of that mode (e.g., ~ 0.14 for the $\{10\bar{1}2\}$ tensile twin). In this early study of the interaction between dislocation climb and twinning, we employ the PTR model which attempts to capture effect of twinning on texture by selectively reorienting grains that surpass a threshold twinning volume fraction (F_T) as defined by Eq. 6

$$F_T = 0.25 + 0.25\frac{F_E}{F_R} \quad (6)$$

When the volume fraction of a grain undergoing twinning exceeds F_T, the grain in question is reoriented according to its most active twin variant. F_E is a measure of the twinning volume fraction calculated by summing the volume fractions of all grains that have been completely reoriented and F_R tracks the total volume fraction of twins which have been invoked and is known as the "real" twinned fraction, as this is akin to the volume fraction that can be measured. The major advantage of this model is computational efficiency, since the number of grains in the calculation remains fixed throughout the simulation. The major disadvantages are that entire grains are reoriented instead of local regions and that the model only accounts for the texture evolution associated with the most active twinning system in each grain.

In this study, the experimentally measured texture is used as an input and boundary conditions appropriate for uniaxial tension parallel to (TD) or 45° away from (45) the sheet transverse direction are imposed. The parameters which were explored include the critical resolved shear strengths (τ) of the prismatic <a>, pyramidal <c + a>, tensile twinning, and basal <a> climb, all relative to the strength of basal <a> slip. The outputs presently under consideration are the r-value and the evolved texture after deformation. The power-law exponents employed in this study were fixed at $n_g = 3$ and$n_c = 3$, for high temperature simulations and $n_g = 20$ for low temperature simulations, though further exploration of the effects of these parameters on the constitutive response and texture evolution is merited.

Experimental Methods

Mg alloy ZK10 sheet with 1 mm thickness was provided by the former Magnesium Elektron North America, part of the Luxfer group, with a nominal composition of 1wt% Zn and 0.5% Zr. The samples were received in the F (as-worked) temper, but examined in the O (soft annealed) temper, which involved annealing at 300 °C for 1 h to fully recrystallize the microstructure. The microstructure of the samples was examined in previous study and found to be comprised of equiaxed grains with a lineal intercept grain size of −10 mμ with only occasional twins from the prior deformation in addition to few precipitates present [6]. The texture is measured using X-ray diffraction from the sheet midplane, both before and after deformation, using a Panalytical X'pert Pro MPD diffractometer, as described previously [17]. The texture analysis and representation were performed using the MTEX toolbox for MATLAB [18]. A preliminary assessment of the twin fractions in deformed samples was made by tracking the orientations on the basal pole figure with tilt values, $\alpha > 50°$ and subtracting them from the orientations collected from the undeformed sample. The process was repeated for $\alpha > 80°$ in order to determine upper and lower bound estimates of the twin fraction.

Tensile samples with an ASTM E-8 sub-sized standard dog bone geometry were prepared from the sheets using electrodischarge machining. The effective gauge section of the samples is approximately 33 mm long by 6 mm wide. The r-values were measured as the ratio of the logarithmic true plastic strains along the width and thickness directions, after deforming the samples to a plastic strain of 0.08–0.12 perpendicular (TD) and 45° (45) to the rolling direction. One test of the accuracy of the strain measurements was to examine volumetric strain implied, and since plasticity is known to be volume constant, only those measurements with implied absolute dilatation of less than ~0.005 were retained in the final analysis.

Tensile tests were performed at temperatures ranging from room temperature up to 350 °C, at strain rates ranging from 10^{-4} to 10^{-3} s^{-1}. The tensile test data were analyzed in terms of the flow stress measured at a plastic strain of ~0.10. The test conditions were chosen in order to obtain a wide range of Zener–Holloman parameter, Z, also known as the temperature-compensated strain rate:

$$Z = \dot{\varepsilon}\exp\left(\frac{Q}{RT}\right) \quad (7)$$

where Q is the activation energy and R is the universal gas constant. Sellars and Tegart [19] suggested a hyperbolic sine function would describe the relationship between rate and flow stress over a wide range of hot working and creep conditions.

$$Z = A\{\mathbf{sinh}(\alpha\sigma)\}^n \quad (8)$$

At low stress (or Z) conditions, this relationship asymptotes to a simple power-law indicative of high temperature creep, whereas it asymptotes to an exponential function of stress at high stresses, which is typical of thermally activated slip. That is, it does a reasonably good job of describing both the power-law and power-law breakdown regimes. The empirical parameters A, α, n, and Q were obtained by least-squares non-linear regression.

Experimental Results

Tensile test results are shown in Fig. 1a. It is clear that increased temperature results in the lowered yield and ultimate tensile stress. In addition, the amount of hardening decreases as the temperature increases. Figure 1b presents the flow stress during uniaxial straining parallel to the TD and between RD and TD, at a tensile plastic strain of ~ 0.10, plotted as a function of the Zener–Holloman parameter, Z. Superimposed on the experimental data is a best-fit Sellars–Tegart function (Eq. 8), with a stress exponent $n = 3.5$. Such a value suggests that the rate-controlling deformation mechanism is dislocation climb at conditions of Z ~ 22 or less. The transition to rate-insensitive plasticity (or power-law breakdown) occurs over the range of $22 \leq Z \leq 35$. At even higher Z levels, the plasticity is rate-insensitive. The focus of this study is placed upon examining the texture evolution and strain anisotropy over these same three Z-value regimes.

Figure 1b also presents the r-values measured at a plastic tensile strain of 0.08–0.12 and plotted as a function of applied Z. In the low Z regime, the r-value is close to 1 (near plastic isotropy) and in the high Z regime, the r-value is ~1.4. There is a slight difference in r-value between sample orientations at the high Z test conditions that is not present at low Z. It is notable that conditions where the constitutive modeling suggests the onset of significant strain accommodation by dislocation climb (i.e., the low Z power-law climb and glide regime), the r-value begins to decrease.

Finally, the crystallographic texture and its evolution after tensile deformation within each of the aforementioned Z-regimes is presented in Fig. 2. Samples tested at high Z conditions show significant activity of prism dislocations, which is manifests as intensity of the ODF in Rodrigues space forming nodes at the top (for 45° samples) or bottom (for TD samples) of the Z axis. In contrast to previous studies on AZ31B, the ZK10 samples in the current study exhibited dramatically lower r-values, which is attributed to reduced prism slip [17]. This also helps to explain the lower yield stresses of these ZK10 sample at ambient temperatures, since the distinct texture permits the softer basal slip mechanism to accommodate greater fraction of the macroscopic strain. Another possible contributor to decreased r-values could be <c + a> slip on pyramidal planes, which was found to be the dominant deformation mechanism of Mg single crystals oriented for $<11\bar{2}0>$ tension or $<0001>$ compression at ambient temperatures in Mg–Li and Mg–Zn [20], Mg–Y [21] and at high temperatures in Mg–Al–Zn [22] and pure Mg [23]. Notably, only single crystals of Mg–Al–Zn did not show a dominance of <c + a> slip under these straining conditions, suggesting that it is the unusual case, even though it is the most common commercial wrought Mg alloy.

The twin volume fractions were estimated to be between 1.5 and 7.5%, based upon the measured deformation textures, though no clear trend with testing direction or test conditions (Z) is apparent. Due to the minor contribution twinning makes to the overall strain accommodation, even at room temperature, it is not considered to be a significant contributor to the low yield strength of this alloy.

As the temperature is increased (and Z-value is correspondingly decreased), the texture evolution undergoes a transition, whereby the intensity forms two fibers along the z axis. In the low Z regime, where the r-value is decreased below 1, the texture appears similar to the undeformed texture, i.e., the deformation texture evolution is greatly slowed at low Z conditions (Fig. 3a). Previous study of Mg alloy AZ31B revealed that this is due to the activation of basal <a> dislocation climb, which does not cause rotations in the texture and, therefore, results in a slowing of the texture evolution which would otherwise occur due to dislocation glide.

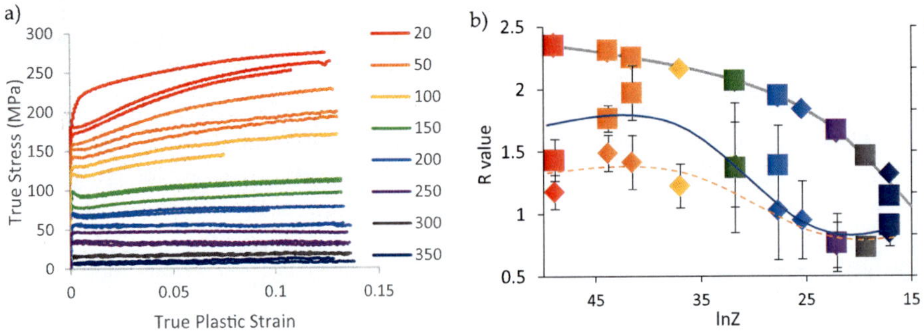

Fig. 1 **a** Stress–strain plots performed from ambient to 350 °C. **b** Plot of log-flow stress vs. log-Z (right y axis). The r-value for samples of textured Mg alloy, ZK10, sheet material tested parallel to the sheet transverse direction (TD) and 45° away from TD (left y axis). The solid (TD data) and dashed (45° data) curves are meant to guide the eye, but suggests three mechanistic regimes at low, intermediate, and high Z-values. (Color figure online)

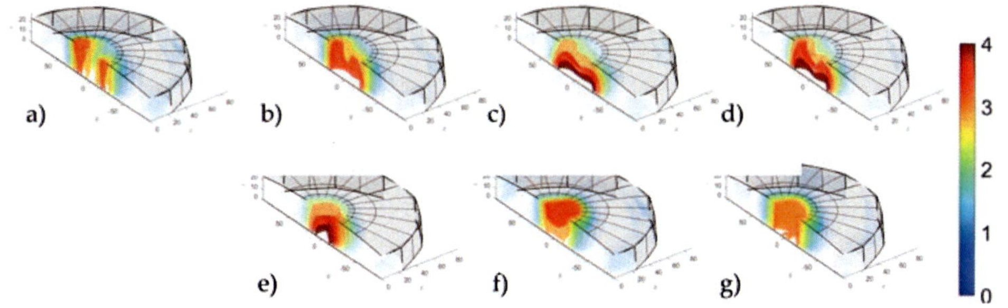

Fig. 2 Crystallographic textures of samples Mg alloy ZK10 sheet material tensile tested parallel to the transverse direction (TD) (**a–d**) and 45° away from it (45) (**e–f**). The textures correspond to (**a**) undeformed and deformed at ln Z = (**b & e**) 48.9, (**c & f**) 31.8, and (**d & g**) 19.3 (**d & h**). Unless provided, all ODFs share the same scale as Fig. 2. (Color figure online)

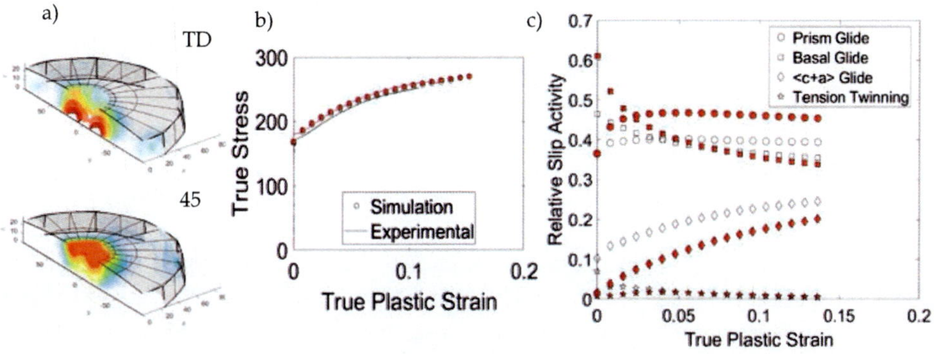

Fig. 3 Low temperature **a** TD and 45° orientations textures. **b** Experimental and simulated stress–strain curves of TD and 45° samples. **c** Relative slip activity plots of simulated tensile tests. Closed symbols and lighter lines represent TD samples and opened symbols and darker lines represent 45° samples. (Color figure online)

Modeling Results

Numerous authors have explored the effects of varying the critical resolved shear stress (CRSS) ratios of basal, prismatic <a>, and pyramidal <c + a> slip can have on the r-values and texture evolutions [e.g., 3]. Here we focus on the effects of climb of basal <a> type dislocations on a twinning prone alloy. Figure 3 shows the optimized set of Voce parameters used to simulate the low temperature tensile tests. It is clear from Fig. 3c that in both sample orientations that the majority of the strain is accommodated through basal and prism glide, which is expected for low temperature deformation. Tensile twinning has a low relative activity, which is related to the low twinning fractions estimated experimentally. Surprisingly, <c + a> glide has significant activity at ambient temperatures in both orientations and may explain the lower r-values of the 45° sample orientation as increased <c + a> glide activity. This is in contrast to the previously mentioned study on AZ31B, where the glide of <c+a> dislocations on second order pyramidal planes was suppressed at low temperatures. This could be due to alloy chemistry as it was shown that Al and Zn additions to single crystal Mg samples increase the CRSS of <c + a> glide and decrease the twinning CRSS [22]. Curiously, it does not seem to have a strong effect on the alloy's ductility as one would predict.

In order to simulate the high temperature experimental samples, large batches of simulations were performed with varying the CRSS values which control tension twinning, prism <a> and second order pyramidal <c + a> glide, along with the critical stress which controls the climb of basal <a> dislocations (all relative to the CRSS of basal <a> glide). Equations 9 and 10 were used to quantify the total residual error between the experiments and simulations, taking into account the plastic anisotropy (r-values) and deformation texture of both sample types.

$$RelErr = TexErr_{TD} + \frac{\left|r_{sim} - r_{exp}\right|_{TD}}{r_{TD\_exp}} + TexErr_{45} + \frac{\left|r_{sim} - r_{exp}\right|_{45}}{r_{45\_exp}} \tag{9}$$

$$TexErr = \sqrt{\int \left(f_{exp}(g) - f_{sim}(g)\right)^2 dg} \tag{10}$$

where $f_{exp}(g)$ and $f_{sim}(g)$ refer to the orientation distribution functions of the experimental and simulated deformation textures, respectively.

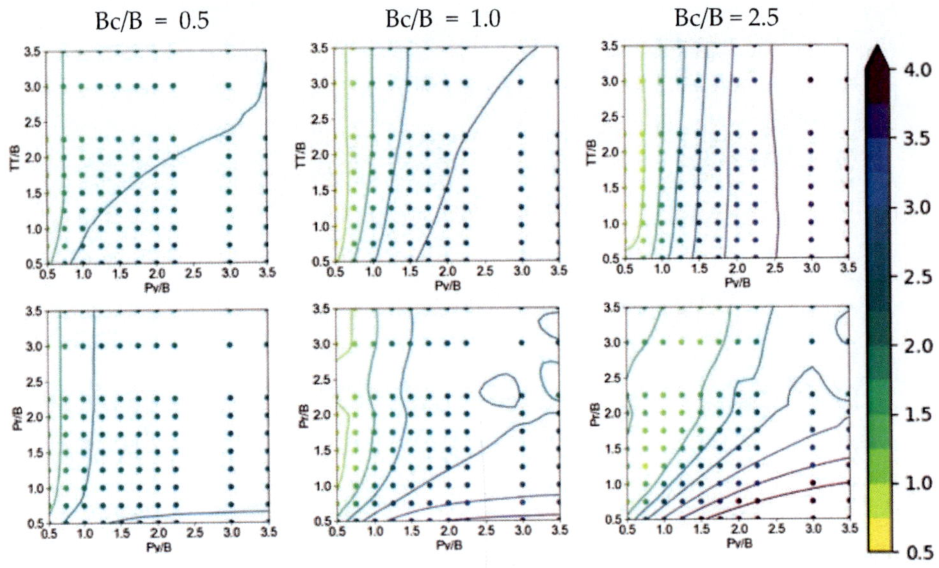

Fig. 4 Top: **a–c** Constant prism CRSS (P/B = 1) and Bottom: **d–f** Constant tension twinning CRSS (TT/B = 1.3). (Color figure online)

Two separate simulation batches were run to elucidate the relative effects of tension twinning as well as dislocation glide and climb slip modes on the high temperature flow. The first batch varied the tension twinning, <c + a> glide, and basal climb critical stresses (Fig. 4a–c). It is clear that climb of basal <a> dislocations is an important mechanism in the high temperature deformation of ZK10, as soft basal climb (Fig. 4a) is required to lower the residual error. As basal climb is deactivated by raising its relative critical stress (Figs. 4b and c), the total error becomes more dependent on the CRSS value of < c + a> slip. The simulations performed here suggest a very low CRSS ratio of <c + a> to basal glide (<1), which is physically unrealistic. This is most likely due to the exceptionally low r-values measured (~0.7), since < c + a> slip is known for lowering the strain anisotropy in textured Mg alloy sheets. Given the fact that the twin volume fraction is less than 7.5% at any temperature, it is not surprising that the CRSS value of twinning does not strongly influence the error. A second batch of simulations was used to highlight the relationship between basal climb and non-basal glide. Again, soft basal climb caused the total error to be nearly independent of prism and <c + a> glide strength (Fig. 4d). Again, when the critical stress ssfor basal climb is increased, soft <c + a> glide is required in order to achieve a lower total error (Fig. 4e and f). Curiously, the CRSS value controlling prism slip plays a minor role in influencing the total residual error in these high temperature simulations.

The optimal CRSS values of twinning, non-basal glide and climb relative to that of basal glide were used in Fig. 5 to illustrate the best-fit set of parameters to describe the high temperature deformation of ZK10. The absolute values of the Voce hardening parameters τ_o and τ_1 were varied to match experimental stress–strain curves. It would appear that <c + a> glide is the dominant mechanism during high temperature deformation, which would account for the exceptionally low r-values. Regardless, climb and glide of basal <a> dislocations still accommodate a significant

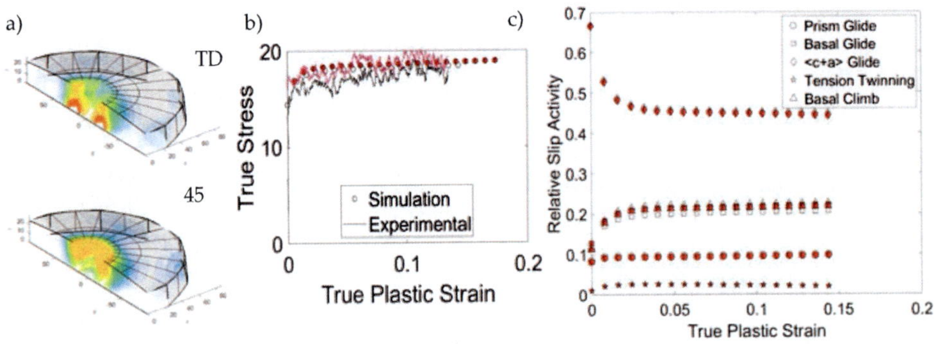

Fig. 5 High temperature-optimized simulations of **a** TD and 45° orientations textures. **b** Experimental and simulated stress–strain curves of TD and 45° samples. **c** Relative slip activity plots of simulated tensile tests. Closed symbols and lighter lines represent TD samples and opened symbols and darker lines represent 45° samples. (Color figure online)

amount of strain (~20% each), which runs counter to traditional beliefs in climb and glide models for creep which stipulate that an overwhelming majority of strain occurs due to dislocation glide.

Conclusions

Measurements of the flow stress, r-values, and texture evolution during TD and 45° tension of basal textured Mg alloy ZK10 sheets have revealed three behavioral regimes corresponding to low ($Z < 22$), intermediate ($22 \leq Z \leq 35$), and high ($Z > 35$) temperature-compensated strain rates. Simple constitutive modeling reveals the behavior within each of these regimes to be characteristic of power-law creep, power-law breakdown, and thermally activated plasticity, respectively. Within the high Z, thermally activated plasticity regime, non-basal slip of <a> dislocations is shown to prevail, since it is required to produce the characteristic node-like textures. However, the low r-values ~1.5 appear to be due to increased <c + a> glide activity, relative to what is observed in alloy AZ31B. The contribution of twinning to in-plane tensile deformation was ultimately revealed to be minimal regardless of Z-value.

A new crystal plasticity model, which incorporates the strain and reorientation characteristics of dislocation climb is used to show that the transitions in both the r-value and texture evolution can be described if one accounts for the climb of basal <a> dislocations. These results continue to emphasize the role which dislocation climb has in strain accommodation and texture evolution, whereas prior work relegated climb to a role of dislocation recovery alone.

Acknowledgements The authors would like to thank the United States National Science Foundation, Division of Materials Research, Metals and Metallic Nanostructures (NSF-DMR-MMN) program, Grant No. 1810197, overseen by program manager Dr. Judith Yang, for financial support of this research. The authors would also like to thank Ricardo Lebensohn of Los Alamos National Laboratory for sharing the version of the VPSC code which was used to perform the simulations reported herein.

References

1. A. Chapuis and J. H. Driver, "Temperature dependency of slip and twinning in plane strain compressed magnesium single crystals," *Acta Mater.*, vol. 59, no. 5, pp. 1986–1994, 2011, https://doi.org/10.1016/j.actamat.2010.11.064.
2. T. Walde and H. Riedel, "Modeling texture evolution during hot rolling of magnesium alloy AZ31," *Mater. Sci. Eng. A*, vol. 443, no. 1–2, pp. 277–284, 2007, https://doi.org/10.1016/j.msea.2006.09.028.
3. A. Chapuis and Q. Liu, "Investigating the temperature dependency of plastic deformation in a Mg-3Al-1Zn alloy," *Mater. Sci. Eng. A*, vol. 725, no. March, pp. 108–118, 2018, https://doi.org/10.1016/j.msea.2018.04.019.
4. A. Jain and S. R. Agnew, "Modeling the temperature dependent effect of twinning on the behavior of magnesium alloy AZ31B sheet," *Mater. Sci. Eng. A*, vol. 462, no. 1–2, pp. 29–36, 2007, https://doi.org/10.1016/j.msea.2006.03.160.
5. K. Y. Xie, Z. Alam, A. Caffee, and K. J. Hemker, "Pyramidal i slip in c-axis compressed Mg single crystals," *Scr. Mater.*, vol. 112, pp. 75–78, 2016, https://doi.org/10.1016/j.scriptamat.2015.09.016.
6. J. Bohlen, M. R. Nürnberg, J. W. Senn, D. Letzig, and S. R. Agnew, "The texture and anisotropy of magnesium-zinc-rare earth alloy sheets," *Acta Mater.*, vol. 55, no. 6, pp. 2101–2112, 2007, https://doi.org/10.1016/j.actamat.2006.11.013.
7. F. Wang and S. R. Agnew, "Dislocation transmutation by tension twinning in magnesium alloy AZ31," *Int. J. Plast.*, vol. 81, pp. 63–86, 2016, https://doi.org/10.1016/j.ijplas.2016.01.012.
8. M. Ritzo, R. Lebensohn, L. Capolungo, and S. and Agnew, "Accounting for the effect of dislocation climb-mediated flow on the anisotropy and texture evolution of Mg alloy, AZ31B," *Int. J. Plast.*, 2021. In preparation.
9. R. A. Lebensohn, C. S. Hartley, C. N. Tome, and O. Castelnau, "Modeling the mechanical response of polycrystals deforming by climb and glide," *Philos. Mag.*, vol. 90, no. 5, pp. 567–583, 2010, https://doi.org/10.1080/14786430903213320.
10. R. A. Lebensohn, R. A. Holt, A. Caro, A. Alankar, and C. N. Tomé, "Improved constitutive description of single crystal viscoplastic deformation by dislocation climb," *Comptes Rendus - Mec.*, vol. 340, no. 4–5, pp. 289–295, 2012, https://doi.org/10.1016/j.crme.2012.02.011.
11. A. Izadbakhsh, K. Inal, R. K. Mishra, and M. Niewczas, "New crystal plasticity constitutive model for large strain deformation in single crystals of magnesium," *Comput. Mater. Sci.*, vol. 50, no. 7, pp. 2185–2202, 2011, https://doi.org/10.1016/j.commatsci.2011.02.030.
12. G. Proust, C. N. Tomé, A. Jain, and S. R. Agnew, "Modeling the effect of twinning and detwinning during strain-path changes of magnesium alloy AZ31," *Int. J. Plast.*, vol. 25, no. 5, pp. 861–880, 2009, https://doi.org/10.1016/j.ijplas.2008.05.005.
13. H. R. W. U.F. Kocks, C. N. Tome, "Texture And Anisotropy: Preferred Orientations in Polycrystals and Their Effect on Materials Properties." pp. 1–676, 1998.
14. A. Arsenlis and D. M. Parks, "Crystallographic aspects of geometrically-necessary and statistically-stored dislocation density," *Acta Mater.*, vol. 47, no. 5, pp. 1597–1611, 1999, https://doi.org/10.1016/S1359-6454(99)00020-8.
15. U. F. Tome, C N, Lebensohn, R A, Kocks, "A model for texture development dominated by deformation twinning: application to zirconium alloys," *Acta Mater.*, vol. 39, no. Ii, pp. 2667–2680, 1991.
16. H. Wang, P. D. Wu, J. Wang, and C. N. Tomé, "A crystal plasticity model for hexagonal close packed (HCP) crystals including twinning and de-twinning mechanisms," *Int. J. Plast.*, vol. 49, pp. 36–52, 2013, https://doi.org/10.1016/j.ijplas.2013.02.016.
17. S. R. Agnew and Ö. Duygulu, "Plastic anisotropy and the role of non-basal slip in magnesium alloy AZ31B," *Int. J. Plast.*, vol. 21, no. 6, pp. 1161–1193, 2005, https://doi.org/10.1016/j.ijplas.2004.05.018.
18. F. Bachmann, R. Hielscher, and H. Schaeben, "Texture analysis with MTEX- Free and open source software toolbox," *Solid State Phenom.*, vol. 160, pp. 63–68, 2010, https://doi.org/10.4028/www.scientific.net/SSP.160.63.
19. C. M. Sellars and W. J. M. Tegart, "Hot Workability," *Int. Metall. Rev.*, vol. 17, no. 1, pp. 1–24, Jan. 1972, https://doi.org/10.1179/imtlr.1972.17.1.1.
20. S. Ando, N. Harada, M. Tsushida, H. Kitahara, and H. Tonda, "Temperature Dependence of Deformation Behavior in Magnesium and Magnesium Alloy Single Crystals," *Key Eng. Mater.*,

vol. 345–346, pp. 101–104, 2007, https://doi.org/10.4028/www.scientific.net/kem.345-346.101.

21. S. Ando, H. Rikihisa, M. Tsushida, and H. Kitahara, "Activities of non-basal slips in deformation of magnesium alloy single and poly crystals," *Mater. Sci. Forum*, vol. 941 MSF, pp. 1242–1247, 2018, https://doi.org/10.4028/www.scientific.net/MSF.941.1242.
22. S. Ando, A. Kodera, K. Fukushima, M. Tsushida, and H. Kitahara, "Tensile deformation of magnesium and magnesium alloy single crystals," *Mater. Sci. Forum*, vol. 783–786, pp. 341–345, 2014, https://doi.org/10.4028/www.scientific.net/msf.783-786.341.
23. S. Ando, M. Tsushida, and H. Kitahara, "Deformation behavior of magnesium single crystal in c-axis compression and a-axis tension," *Mater. Sci. Forum*, vol. 654–656, October, pp. 699–702, 2010, https://doi.org/10.4028/www.scientific.net/MSF.654-656.699.

Interactions of a Basal Edge Dislocation with Vacancies and Interstitials in Magnesium

Defei Li, Jing Tang, Zikun Li, Xiaobao Tian, Yan Li, and Haidong Fan

Abstract

As the lightest structural metal, magnesium (Mg) and its alloys have an important application on aircrafts, and would work under an irradiation condition, i.e. in outer space. So it is significant to study the microstructure and mechanical property of irradiated magnesium. In this work, we performed molecular dynamics simulations on the interactions of basal edge <*a*> dislocations with interstitials and vacancies. We found that both point defects have a blocking effect on dislocation motion. However, the blocking effect of interstitials is much stronger than that of vacancies. This is due to the different interaction types. The interstitials are absorbed by the basal dislocation, so the stronger blocking effect is a result of the short-range interaction, while the weaker blocking effect of vacancies is induced by the long-range interaction only since vacancies cannot be absorbed. Current work is useful for understanding the irradiation effects in magnesium and its alloys.

Keywords

Magnesium • Basal dislocation • Vacancies and interstitials • Hardening effect • Molecular dynamics simulation

D. Li
State Key Laboratory of Hydraulics and Mountain River Engineering, Sichuan University, Chengdu, China

D. Li · J. Tang · Z. Li · X. Tian · Y. Li · H. Fan (✉)
Department of Mechanics, Sichuan University, Chengdu, 610065, China
e-mail: hfan85@scu.edu.cn; haidongfan8@foxmail.com

Introduction

As the lightest structural metal, magnesium is attracting a lot of attention in recent years for its potential use in the aerospace, automotive, electronics, and defense applications [1–3]. On aircrafts in outer space, magnesium and magnesium alloys would work under an irradiation condition, where dense irradiation-induced defects are introduced in the materials, leading to the degeneration in the mechanical property and reduction of the service life [4]. So it is significant to study the microstructure and mechanical property of irradiated magnesium.

In the irradiated metals, high-energy particles knock the crystalline lattice of metallic materials, and leave dense point defects by collision cascade, such as vacancies and interstitials [5]. During plastic deformation, the point defects interact with dislocations and have a hardening effect on the mechanical property [6]. Clouet studied the interactions between vacancies and edge dislocations in face-centered cubic (FCC) metals (Al, Au, Cu, Ni) [7]. It was shown that the vacancy is attractive by the dislocation core above the slip plane (in the compressive region), while repulsive below the slip plane. In body-centered cubic (BCC) iron, a vacancy is repulsed while an interstitial is attracted by a screw dislocation, but the repulsive effect is weaker [8]. For edge dislocation in iron, the interaction between a vacancy and dislocation is far weaker than that with an interstitial [9]. In Fe–0.3Cu [10] and Fe–9Cr [11] alloys, screw dislocations are able to reduce the dimensions and density of vacancy-like clusters by combining interstitial- and vacancy-like defects. In aluminum, intrinsic mechanism is preferential for diffusion along a screw dislocation, whereas vacancy mechanism is predominant along an edge dislocation [12]. After the absorption of point defects, the edge dislocation climbs, while screw dislocation cross-slips, leading to a hardening effect on the mechanical property [13].

From the discussions mentioned above, we can see that the interactions between dislocations and irradiation-induced

V. M. Miller et al. (eds.), *Magnesium Technology 2021*, The Minerals, Metals & Materials Series,
https://doi.org/10.1007/978-3-030-65528-0_7

point effects were studied mostly in FCC and BCC metals, but were rarely investigated in magnesium and magnesium alloys. Therefore, in this work, we performed molecular dynamics (MD) simulations to study the interactions of interstitials and vacancies with basal dislocations, as well as their hardening effects on the mechanical behavior of magnesium.

Methodology

In the current work, the interactions of a basal edge <*a*> dislocations with interstitials and vacancies are studied using MD simulations. LAMMPS is used to conduct all the MD simulations [14]. In order to model the inter-atomic interaction, the Mg embedded atom method (EAM) potential that was developed by Liu et al. is utilized [15]. All atomic structure in the paper are visualized using software OVITO [16]. A simulation cell with the sizes of 20.5 nm × 60.8 nm × 60 nm, as shown in Fig. 1, contains initially a basal edge dislocation and interstitials/vacancies. Periodic boundary conditions are imposed along x- and y-directions, while free surface boundary conditions along z-direction.

The edge dislocation is realized by deleting two half-planes of atoms. The interstitials are introduced by creating atoms, and the vacancies are introduced by deleting atoms randomly. The concentration of point defects is varied from 0 to 8‰. At the beginning of all simulations, an energy minimization step is performed in order to obtain the stable configurations of dislocation and point defects in a stress-free condition. Then, a pure shear strain ε_{yz} is imposed on the free surfaces at a shear strain rate of $1.67 \times 10^8 s^{-1}$ to move the dislocation towards the point defects [17, 18].

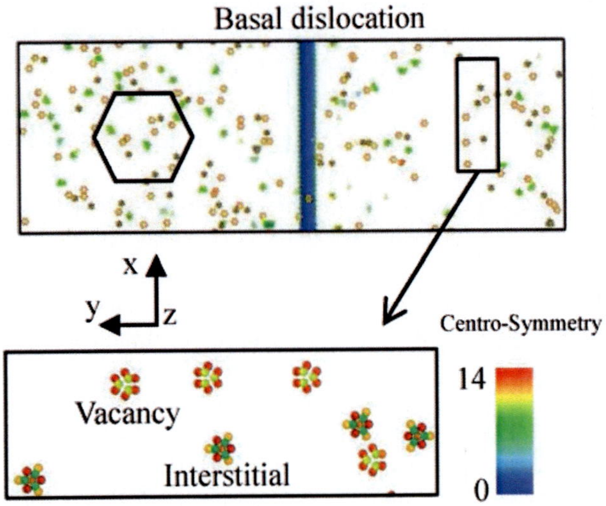

Fig. 1 Simulation cell showing a basal edge dislocation and interstitials/vacancies. (Color figure online)

Results and Discussion

The shear stress–strain curves as predicted from the current MD simulations for different concentrations of vacancies and interstitials are shown in Fig. 2. As shown in Fig. 2a in the case of interstitials and vacancies, as the strain increases, the stress increases linearly in the elastic stage, and increases nonlinearly in the plastic stage. However, as compared to the simulations without point defects, the stress level in the plastic stage increases with the increasing concentration of point defects, indicating a blocking effect of the point defects on dislocation motion. In order to see the different blocking effects of interstitials and vacancies, additional simulations were performed for interstitials only and vacancies only, as shown in Fig. 2b, c. By a comparison between the two cases, we can see that the stress levels of interstitials only in the plastic stage are much higher than those of vacancies only, suggesting a stronger blocking effect of interstitials than that of vacancies.

In order to see the blocking effects of these point defects quantitatively, we now study the glide resistance stress of the dislocation. In the simulations, the dislocation starts gliding at the yielding point. So the yielding shear stress can be treated as the dislocation glide stress or critical resolved shear stress (CRSS), here which is the shear stress at a plastic strain of 0.5‰. Then the blocking stress of the point defects on dislocation is calculated as the ΔCRSS, which is the increment of CRSS with respect to that without point defects. The blocking stress is plotted in Fig. 3 for the three cases studied. We can see that the blocking stress as a function of concentrations increases significantly for the case of interstitials only, while weakly for vacancies only (even though the concentration increases to 8‰), again confirming that the blocking effect of interstitials is stronger. In addition, the blocking effect of interstitials and vacancies is intermedium, which is a mixed blocking effect of interstitials and vacancies.

In order to explain the different blocking effects of interstitials and vacancies on dislocation motion, we plotted the interaction detail in Fig. 4. Here the point defects of different numbers (i.e. 1, 5, 10, 20, 50) are distributed on the slip plane of the dislocation, then we can see the interaction clearly. For the interstitials, i.e. in the simulation of only one

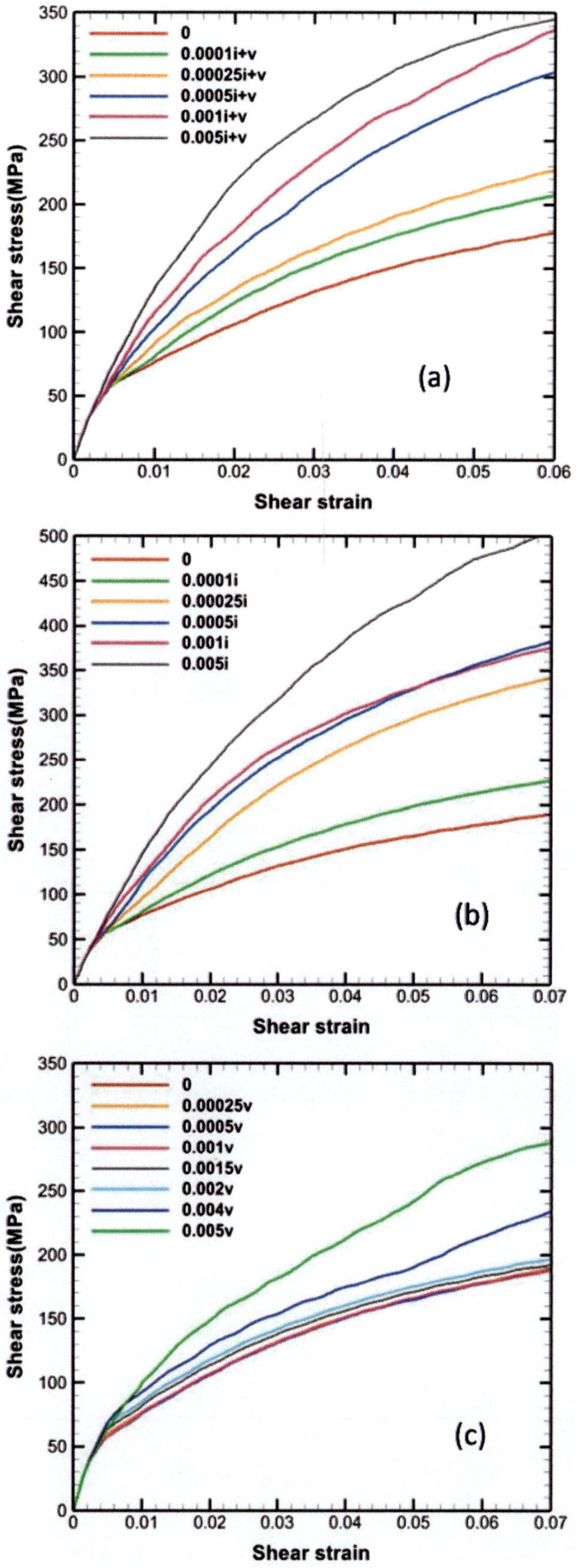

Fig. 2 Shear stress–strain curves at different point defect concentrations. **a** interstitials and vacancies (i + v); **b** interstitials only (i); **c** vacancies only (v). (Color figure online)

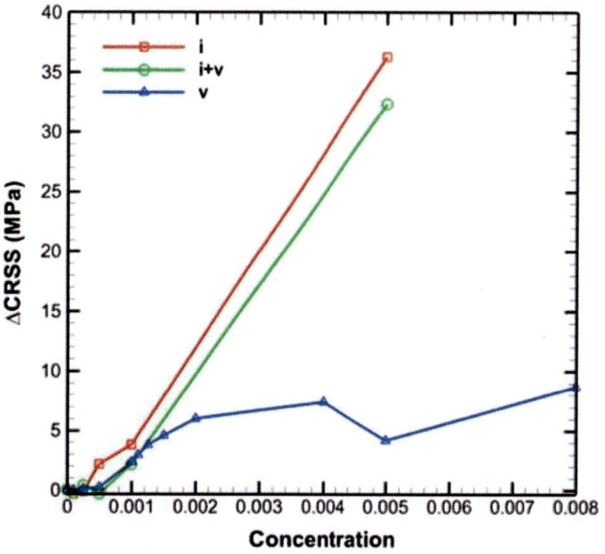

Fig. 3 The blocking effects of interstitials and vacancies as a function of the concentration. (Color figure online)

interstitial, the interstitial is observed to be absorbed by the dislocation, forming a jog on the dislocation. The dislocation is able to glide further while dragging the jog. As the interstitial number increases, i.e. 5–20 interstitials, the dislocation absorbs all the interstitials on the slip plane and a number of jogs are formed on the dislocation line, leading to the dislocation gliding difficulty. For the case of 50 interstitials, the dislocation still absorbs interstitials, but four interstitials are released from the jogged dislocation, which indicates that the capacity of interstitials tends to a saturation. In contrast, the dislocation is observed to absorb no vacancies at all, as shown in Fig. 4. However, the dislocation is impeded as well since the dislocation line becomes un-straight. This is because of the long-range interactions between the dislocation and vacancies.

Conclusions

Molecular dynamic simulations were employed to investigate the interactions of a basal edge <*a*> dislocation with interstitials and vacancies as well as their blocking effect on dislocation motion in magnesium. It was shown that both interstitials and vacancies have a blocking effect on dislocation motion, while the blocking effects are different as the point defect concentration increases. The blocking effect of interstitials is much stronger than that of vacancies. Such different blocking effects origin from the different interaction types. The interstitials are absorbed by the edge dislocation (short-range interactions), leading to a stronger blocking effect, while vacancies are not absorbed, leading to a weaker blocking effect by long-range interactions only.

Fig. 4 Interaction of dislocation with interstitials and vacancies. (Color figure online)

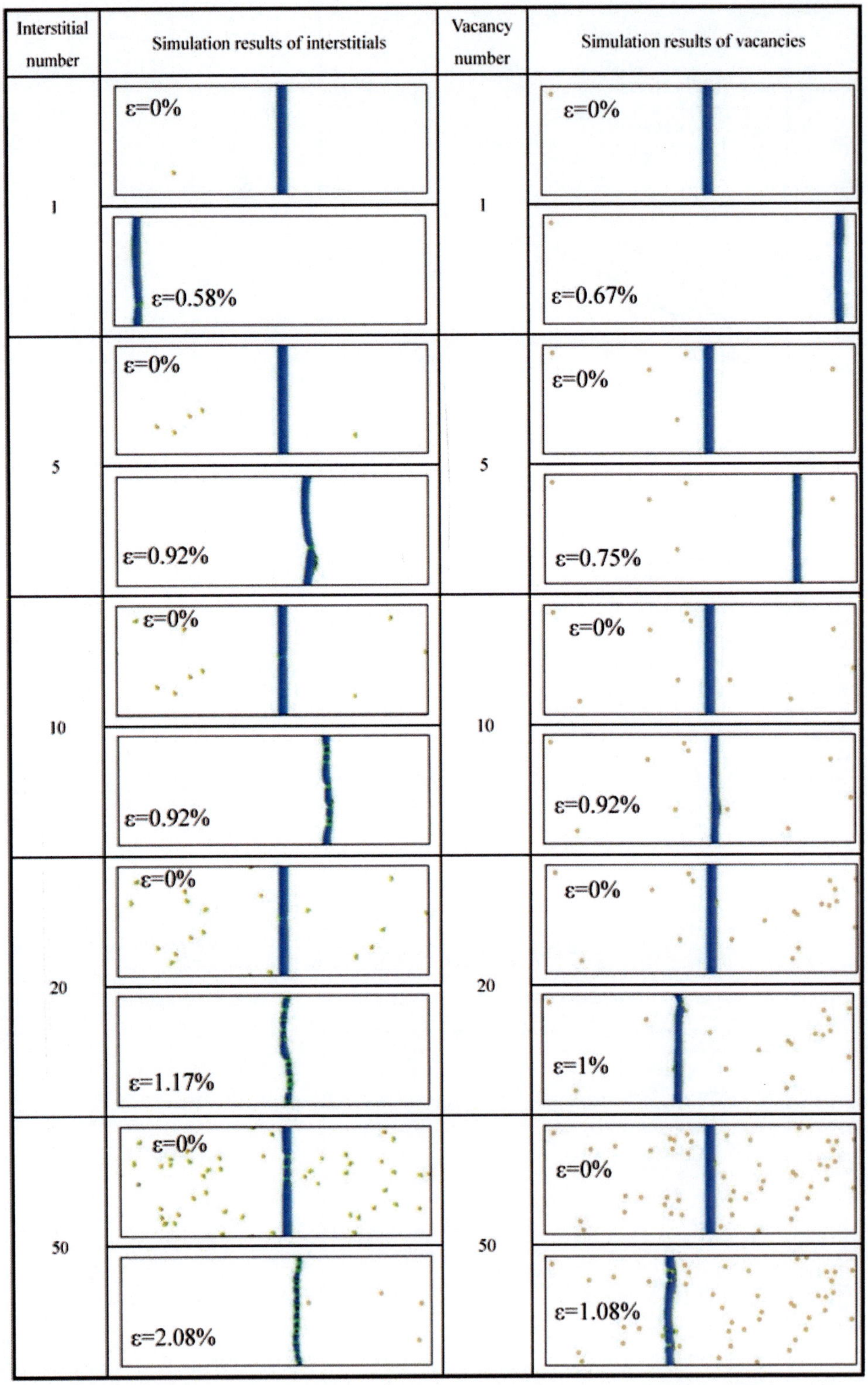

Acknowledgements The financial support from Sichuan Province Science and Technology Project (2020JDJQ0029) and National Natural Science Foundation of China (12072211, U1730106) is acknowledged.

References

1. Stalmann A, Sebastian W, Friedrich H, Schumann S, Dröder K, (2001) Properties and processing of magnesium wrought products for automotive applications. Advanced Engineering Materials. 3 602–608
2. Luo AA, (2013) Magnesium casting technology for structural applications. Journal of Magnesium and Alloys.1 2–22
3. Agnew SR, Duygulu Ö, (2005) Plastic anisotropy and the role of non-basal slip in magnesium alloy AZ31B. International Journal of Plasticity. 21 1161–1193
4. Fastenau RHJ, Baskes MI, (1983) A combined atomistic and monte carlo simulation of point defect–dislocation interactions. Physica Status Solidi. 75 323–334
5. Pál LI, Németh G, (1959) A statistical theory of lattice damage in solids irradiated by high-energy particles. Il Nuovo Cimento 12 293–309
6. Birnbaum HK, (1963) Formation of point defects during plastic deformation and their subsequent annealing behavior. Journal of Applied Physics. 34 2175–2185
7. Clouet E, (2006) The vacancy–edge dislocation interaction in fcc metals: A comparison between atomic simulations and elasticity theory. Acta Materialia. 54 3543–3552
8. Hayward E, Deo C, Uberuaga BP, Tomé CN, (2012) The interaction of a screw dislocation with point defects in bcc iron. Philosophical Magazine. 92 2759–2778
9. Shastry V, Diaz de la Rubia T, (1998) Interaction between point defects and edge dislocation in BCC iron. Journal of Engineering Materials and Technology. 121 126–128
10. Liu H, Wang X, Cui W, Li Q, Xu B, Liu W, Shu G, (2019) The effect of dislocations on irradiation-induced vacancy-like defects in FeCu alloy and reactor pressure vessel steel. Journal of Nuclear Materials. 524 80–89
11. Haley JC, Liu F, Tarleton E, Cocks ACF, Odette GR, Lozano-Perez S, Roberts SG, (2019) Helical dislocations: Observation of vacancy defect bias of screw dislocations in neutron irradiated Fe-9Cr. Acta Materialia. 181 173–184
12. Purja Pun GP, Mishin Y, (2009) A molecular dynamics study of self-diffusion in the cores of screw and edge dislocations in aluminum. Acta Materialia. 57 5531–5542
13. Voskoboinikov RE, (2013) Interaction of collision cascades with an isolated edge dislocation in aluminium. Nuclear Instruments & Methods in Physics Research. 303 125–128
14. Plimpton S, (1995) Fast parallel algorithms for short-range molecular dynamics. Journal of Computational Physics. 117 1–19
15. Liu XY, Ohotnicky PP, Adams JB, Lane Rohrer C, Hyland RW, Jr (1997) Anisotropic surface segregation in A1-Mg alloys. Surface Science. 373 357–370
16. Stukowski A, (2010) Visualization and analysis of atomistic simulation data with OVITO — the Open Visualization Tool. Modelling and Simulation in Materials Science and Engineering. 18 015012
17. Fan H, El-Awady JA, Wang Q, (2015) Towards further understanding of stacking fault tetrahedron absorption and defect-free channels – A molecular dynamics study. Journal of Nuclear Materials. 458 176–186
18. Fan H, El-Awady JA, (2015) Towards resolving the anonymity of pyramidal slip in magnesium. Materials Science and Engineering A. 644 318–324

Three-Dimensional Interaction of $\{10\bar{1}2\}$ Twin with Tilt Boundaries in Mg: Twin and Dislocation Transmission

Khanh Dang, John Graham, Carlos N. Tomé, and Laurent Capolungo

Abstract

While both dislocations and deformation twins accommodate plastic shear, the former are linear and the latter are three-dimensional domains bounded by complex interfaces. Therefore, the transmission of twins is much more complex than that of dislocations. In this work, we characterize the reactions and possible scenarios of associated interactions between twins and tilt boundaries using both atomistic and phase-field simulations. We find that the interaction is a competition between slip and twin transmission, depending on the geometrical alignment of the slip and/or twinning systems, resolved shear stress, and short-range interactions between intrinsic defects within the twins and grain boundaries. More importantly, we find that lateral twin transmission is easier than the forward twin transmission. We use a phase-field model to systematically investigate the role played by interfacial properties of the twin, such as facet energies and mobilities, on lateral and forward transmission into neighboring grains.

Keywords

Twinning • Twin GB interaction • HCP metals • Atomistic simulations • Phase field simulations

In polycrystals, twin growth is often influenced by its interactions with other structural defects such as dislocations [1], other twins [2], and grain boundaries (GBs). The interaction between dislocation and GBs has been extensively studied both experimental, theoretically and computationally. On the other hand, twin-GB interactions are typically analyzed by [3–11] adopting and adapting criteria for dislocation transmission across GB interactions, which may not be totally appropriate. While both dislocations and twins accommodate plastic shear, they are fundamentally different. The former are line defects while the latter are 3-dimensional domains bounded by interfaces/facets. As a consequence, the propagation of twins and their interactions with GBs are much more complex than those of dislocations. Further, insight about the structure and mobility of twin facets is essential to understand twin morphologies, growth and eventually transmission or lack thereof across grain boundaries.

Therefore, the goal of this paper is to characterize the reactions, possible transmission scenarios, and controlling factors associated with the twin interaction with grain boundaries. In this work, the forward and lateral interactions between $\{10\bar{1}2\}$ twin and tilt grain boundaries in Mg has been investigated using the combined approach of molecular dynamics (MD) and phase-field (PF) simulations. The outcomes of the interactions are summarized in Fig. 1 where the twin can get completely absorbed, transmit through the GB, or dislocations/stacking faults can be emitted from the GBs. Atomistic simulations of the interaction between the $\{10\bar{1}2\}$ twin and tilt boundaries in magnesium are performed using the classical atomistic simulation code LAMMPS [12]. The modified embedded atom method (MEAM) interatomic potential developed by Wu et al. [13], which is a reparameterization of a prior MEAM potential by Kim et al. [14], is used to model Mg systems due to its accurate reproduction of the lattice parameters, cohesive energy, and defect energies. Open Visualization Tool (OVITO) and its Dislocation Extraction Algorithm (DXA) is used to visualize and analyze the results from atomistic simulations [15, 16].

Figure 2 shows the configurations that are used to study the forward and lateral interactions between $\{10\bar{1}2\}$ twin and tilt grain boundaries in Mg. Specifically, one grain is oriented such that the X, Y, and Z axes are along η ($[\bar{1}011]$), K($\{10\bar{1}2\}$), and λ ($[\bar{1}2\bar{1}0]$) directions, respectively. The 2nd grain is rotated by θ^o about the certain axes relative to the 1st grain. The GB plane is always positioned normal to either η

K. Dang · J. Graham · C. N. Tomé · L. Capolungo (✉)
MST-8 Los Alamos National Laboratory, Los Alamos, NM 87544, USA
e-mail: laurent@lanl.gov

V. M. Miller et al. (eds.), *Magnesium Technology 2021*, The Minerals, Metals & Materials Series,
https://doi.org/10.1007/978-3-030-65528-0_8

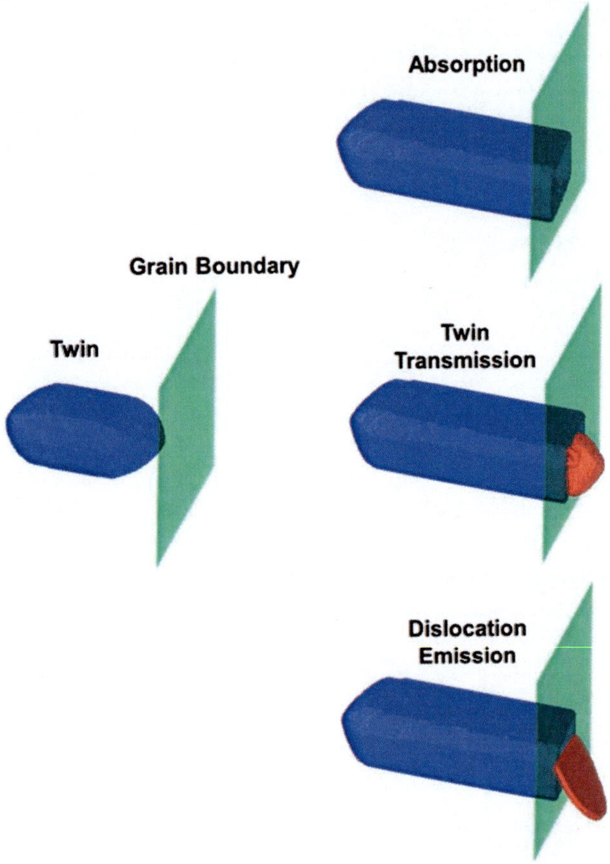

Fig. 1 Possible outcomes when twin interact with a grain boundary. (Color figure online)

or λ to allow perfect forward and lateral interactions, respectively, instead of mixed interactions that commonly used in previous MD studies with symmetric tilt grain boundaries [17]. Depending on the axis of rotation, different types of tilt grain boundaries can be studied. Here, two types of tilt grain boundaries are considered. Type I tilt boundaries are boundaries whose axes of rotation are λ and η for the forward and lateral interactions, respectively. On the other hand, type II tilt boundaries have axes of rotation parallel to the K direction. This allows a more direct comparison between lateral and forward interactions of the twin with the GBs since the geometric factor and global resolved shear stress of both cases are identical. For each configuration, five different θ angles are investigated to understand the role of misorientation angle to the outcome of the interactions as shown in Fig. 2.

Results from this study show that in addition to the geometric alignment parameters, several factors can influence the outcomes of the twin-GB interactions. First, both global resolved shear stress on the slip system and twinning modes of the neighboring grain plays a major role in the outcome of the interactions. Compared to the twinning event (transmission or nucleation), basal dislocation emission has much lower critical shear stress for activation, which is in qualitative agreement with experimental results on micropillars [18]. There are differences in the outcomes of forward and lateral interactions between twin and GBs. Using phase-field simulations, these differences are likely

Fig. 2 Atomistic simulations performed in this work. (Color figure online)

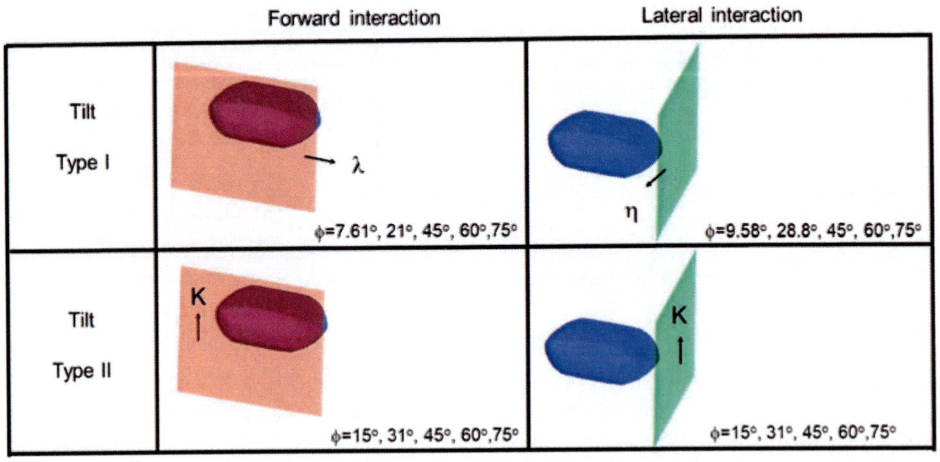

due to (1) the short-range interaction between the intrinsic defects within the twin facets and the grain boundary as well as (2) the local resolved shear stress.

References

1. M. Gong, G. Liu, J. Wang, L. Capolungo, C.N. Tomé, Acta Mater. 155 (2018) 187–198.
2. M. Gong, S. Xu, Y. Jiang, Y. Liu, J. Wang, Acta Mater. 159 (2018) 65–76.
3. I.J. Beyerlein, L. Capolungo, P.E. Marshall, R.J. McCabe, C.N. Tomé, Philos. Mag. 90 (2010) 2161–2190.
4. I.J. Beyerlein, R.J. McCabe, C.N. Tomé, J. Mech. Phys. Solids. 59 (2011) 988–1003.
5. A. Fernández, A. Jérusalem, I. Gutiérrez-Urrutia, M.T. Pérez--Prado, Acta Mater. 61 (2013) 7679–7692.
6. A. Khosravani, D.T. Fullwood, B.L. Adams, T.M. Rampton, M. P. Miles, R.K. Mishra, Acta Mater. 100 (2015) 202–214.
7. X. Hong, A. Godfrey, W. Liu, Scr. Mater. 123 (2016) 77–80.
8. M. Arul Kumar, I.J. Beyerlein, R.J. McCabe, C.N. Tomé, Nat. Commun. 7 (2016) 13826.
9. J. Kacher, J.E. Sabisch, A.M. Minor, Acta Mater. 173 (2019) 44–51.
10. X. Song, J. Luo, Z. Shi, L. Zhuang, Y. Qiao, J. Zhang, Mater. Des. 192 (2020) 108745.
11. C. Guo, R. Xin, J. Xu, B. Song, Q. Liu, Mater. Des. 76 (2015) 71–76.
12. S. Plimpton, J. Comput. Phys. 117 (1995) 1–19.
13. Z. Wu, M.F. Francis, W.A. Curtin, Model. Simul. Mater. Sci. Eng. 23 (2015) 015004.
14. Y.-M. Kim, N.J. Kim, B.-J. Lee, Calphad. 33 (2009) 650–657.
15. A. Stukowski, Model. Simul. Mater. Sci. Eng. 18 (2010) 015012.
16. A. Stukowski, V. V Bulatov, A. Arsenlis, Model. Simul. Mater. Sci. Eng. 20 (2012) 085007.
17. Z. Li, B. Xu, W. Liu, Comput. Mater. Sci. 154 (2018) 147–151.
18. Y. Liu, N. Li, M. Arul Kumar, S. Pathak, J. Wang, R.J. McCabe, N.A. Mara, C.N. Tomé, Acta Mater. 135 (2017) 411–421.

Thermally Activated Nature of Basal and Prismatic Slip in Mg and Its Alloys

Mohammed A. Shabana, Jishnu J. Bhattacharyya, Marek Niewczas, and Sean R. Agnew

Abstract

Throughout the literature, a large discrepancy exists among the activation volumes reported for Mg and its alloys. The present work surveys the reported values for basal and prismatic <a> slip of pure and alloyed Mg single crystals as well as polycrystals. A focus is placed on recent results obtained for rare earth element solutes, Sc and Y. The measured values are discussed in light of a recently developed predictive model for thermally activated basal-solute interaction in Mg alloys. It is found that if the single crystal activation volumes for basal slip in solid solution alloys are computed using the total stress instead of a presumed "thermal component" of the stress, i.e. admitting that thermal fluctuations can aid in overcoming any obstacles present in those materials, then the experimental results are in much better accordance with the theoretical predictions. Possible implications of the combined activities of different deformation modes on the activation volume of pure and alloyed Mg polycrystals are briefly introduced. Finally, using polycrystal elasto-viscoplasticity modelling, it is shown that under conditions relevant to tests performed on polycrystalline, solute-containing binary Mg alloys, basal slip can be the dominant deformation mode at 0.2% offset strain at which the initial activation volume is often assessed.

M. A. Shabana (✉) · J. J. Bhattacharyya · S. R. Agnew
Department of Materials Science and Engineering, University of Virginia, Charlottesville, VA 22904, USA
e-mail: mas5am@virginia.edu

J. J. Bhattacharyya
e-mail: jjb4cp@virginia.edu

S. R. Agnew
e-mail: agnew@virginia.edu

M. Niewczas
Department of Materials Science and Engineering, McMaster University, Hamilton, ON L8S 4L8, Canada
e-mail: niewczas@mcmaster.ca

Keywords

Strain rate sensitivity • Thermal activation • Activation volume • Solute-strengthening • Haasen plot

Introduction

One of the established frameworks for investigating the thermodynamics and kinetics of plasticity is to seek relationships between applied uniaxial stress (σ) and the plastic shear strain rate ($\dot{\varepsilon}_p$). We begin with the Orowan equation

$$\dot{\varepsilon}_p = \rho_m b \bar{v} \tag{1}$$

which involves ρ_m the mobile dislocation density, b the magnitude of the Burgers vector of those dislocations, and $\bar{v}$ the average dislocation velocity. For a mobile dislocation to sweep an area, it has overcome obstacles, such as the Peierls barrier, solute atoms, or forest dislocations. Thus, $\bar{v}$ is described by an Arrhenius relationship in which the activation energy barrier $\Delta G(\tau)$ is taken to be a function of the resolved shear stress (τ) on the slip plane.

$$\bar{v} = v_0 \lambda e^{-\left(\frac{\Delta G(\tau)}{kT}\right)} \tag{2}$$

where v_0 is a microscopic attempt frequency, λ is an effective dislocation displacement, k is the Boltzmann's constant, and T is temperature. The shape of the free energy–stress profile is often assessed in terms of the first-order Taylor series expansion,

$$\Delta G(\tau) = \Delta G^0 + \tau \frac{\partial \Delta G(\tau)}{\partial \tau} \tag{3}$$

where the slope of the free energy profile can be thought of as an activation "volume" (ΔV), which can be related to the area swept by the dislocation during an activation event by dividing the Burgers vector $\left(\Delta a = \frac{\Delta V}{b}\right)$.

V. M. Miller et al. (eds.), *Magnesium Technology 2021*, The Minerals, Metals & Materials Series,
https://doi.org/10.1007/978-3-030-65528-0_9

$$\Delta V = -\frac{\partial \Delta G(\tau)}{\partial \tau} \quad (4)$$

The scale of this activation volume (or area) has come to be associated with various obstacles that may control the relationship between applied stress and strain rate. Experimentalists assess this activation volume using either strain rate jump or stress relaxation tests. The strain rate jump test permits assessment of the steady state Strain Rate Sensitivity (SRS) parameter (also denoted m) through the measured instantaneous stress change ($\Delta\sigma$) between two different strain rates using the methodology described by Basinski [1] and Wagoner [2]. This is possible due to the fact that m is expressed in the form of

$$m = \frac{\partial \ln\sigma}{\partial \ln\dot{\varepsilon}_p} = \frac{\Delta\sigma}{\sigma\Delta\ln\dot{\varepsilon}_p} \quad (5)$$

In the case of stress relaxation testing, the SRS can still be determined since the plastic strain rate during stress relaxation is equivalent to the negative ratio between the time derivative of the stress ($\dot{\sigma}$) and the effective elastic modulus (E_{eff}) of the sample and machine, $\dot{\varepsilon}_p = -\dot{\sigma}/E_{eff}$. As described by Caillard and Martin [3], the stress rate can be obtained by fitting the recorded stress change with time with a logarithmic function and then taking the time derivative of the fitted expression. In either case, differentiating Eq. (1) to obtain the SRS and simplifying the resulting differential leads to

$$\frac{\partial \ln\dot{\varepsilon}_p}{\partial \ln\tau} = -\frac{\tau}{kT}\frac{\partial \Delta G(\tau)}{\partial \tau} = \frac{\tau\Delta V}{kT} \quad (6)$$

If one performs experiments on single crystals, the relationship between the applied stress state and the resolved shear stress (τ) on the active slip plane is readily obtained through the Schmid law. For polycrystalline cubic metals, it has frequently been adequate to assume conditions of "polyslip" within each of the grains during macroscopic plasticity, and hence, τ is reasonably related to σ by the upper bound approximation, the Taylor factor (M) by $\sigma = M\tau$. Under these conditions, the relationship between the rate sensitivity and the activation volume may be simply expressed as

$$\Delta V = \frac{MkT}{\sigma m} \quad (7)$$

and the Haasen plot representation of this relation simply obtained by substituting Eq. (5) into the inverse of Eq. (7) as

$$\frac{1}{\Delta V} = \frac{\Delta\sigma}{MkT\Delta\ln\dot{\varepsilon}_p} \quad (8)$$

We do not dwell on the complications associated with parsing total and mobile dislocation densities or with the detailed role of machine compliance, as these have been treated extensively elsewhere, [3] and are common to all such analyses. Rather, we discuss some of the specific complications which arise when seeking to perform this type of analysis for non-cubic metals, such as Mg and its alloys. For example, it is important to note that in non-cubic materials e.g. Mg polycrystals, a variety of slip/twin modes may be active, and one needs to be careful when using the Taylor factor M in the traditional sense, as introduced in Eqs. (7) and (8). An alternative has recently been proposed by some of the authors, in which crystal plasticity modeling is used to obtain an effective Taylor factor [4]. We begin with a review of relevant single crystal and polycrystal data and theory, seeking to provide some clarity to an otherwise confusing collection of results and theories. We then propose paths forward for researchers to relate insights from theory to practice.

Activation Volumes in Mg and Its Alloys

Table 1 summarizes all the values of activation volume at yield for the different slip conditions that are discussed throughout this section. It also contains values that were computed using the Leyson-Curtin model [5] and the theory from Yasi et al. [6] for non-basal (prismatic) slip. By analyzing the values of ΔV for the different cases presented in Table 1, it is convenient to categorize the activation volume values into three groups: **high** for basal slip and **low** for prismatic slip in pure single crystals, and **intermediate** for basal slip in alloys and both pure and alloyed polycrystals.

Basal Slip in Pure and Alloyed Mg Single Crystals

Throughout the literature, the activation energy is often taken to be a function of the "thermal stress," $\tau^* = \tau - \tau^G$, since some long-range barriers induce a level of back-stress,τ^G, on dislocations that cannot be overcome by thermal activation. Examples of obstacles which are expected to induce such an "athermal" barrier to dislocation motion include grain boundaries and large, unshearable particles. On the other hand, researchers have often observed an "athermal plateau" in the flow stress above a certain temperature, above which the flow stress only decreases in so much as the relevant shear moduli of the material decrease with temperature. Because of this, researchers have reported activation volumes as a function of the thermal stress,τ^*, often with rather complex deconvolution schemes [16, 17] used to extract this value from the applied stress,τ. Furthermore,

Table 1 A summary of the typical values of activation volume at yield for the different slip conditions examined in Activation Volumes in Mg and Its Alloys

Test Orientation	Mg	Species	$\Delta V(b^3)$	
			78 K	**298 K**
Single Crystal Basal Slip	Pure	–	2000 [7] 400 [8]	900 [8]
	Alloyed	Zn (0.15–0.45 at %)	240–430 [7] 110–230[a]	–
		Al (0.237–1.63 at %)	70–220[a]	–
		Y (0.5 at%)	75[a]	299 [a]
		Y (1 at%)	55[a]	60 [9], 188 [a]
		Dy (0.5 at%)	76[a]	313 [9], 303 [a]
Single Crystal Prismatic <a> Slip	Pure	–	5 [10]	50 [10], 30 [11] 30 [b]
	Alloyed	Zn (0.258–0.45 at %)	15–20 [10]	15–35 [10]
		Li (7.9, 12.9 at%)	25–50 [12]	100 [12]
Polycrystal	Pure	–	300 [13] (at 87 K) 261 [14]	200 [13] (at 197 K) 132 [14], 146 [15]
	Alloyed	Sc (0.2–0.4 at%)	122–193 [14]	383–531 [14]
		Y (0.3–1.3 at%)	41–245 [15]	137–174 [15]

Note The values with (a) sign, were obtained using the Leyson-Curtin model [5]. The values with (b) sign, were obtained using the theory from Yasi et al. [6]

researchers have rather arbitrarily assigned mechanisms responsible for the athermal stress, such as forest dislocations, [7] which are otherwise known to be obstacles potentially overcome by thermal activation.

Leyson et al. [5] have recently proposed a theory for the thermally activated flow of basal oriented single crystals, which obviates the need to distinguish between the thermal and athermal portions. The origin for the apparent athermal plateau is explained in their paper as due to broad dislocation core spreading that results in two different activation energies, one of which is large (of order ~ 10 eV). This leads to a transition in the dominant activation energy configuration with increasing temperature; a transition that is indicated by a "plateau" at the high temperature regime in the strength vs temperature plot. The high energy barrier results in a very weak dependence of strength upon temperature in the higher temperature regime. Per Leyson's proposal, reports of the activation volumes, in Mg alloy single crystals oriented for basal slip, have been re-examined without the assumption of an "athermal" stress and these activation volumes are reported in Table 1.

Conrad et al. [17] derived activation volumes for basal slip (ΔV^{basal}) of pure Mg single crystals. Akhtar and Teghtsoonian [7] performed a similar analysis and concluded that pure Mg single crystals have $\Delta V^{basal} \approx 2 \times 10^4 b^3$ after they remove an athermal stress component. These values are similar to those observed by numerous researchers who have examined pure FCC metal single crystals, including a recent study by Bochniak [18]. However, the value at yield is approximately $2000\,b^3$ if one employs the entire stress (i.e.τ, without subtracting off an athermal component τ^G). A recent study by Bhattacharya and Niewczas [8] reinvestigated the thermally activated plasticity of single crystal pure Mg oriented for basal single-slip, using strain rate jump tests and found the activation volume, $\Delta V^{basal} \approx 900\,b^3$, at yield under ambient temperature conditions. The precise reason for the discrepancy between one data set and the next is not known, but two possibilities are variations in the grown-in dislocation density (which remained even after pre-annealing) and differences in the impurity level within the nominally pure Mg samples. At higher strain levels and higher associated dislocation densities, the activation volumes are observed to similarly decrease. Ultimately, our interest here is in alloys in which the reported variations are not nearly large, so the discrepancies in the activation volumes of pure single crystals oriented for basal slip is not alarming.

Miura et al. report a value of $2000\,b^3$ for Mg-0.5%Dy after assuming the presence of a thermal and athermal component to the applied stress [9]. However, accounting for the total stress reveals $\Delta V = 313\,b^3$ at RT. This value is in agreement with the one obtained by using the

Leyson-Curtin model [19, 20] for Mg-0.5%Dy ($\Delta V \approx 303\,b^3$) at ambient. However, a discrepancy arises when this concept is applied to Mg-1%Y; the reported activation volume is about $1200b^3$ [9], and it becomes about $60b^3$ when the total stress is employed in the ΔV calculation. The Leyson-Curtin model prediction is much larger. We note that the theoretically predicted ΔV for Mg-0.5%Y and Mg-0.5%Dy were found to be the same at the cryogenic and ambient temperatures shown in Table 1. Finally, it is worth noting that the Leyson-Curtin model was also used to predict the behavior of single crystal Mg–Al and Mg-Zn alloys by Tehranchi and Curtin [21], and the activation volumes implied by their predictions are in the low hundreds of b^3 for both Mg–Al and Mg-Zn. These predictions are only a factor of two off from the experimental observations, and such predictions are admittedly challenging, relative to predicting the flow stress values [7].

Thus, for basal slip, all the alloys reported in Table 1 (except for Mg-1%Y) show corrected activation volume values consistent with the Leyson-Curtin model when the total stress is used. The presently observed general agreement between the Leyson-Curtin model and existing experimental observations provides a new lens through which to reconsider Basinski's observation of "stress-equivalency" in a wide range of FCC metal alloys and HCP Mg alloys oriented for basal slip [22]. Note however that previous researchers observed that the dislocation density in the material increased as the solute content was increased [7]. It is interesting that the contribution of this increased dislocation density to the apparent "athermal" plateau did not need to be taken into account to obtain agreement with experiments performed on Mg-Zn, Mg–Al, and Mg-0.5%Dy [21]. The effects of dislocation–dislocation interactions in the presence of solute are the subject of ongoing study, and it is speculated that they may help to explain the observed discrepancy for the case of Mg-1%Y.

Prismatic Slip in Pure and Alloyed Mg Single Crystals

A few studies have focused on the thermally activated nature of prismatic slip of <a> dislocations in pure Mg. One of the most often cited studies was performed by Ward-Flynn et al. [11], whose data suggest an activation volume, $\Delta V^{prism} = 30b^3$, near ambient temperatures, as reported by Couret and Caillard [23] and Yasi et al. [6]. Couret and Caillard as well as Püschl [24] go on to report the possibility of quite small activation volumes ($\Delta V^{prism} \approx 10b^3$) for single crystal prismatic slip in pure Mg. More importantly, they highlight the fact that a single mechanism of prismatic "cross" glide, which they denote as the double-jog (or double-kink) mechanism occurs over a wide range of temperatures from 50 to 650 K as opposed to transitioning to a single bow-out mechanism at low temperatures and high stresses. This is rationalized as due to the fact that the stable configuration for <a> dislocations in Mg is dissociated within the basal plane, regardless of the plane of motion [23].

Akhtar and Teghtsoonian [10] examined the effect of solid solution alloying on the thermally activated response of single crystals oriented for prismatic slip and found ΔV^{prism} to be in the range of $15b^3 - 35b^3$ at RT for Mg-Zn alloys of different concentrations and $\sim 50b^3$ for pure Mg. Below ambient temperatures, ΔV^{prism} does not seem to change much ($\sim 15b^3 - 20b^3$) for the alloys while it becomes $\sim 5b^3$ for pure Mg. Ahmadieh et al. [12] investigated the same for single crystal Mg-Li alloys and found $\Delta V^{prism} \approx 100b^3$ around ambient which decreases to small tens at lower temperatures. These observations make sense in light of the fact that prismatic slip is a hard slip mode in which a large pseudo-Peierls barrier is overcome via the aforementioned double-kink mechanism [6, 23, 24]. Conceptually, the addition of solutes can either increase or decrease the activation volume because the softening observed at low temperatures is associated with solute-induced easing of kink *formation* and solute hardening observed at higher temperatures is associated with an increased difficulty of kink *motion* [6].

Pure and Alloyed Mg Polycrystals

For polycrystalline pure Mg, a study by Sastry et al. [13] examined the activation volumes at 87 and 197 K. The values were extrapolated from the reported activation volume plot and were found to lie between $200b^3$ and $300b^3$ for fine-grained (3 μm) pure Mg polycrystals. Sastry also examined the case of 30-μm grains, and the activation volumes were much higher than in the fine-grained case. Other studies investigated other commercial polycrystalline Mg alloys including the ones made by Lukáč and Trojanová [25] and Trojanová et al. [26] for AZ31 and AZ63, respectively.

There are several published works by Silva et al. [14, 27, 28] which discuss Mg-Sc polycrystals. It is helpful to refer to the Mg-Sc phase diagram by Kang et al. [29] which demonstrates the limited solubility of Sc solute atoms in Mg. In fact, the phase diagram indicates that Sc is nearly insoluble in Mg at ambient temperatures. Silva et al. [28] reported the presence of precipitates—scarce but present—noting that it was difficult to produce Mg-Sc solid solutions. Silva [14] reported activation volumes in the small hundreds of b^3 for alloys with 0.2, 0.3, and 0.4 at% Sc. In another study, Jia [15] reports on the thermally activated plasticity of Mg-Y solid solution alloyed polycrystals. Here, the activation volume data are reported to be in the order of tens to small hundreds of b^3 as shown in Table 1. Both Mg-Sc and Mg-Y are further discussed in the terms of the so-called Haasen plot.

Haasen Plot for Binary Mg Alloy Polycrystals

A plot of the inverse activation volume as a function of the flow stress is called a "Haasen plot" in the literature. This plot is developed on the premise of the Cottrell-Stokes Law, which states that the (CS) ratio of flow stresses of a material at two different temperatures, for a specified strain rate, does not depend on the level of strain hardening [30]. Peter Haasen [31] mentioned that the Cottrell-Stokes law would still be valid with additional contribution to the flow stress and determined that a difference between flow stresses (instead of the CS ratio) would present a linear relationship with flow stress. Since a change in temperature leads to changes in the elastic modulus and the dislocation core configuration, an analogous understanding of the Cottrell-Stokes Law was developed by employing a strain rate change at a specified temperature instead of a temperature change at a specified strain rate [32]. This change in flow stress with respect to a change in strain rate corresponds to our definition of SRS. Thus, the stress change and the SRS can be related to the inverse of the activation volume by Eq. (8). The Haasen plots obtained from up-jump tests of Mg-Sc and Mg-Y at 78 K are presented in Fig. 1. We note here that Silva [14] and Jia [15] performed both up-jump and down-jump tests, showing similar trends, so the up-jump data are deemed representative.

Two things are evident from the figure. First, for pure, polycrystalline Mg as well as for all the alloys, the data are linear and secondly, it appears that neither Sc nor Y additions change the slope of the Haasen plot, with respect to that of the pure Mg. A constant SRS (i.e. the Haasen plot slope) with straining suggests that the generalized Cottrell-Stokes law holds, and the activation volume, ΔV decreases with strain hardening as $1/(\sigma - \sigma_0)$. The intercept of the Haasen plot provides valuable information regarding the solute interaction with dislocations at yield and the slope of the Haasen plot provides insight about the effect of solutes on dislocation–dislocation interaction. The results are consistent with a simple linear superposition law for the inverse activation volumes of the two obstacles to dislocation motion (solute and forest dislocations).

Figure 2a shows a plot of the intercept of the Hassen plot (Fig. 1) as a function of solute content. Interestingly, the intercepts for both Sc and Y solutes appear to linearly increase with solute addition. However, it is hard to be certain for Sc due to its limited solubility in Mg. For both Sc and Y, the data from the alloys would extrapolate close to the origin, indicating that the activation volume is large (as expected) for the pure material (i.e. zero solute content). The actual measurement from the pure polycrystalline Mg samples, on the other hand, has an intercept value of ~0.008, which may be due to impurities present in the alloys (and we note that commercial purity 99.8% material was employed in the study), or it could be due to the composite effect of simultaneous activity of basal and prismatic slip within the polycrystal, as discussed in the polycrystal plasticity modelling Polycrystal Modeling below.

Figure 2b shows the SRS obtained from the Haasen plot as a function of solute content. This plot shows that the SRS values lie within 0.006 and 0.008, for both Sc and Y. In other words, it appears insensitive to solute content. The 1.30 at% Y data may be an outlier, so it is difficult to conclude whether the SRS actually increases at higher solute content or if this is related to some other factor such as the fact that the grain size of this sample was much smaller for this sample than in the other Mg-Y samples [15]. Recall that Sastry noted a decrease in ΔV (increase in SRS) for finer grain sizes in pure Mg polycrystals. The similarity in the Haasen plot intercept and the SRS for Sc and Y solutes at 78 K, suggests that they are not that different with regards to thermal activation despite the fact that Y is predicted to have much stronger interaction energies than Sc [33] and again raises questions concerning "stress equivalency" [22]. It is important to note that the Haasen plot analysis at ambient temperatures leads to a very different result; both Mg-Sc [14] and Mg-Y [15] show a strong decrease in slope at RT, even reaching negative values which is likely associated with dynamic strain aging (DSA), and this is the focus of an ongoing study.

Polycrystal Modeling

Crystal plasticity modeling enables parsing the contributions from different deformation modes in non-cubic metals and alloys like Mg, where multiple modes can potentially be active. Out of several homogenization schemes often employed in crystal plasticity models, an Elasto-ViscoPlastic Self-Consistent (EVPSC) polycrystal plasticity code is used in the present work to demonstrate the effect of having multiple modes active in a polycrystal. The description of the model is detailed elsewhere [34–37], hence is not repeated here for brevity. A key point to keep in mind is that the viscoplastic constitutive response is described by a power law flow rule: $\dot{\varepsilon}_p = \dot{\gamma}_0 \sum_i m^\alpha \left(\frac{m^\alpha : \sigma}{\tau^\alpha}\right)^n$ where, $\dot{\gamma}_0$ is the reference shear rate, m^α is the Schmid tensor, σ is the grain level stress, τ^α is the current threshold stress of slip mode i, and n is the

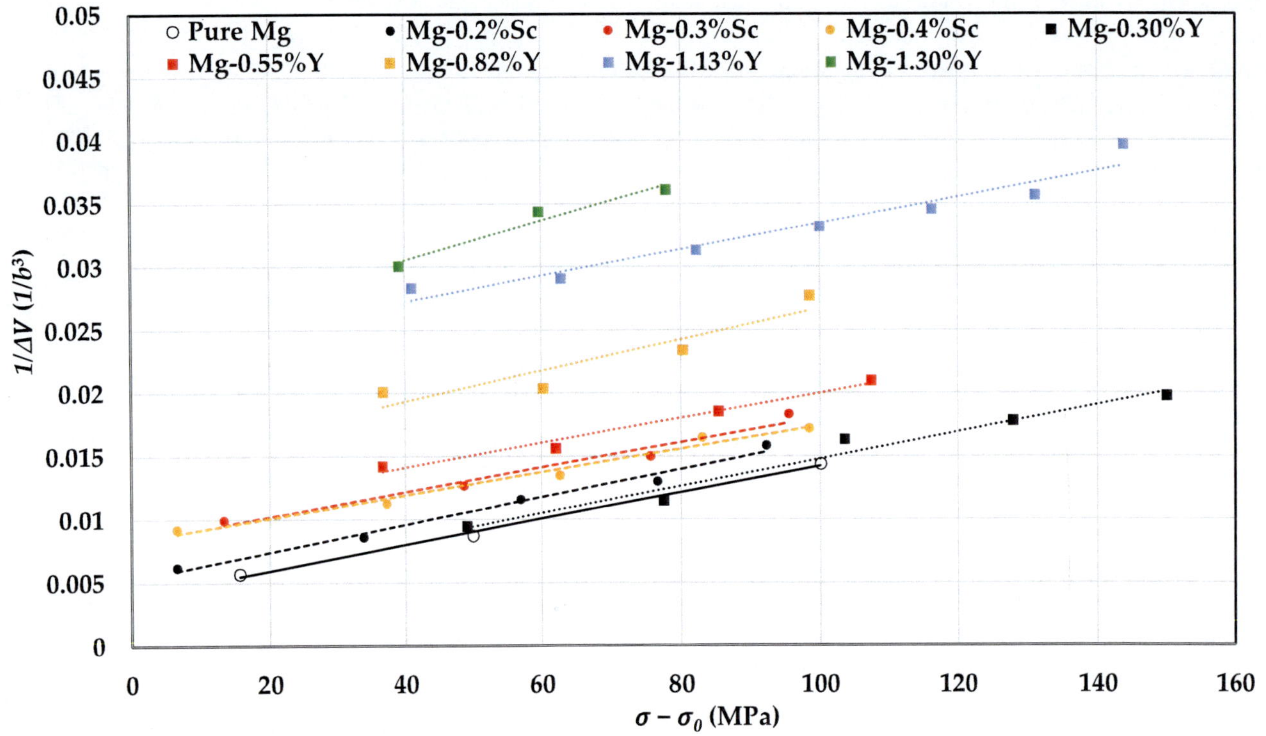

Fig. 1 Combined Haasen plots for different concentrations of Sc and Y in Mg, obtained from extracted data of tensile rate jump tests at 78 K [14, 15]. (Color figure online)

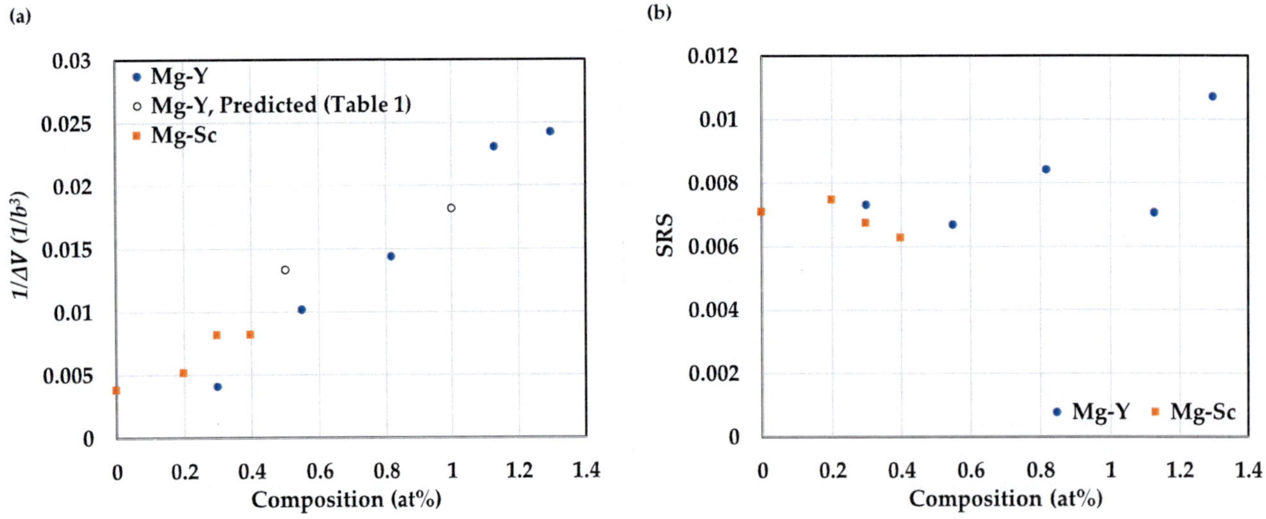

Fig. 2 78 K Haasen plot intercepts (a) and slopes (SRS) (b) for Mg-Sc and Mg-Y superimposed on each other. (Color figure online)

stress exponent (i.e., the inverse of Eq. (5)). For the present work, the stress exponents reported by Wang et al. [38] for Mg alloy AZ31B are used, i.e. the n value for basal, prismatic, <c + a> slip, and extension twinning are 100, 80, 33 and 120, respectively.

The stress–strain response and the corresponding relative activities for a uniaxial test along the rolling direction (RD) of a moderately basal-textured Mg polycrystal sheet is shown in Fig. 3. The texture employed for this simulation has a maximum strength of ~4 m.r.d, typical of low solute-containing Mg-Y alloys [39]. For the purpose of demonstration, the simulations were carried out using Voce parameters that are typical of Mg-Y alloys e.g. where the ratios of the critical resolved shear stresses (CRSS) for basal,

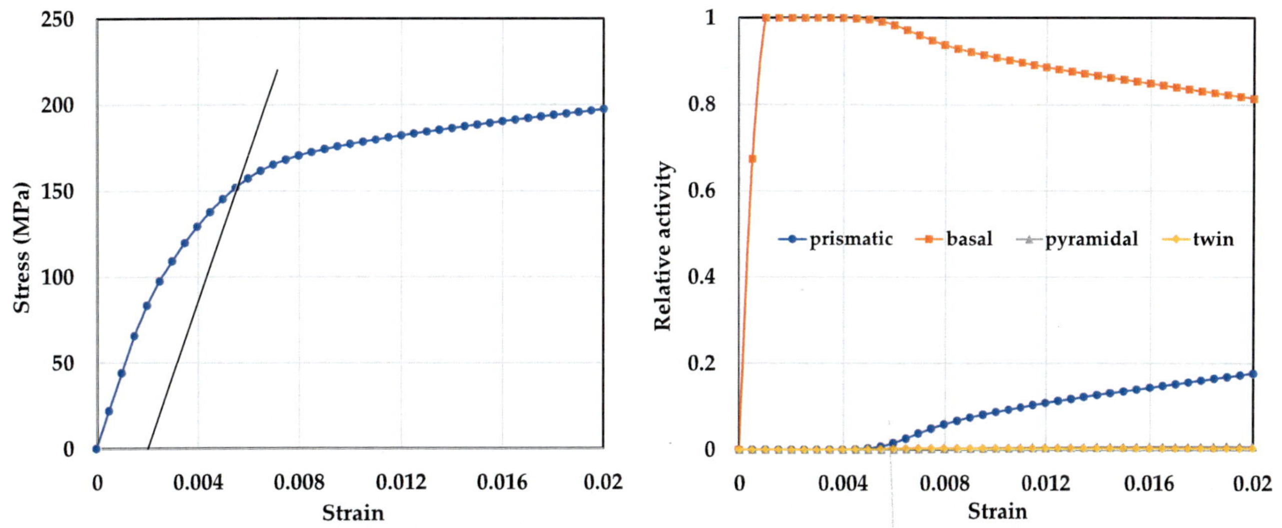

Fig. 3 EVPSC simulated stress–strain response showing the 0.2% offset line which is typically used to determine the yield stress of the material, and corresponding relative activities, showing that basal slip is the dominant deformation mode up to a strain of 0.006, after which the prismatic slip activates. (Color figure online)

prismatic, <c + a> slip, and extension twinning are 1:5:6:4, respectively [40]. On the stress–strain curve, the 0.2% offset line, typically used to determine the yield stress, is shown. The yield stress thus obtained is ~150 MPa, and the strain value corresponding to this stress is 0.0055. It is evident from the relative activity plot that up to a strain-level of ~0.006, basal slip is the dominant deformation mode. At a strain of 0.006, the prismatic slip is activated, which results in a reduction in the basal slip activity. This suggests that the activation volume at the 0.2% offset yield condition would resemble that of the basal slip rather than the prismatic slip for these polycrystals. This is indeed observed in Table 1 and Fig. 2, where, depending on the exact composition, the polycrystal activation volume values for Y containing alloys are found to be 41–245 b^3. These values are consistent with the theoretical predictions of the Leyson-Curtin model for basal slip in Mg-Y alloys shown in Table 1.

Conclusions

1. The activation volume predictions of the Leyson-Curtin model for basal slip in solute-containing Mg alloys are in general agreement with that reported in the literature, provided the total stress instead of the "thermal" stress is used to recalculate the experimental activation volume values.
2. The generalized Cottrell-Stokes law holds for the binary solid solution Mg alloys with the rare earth elements, Sc and Y, tested at 78 K and the inverse of their activation volumes at yield show a linear relationship with composition.
3. For both Mg-Sc and Mg-Y polycrystals, the SRS values at 78 K lie within 0.006 and 0.008, over the range of compositions investigated, and it appears that the SRS is insensitive to the amount of solute present.
4. The activation volumes at yield lie between 41 and 245 b^3 for Mg-Y polycrystals, with Y solute content ranging from 0.3 to 1.3 at.%, and these values are consistent with the theoretical predictions of the Leyson-Curtin model for basal slip in Mg-Y alloys.
5. Polycrystal plasticity simulations reveal that basal slip is the dominant deformation mode at the 0.2% offset strain, in a moderately textured (~4X random) Mg- Y alloy sheet, thereby helping to explain the similarity between the thermal activation of single crystals oriented for basal slip and polycrystals.

Acknowledgements The research at University of Virginia was sponsored by the Department of Energy, Basic Energy Sciences, Mechanical Behavior, and Radiation Effects Program led by Dr. John Vetrano, Grant # DE-SC0018923. Also, helpful discussions with W.A. Curtin (EPFL) and A. Kula (AGH University of Science and Technology) are gratefully acknowledged.

References

1. Basinski, Z. S., Jackson, P. J., & Duesbery, M. S. (1977). Transients in steady-state plastic deformation produced by changes of strain rate. *Philosophical Magazine*, *36*(2), 255-263.
2. Wagoner, R. H. (1984). Strain-rate sensitivity of zinc sheet. *Metallurgical and Materials Transactions A*, *15*(6), 1265-1271.

3. Caillard, D., & Martin, J. (2003). Experimental characterization of dislocation mechanisms. *Thermally activated mechanisms in crystal plasticity, Pergamon Materials Series, Elsevier Science, Cambridge, UK*, 15-18.
4. Bajikar, V., Bhattacharyya, J. J., Peterson, N., & Agnew, S. R. (2019). Thermally Activated Slip in Rare Earth Containing Mg-Mn-Ce Alloy, ME10, Compared with Traditional Mg-Al-Zn Alloy, AZ31. *JOM*, *71*(6), 2040-2046.
5. Leyson, G. P. M., Hector Jr, L. G., & Curtin, W. A. (2012). First-principles prediction of yield stress for basal slip in Mg–Al alloys. *Acta Materialia*, *60*(13-14), 5197-5203.
6. Yasi, J. A., Hector Jr, L. G., & Trinkle, D. R. (2011). Prediction of thermal cross-slip stress in magnesium alloys from direct first-principles data. *Acta Materialia*, *59*(14), 5652-5660.
7. Akhtar, A., & Teghtsoonian, E. (1969). Solid solution strengthening of magnesium single crystals—I alloying behaviour in basal slip. *Acta Metallurgica*, *17*(11), 1339-1349.
8. Bhattacharya, B., & Niewczas, M. (2011). Work-hardening behaviour of Mg single crystals oriented for basal slip. *Philosophical Magazine*, *91*(17), 2227-2247.
9. Miura, S., Imagawa, S., Toyoda, T., Ohkubo, K., & Mohri, T. (2008). Effect of rare-earth elements Y and Dy on the deformation behavior of Mg alloy single crystals. *Materials Transactions*, 0804070382–0804070382.
10. Akhtar, A., & Teghtsoonian, E. (1969). Solid solution strengthening of magnesium single crystals—ii the effect of solute on the ease of prismatic slip. *Acta Metallurgica*, *17*(11), 1351-1356.
11. Flynn, P. W., Mote, J. E. D. J., & Dorn, J. E. (1961). On the thermally activated mechanism of prismatic slip in magnesium single crystals. *Transactions of the Metallurgical Society of AIME*, *221*(6), 1148-1154.
12. Ahmadieh, A., Mitchell, J., & Dorn, J. (1964). Lithium alloying and dislocation mechanisms for prismatic slip in magnesium. (No. UCRL-11417). California. Univ., Berkeley. Lawrence Radiation Lab.
13. Sastry, D. H., Prasad, Y. V. R. K., & Vasu, K. I. (1970). The rate-controlling dislocation mechanism for plastic flow in polycrystalline magnesium. *Current Science*, 97–100.
14. Catherine J. Silva. (2014). Effect of Sc Addition on the Mechanical Properties of Mg-Sc Binary Alloys. M.S. thesis, McMaster University.
15. Jia, X. (2013). Solid solution strengthening and texture evolution in Mg-Y alloys. M.S. thesis, McMaster University.
16. Conrad, H., & Robertson, W. D. (1957). Effect of temperature on the flow stress and strain-hardening coefficient of magnesium single crystals. *JOM*, *9*(4), 503-512.
17. Conrad, H., Hays, L., Schoeck, G., & Wiedersich, H. (1961). On the rate-controlling mechanism for plastic flow of Mg crystals at low temperatures. *Acta Metallurgica*, *9*(4), 367-378.
18. Bochniak, W. (1995). Mode of deformation and the Cottrell-Strokes law in FCC single crystals. *Acta metallurgica et materialia*, *43*(1), 225-233.
19. Leyson, G. P. M., & Curtin, W. A. (2016). Solute strengthening at high temperatures. *Modelling and Simulation in Materials Science and Engineering*, *24*(6), 065005.
20. Leyson, G. P. M., & Curtin, W. A. (2016). Thermally-activated flow in nominally binary Al–Mg alloys. *Scripta Materialia*, *111*, 85-88.
21. Tehranchi, A., Yin, B., & Curtin, W. A. (2018). Solute strengthening of basal slip in Mg alloys. Acta Materialia, *151*, 56-66.
22. Basinski, Z. S., Foxall, R. A., & Pascual, R. (1972). Stress equivalence of solution hardening. *Scripta Metallurgica*, *6*(9), 807-814.
23. Couret, A., & Caillard, D. (1985). An in situ study of prismatic glide in magnesium—I. The rate controlling mechanism. *Acta Metallurgica*, *33*(8), 1447-1454.
24. Püschl, W. (2002). Models for dislocation cross-slip in close-packed crystal structures: a critical review. *Progress in Materials Science*, *47*(4), 415-461.
25. Lukáč, P., & Trojanová, Z. (2011). Stress relaxation in an az31 magnesium alloy. In *Key Engineering Materials* (Vol. 465, pp. 101–104). Trans Tech Publications Ltd.
26. Trojanová, Z., Máthis, K., Lukáč, P., Németh, G., & Chmelík, F. (2011). Internal stress and thermally activated dislocation motion in an AZ63 magnesium alloy. *Materials Chemistry and Physics*, *130*(3), 1146-1150.
27. Silva, C. J., Kula, A., Mishra, R. K., & Niewczas, M. (2018). The effect of Sc on plastic deformation of Mg–Sc binary alloys under tension. *Journal of Alloys and Compounds*, *761*, 58-70.
28. Silva, C. J., Kula, A., Mishra, R. K., & Niewczas, M. (2017). Mechanical properties of Mg-Sc binary alloys under compression. *Materials Science and Engineering: A*, *692*, 199-213.
29. Kang, Y. B., Pelton, A. D., Chartrand, P., & Fuerst, C. D. (2008). Critical evaluation and thermodynamic optimization of the Al–Ce, Al–Y, Al–Sc and Mg–Sc binary systems. *Calphad*, *32*(2), 413-422.
30. Cottrell, A. H., & Stokes, R. J. (1955). Effects of temperature on the plastic properties of aluminium crystals. *Proceedings of the Royal Society of London. Series A. Mathematical and Physical Sciences*, *233*(1192), 17–34.
31. Haasen, P. (1958). Plastic deformation of nickel single crystals at low temperatures. *Philosophical Magazine*, *3*(28), 384-418.
32. Nabarro, F. R. N. (1990). Cottrell-Stokes law and activation theory. *Acta Metallurgica et Materialia*, *38*(2), 161-164.
33. Yasi, J. A., Hector Jr, L. G., & Trinkle, D. R. (2012). Prediction of thermal cross-slip stress in magnesium alloys from a geometric interaction model. *Acta Materialia*, *60*(5), 2350-2358.
34. Molinari, A., Ahzi, S., & Kouddane, R. (1997). On the self-consistent modeling of elastic-plastic behavior of polycrystals. *Mechanics of Materials*, *26*(1), 43-62.
35. Mercier, S., & Molinari, A. (2009). Homogenization of elastic–viscoplastic heterogeneous materials: Self-consistent and Mori-Tanaka schemes. *International Journal of Plasticity*, *25*(6), 1024-1048.
36. Wang, H., Wu, P. D., Tomé, C. N., & Huang, Y. (2010). A finite strain elastic–viscoplastic self-consistent model for polycrystalline materials. *Journal of the Mechanics and Physics of Solids*, *58*(4), 594-612.
37. Bhattacharyya, J. J., Sasaki, T. T., Nakata, T., Hono, K., Kamado, S., & Agnew, S. R. (2019). Determining the strength of GP zones in Mg alloy AXM10304, both parallel and perpendicular to the zone. *Acta Materialia*, *171*, 231-239.
38. Wang, H., Wu, P., Kurukuri, S., Worswick, M. J., Peng, Y., Tang, D., & Li, D. (2018). Strain rate sensitivities of deformation mechanisms in magnesium alloys. *International Journal of Plasticity*, 107, 207-222.
39. Kula, A., Jia, X., Mishra, R. K., & Niewczas, M. (2017). Flow stress and work hardening of Mg-Y alloys. *International Journal of Plasticity*, *92*, 96-121.
40. Stanford, N., Cottam, R., Davis, B., & Robson, J. (2014). Evaluating the effect of yttrium as a solute strengthener in magnesium using in situ neutron diffraction. *Acta Materialia*, *78*, 1-13.

Mechanisms and Machine Learning for Magnesium Alloys Design

Zongrui Pei

Abstract

We will show our extensive high-throughput studies for magnesium alloys through both the dislocation and twinning mechanisms. Possible descriptors for the mechanisms are explored and a united picture is demonstrated, which is consistent with available experiments. There are two major contributions of this work, i.e., (i) The relationship between two well-acknowledged deformation mechanisms based on dislocations is clarified; (ii) Machine-learning models show that it is possible to design ductile magnesium alloys without the prior knowledge of deformation mechanisms.

Keywords

Deformation mechanisms • Ductility • Machine learning • Magnesium alloys

Introduction

Traditional alloy design strategy is trial-and-error, which is time-consuming and economically expensive. With the rise of density functional theory and machine learning, high-throughput materials design gradually becomes an important alternative to the traditional strategy. High-throughput materials design can be arguably classified into two groups [1]: (i) descriptors-based high-throughput calculations and (ii) machine-learning models. Identifying descriptors usually requires us to firstly find the critical mechanisms determining the physical properties, e.g., the plasticity of magnesium (Mg). Mechanisms that play either directly (causal relations) or indirectly (correlative relations) roles in determining the physical properties can be good candidates for descriptors. The effectiveness of descriptors lies in the strong correlative relations, and both correlation and anticorrelation are of equal effectiveness. This is an excellent property of descriptors that makes them work effectively. In contrast to correlations, causal relations are usually more difficult and expensive to identify. The causal relations between descriptors and physical properties are more important for understanding the physics in a problem than for materials design. Even so, identifying effective descriptors by experiments and/or simulations can still take unacceptable long time. There is an urgent need to find systematic methods to construct effective and computationally inexpensive descriptors without the need for understanding the physics. Excitingly, there are some proceedings in this direction, e.g., the emergence of so-called SISSO algorithms [2] and its application to find new descriptors [3].

Different from descriptor-based methods, machine learning (ML) does not require a thorough understanding of the mechanisms at all. It connects the input information/parameters with the mechanical properties numerically, which can be done in minutes, given sufficient experimental/simulation data is available. Recently, ML studies are boosted by the emergence and growth of quantum–mechanical databases, e.g., Nomad [4], Aflow [5], OQMD [6], Materials Project [7], Materials Cloud [8], just to name a few.

The rise of ML attracts many materials scientists to explore its potency in materials science applications. ML is dominantly used as a mathematical tool, a black-box tool that follows a similar mechanism as human–environment interaction [9]. Actually, the ML methods can contribute to understand the transparent physics in materials science as well [10]. This exciting direction is worth more efforts to explore.

Z. Pei
Oak Ridge National Laboratory, Oak Ridge, TN 37830, USA

Present Address:
Z. Pei (✉)
National Energy Technology Laboratory, 1450 Queen Ave SW, Albany, OR 97321, USA
e-mail: peizongrui@gmail.com

V. M. Miller et al. (eds.), *Magnesium Technology 2021*, The Minerals, Metals & Materials Series,
https://doi.org/10.1007/978-3-030-65528-0_10

The above ideas will be demonstrated by examples of Mg alloys. Mg and its alloys are among the lightest metallic structural materials available on the earth. Given the high specific strength and the high reserve of Mg in the earth crust, Mg alloys are promising materials for automobile and aerospace industries. However, Mg alloys suffer from limited ductility that hinders their wide applications in these sectors. Designing ductile Mg alloys with affordable price is one of the central topics of the Mg community. In this proceeding, we will demonstrate how mechanisms and machine learning can contribute to the design of Mg alloys with enhanced performance.

Results and Discussion

The I_1SFE and YSI Descriptors

Researchers at Max-Planck Institute for Iron Research identified an effective connection between the enhanced ductility of binary Mg alloys and reduced stacking fault energies (SFEs) based on transmission electron microscopy (TEM) analysis and density functional theory (DFT) calculation [11–15]. They used pure Mg and Mg-3wt.%-Y as samples in the tensile tests and TEM analysis. Mg-3wt.%-Y shows a five-time better ultimate tensile strain than pure Mg. TEM observation and DFT calculations confirmed that a small amount of yttrium renders basal <a> dislocations [16] and <c + a> dislocations [12, 15] more active.

The importance of <c + a> dislocations has been extensively discussed since 1981 [17–19], because they are the only dislocations that can supply the out-of-basal-plane strain in additional to the <c> dislocations. The out-of-basal-plane strain is required by the von Mises' law for arbitrarily adjustable deformation. In Ref. [11, 15], the yttrium-promoted activity of <c + a> was identified to be strongly correlated with stacking faults (SF), i.e., the present of intrinsic I_1 SFs. Both the dislocations and SFs are more active in Mg-Y alloys than pure Mg. The connection between I_1 SF and {11–22} <11–23> dislocation is that the latter can dissociate to generate the former bounded by two <c + a> partial dislocations. With this connection the authors argued that I_1 SFs may act as the nucleation source of the important <c + a> dislocations. Not much attention was paid to this correlation prior to the studies. Although further investigations are needed to clarify the <c + a> mobility, the I_1 SFE is arguably one of the best descriptors identified so far to screen for ductile binary/ternary Mg alloys. For example, the I_1 SFE criterion identifies only the rare-earth (RE) elements can ductilize Mg among all hcp and double hcp (dhcp) structure elements, and the group of non-RE ones do not have such an effect. Experiments on several Mg-RE alloys indeed supported the predictions [13]. Pei et al. extended the searching of solutes to the whole Periodic Table of Elements and confirmed that there are almost no new elements can independently ductilize Mg in addition to the RE ones [20]. It is worth mentioning that some alkaline earth elements are around the boundary of the two groups according to our criterion, so their behavior cannot be predicted. Interestingly, lithium, one of the unpredictable elements, was experimentally confirmed to improve the ductility of Mg [19].

To identify affordable Mg alloys without RE elements, Pei et al. continued searching for two elements (i.e., solute pairs) among the 76 promising elements after excluding rare gas, radioactive elements, etc. This yields combinations of 76(76–1)/2 = 2850. A searching space of this size makes DFT calculations of I_1 SFE unpracticable. They proposed a new descriptor based on the correlations between SFE and elemental properties, i.e., elemental volume (V), electronegativity (ν), and bulk modulus (B). The three properties were identified as the strongest correlated ones with SFEs. The new descriptor, so-called yttrium similarity index (YSI) measures the distance/similarity between an element/element pair and yttrium in the three-dimensional space. As an extensively studied element in Mg, yttrium is naturally chosen as the reference. Mathematically, YSI is expressed by [20]

$$\mathrm{YSI} = 1 - \left(\sum\nolimits_i c_i\left(\alpha_i - \alpha_i^Y\right)\right)^{1/2} \quad (1)$$

where α_i represent the three elemental properties of V, ν and B for one element or element pair, and α_i^Y are the same properties of yttrium. Coefficients c_i are determined by linear regressions. According to the correlation between YSI and I_1 SFE, when YSI > 0.84, the element or element pair are promising to ductilize Mg. The higher the YSI, the more promising the candidate. For element pairs, concentration-averaged physical properties are used. The predictions of YSI for ternary systems are shown in Fig. 1. Yellow regions are the most promising element pairs. Interestingly, there are still non-RE pairs with YSI as high as 0.95. After applying the screening criteria of high element solubility and low cost, we have only one pair left, i.e., Al-Ca, pointing to Mg–Al-Ca alloys. The affordable Mg–Al-Ca alloys were experimentally confirmed to be much more ductile than pure Mg without loss of strength [21], demonstrating the effectiveness of YSI and I_1 SFE.

The Relation Between Two Ductility Mechanisms

One of the concerns on <c+a> dislocations is their mobility reduction due to dislocation dissociation into the basal plane. Focusing on the screw <c + a> dislocations, Wu et al. proposed a different mechanism that may improve the mobility of the <c + a> dislocations. They argued that if the <c +

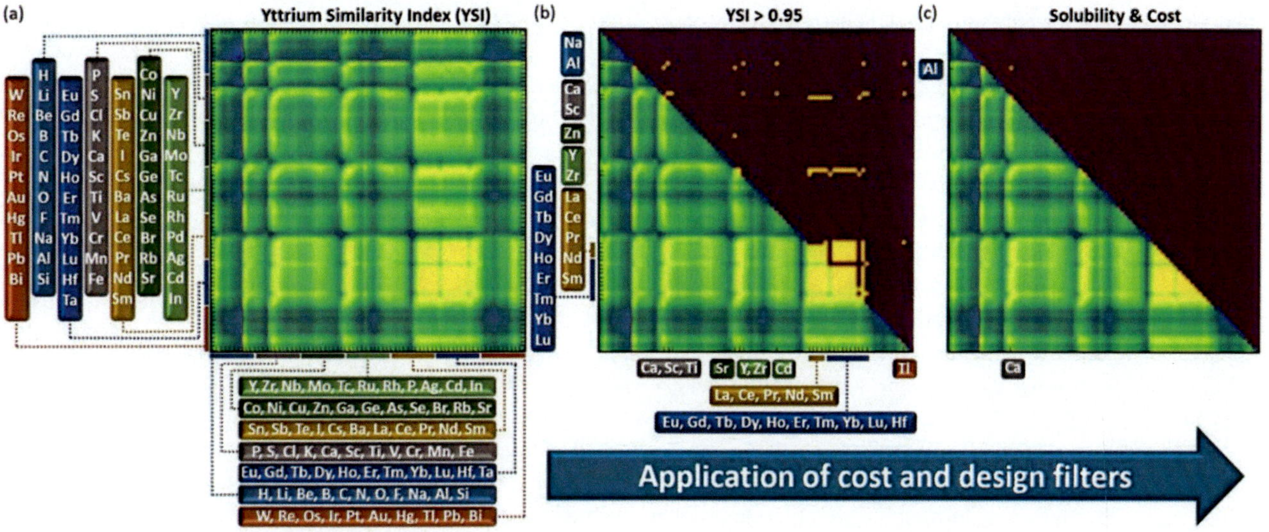

Fig. 1 High-throughput screening of ternary Mg alloys using the yttrium similarity index, YSI (Eq. 1) for the 2850 solute pairs and visualized in the form of a symmetric matrix **a** with yellow regions for high similarity and blue low ones. Solute pairs that have high indices (YSI > 0.95) are shown in the upper triangular part in (**b**). Applying a cost and solubility filter only a single pair of Al-Ca remains (**c**). Details are referred to Refs. [20, 21] and the figure is reproduced from Ref. [21]. (Color figure online)

a> screw dislocation cross-slip from the pyramidal II plane to pyramidal I plane, the high mobility may retain. RE elements like yttrium may be beneficial to this process [22]. Like the mechanism introduced in Sect. 2.1, this mechanism needs more studies to clarify their actual role in the deformation process. There are questions that cannot be answered at the moments. For example, it is unknown whether <c + a> screw dislocations are the deformation carrier or not. Since the <c + a> screw dislocations can also dissociate into the pyramidal I plane, it is not clear if this dislocation configuration is indeed as beneficial to dislocation mobility as is expected, although there are some implications.

Clarifying the controversies in the two ductility mechanisms is a long way to go. But understanding the origins of their correlated effectiveness can be answered with our current results immediately, which is an interesting problem in the scope of materials informatics. To this end, we firstly present some equations of Wu et al. [22] as well as our newly derived equations in Ref. [23].

The Wu et al. mechanism requires the cross-slip rate of <c + a> screw dislocation from pyramidal II ({11–22}) to pyramidal I ({10–11}) plane is much faster than dislocation dissociation into the basal plane, i.e.,

$$\nu_0\left(\frac{L}{l_{XS}}\right)\exp\left(-\frac{\Delta G_{xs}}{k_BT}\right) \gg \nu_0\left(\frac{L}{l_{PB}}\right)\exp\left(-\frac{\Delta G_{PB}}{k_BT}\right), \quad (2)$$

where $\Delta G_{xs}, \Delta G_{PB}$ are the energy barriers of the cross slip and dissociation from the pyramidal II plane into the basal plane. L is the total length of a screw dislocation consisting of multiple kinks of a typical length l, and ν_0, k_B, T are the attempt frequency, Boltzmann's constant and temperature. One possible strategy to increase the cross-slip rate is to reduce its energy barrier that is comprised of four terms

$$\Delta G_{XS} = \Delta G_{XS,i} + \Delta E^{\mathrm{I-II}} l_{XS} + \Gamma \Delta s - \Delta \tau b A. \quad (3)$$

Only $\Delta E^{\mathrm{I-II}}$ in the second term can be tuned through alloying, while the other three terms are unchangeable upon alloying. This quantity involves the calculations of stacking fault energies with solutes for both pyramidal planes. Such calculations are required for solutes at multiple sites near the SFs and within the interaction ranges of solutes and SFs. To efficiently evaluate the alloying effect on this mechanism, we need to find a reliable and quick way to check how the solutes change $\Delta E^{\mathrm{I-II}}$. After some tedious mathematical derivation in Ref. [23], we arrive at a rather clean and simple equation for solute X,

$$\Delta E^{\mathrm{I-II}}(X) = -3BV_0(C_{\mathrm{I}} - C_{\mathrm{II}})\delta_X, \quad (4)$$

where B, V_0 are the bulk modulus and atomic volume of the matrix Mg, $C_{\mathrm{I}} - C_{\mathrm{II}}$ is the geometric constant associated with the two pyramidal planes, and δ_X is the lattice misfit parameter of solute X in Mg. The 11 solutes considered by Wu et al. show the constant $C_{\mathrm{I}} - C_{\mathrm{II}} > 0$, so the sign of $E^{\mathrm{I-II}}(X)$ is equivalent to the sign of $-\delta_X$, i.e.,

$$\mathrm{sign}\left(E^{\mathrm{I-II}}(X)\right) = \mathrm{sign}(-\delta_X). \quad (5)$$

The lattice misfit parameter can be used as a descriptor of the mechanism introduced by Wu et al. It is worth mentioning that the relations of Eqs. (4) and (5) are generally valid for planar defects. Similar relation is found also for the segregation energies of 23 solutes around two profuse twin boundaries after checking the 92 DFT calculations in Ref. [24].

With the approximations introduced above, we are now able to compare the two mechanisms by their descriptors on the same footing. The mechanisms I (source mechanism) is described by SFE misfit parameter ϵ_{SFE}, and mechanism II (mobility mechanism) by lattice misfit δ_X of the solute X in Mg matrix. Both quantities can be calculated using DFT methods with the same computational conditions. For consistency, we adopt only the data generated by Yasi et al. [25]. As is shown in Fig. 2, the two descriptors are strongly correlated, with a favorable Pearson's r = 0.86, explaining the origin of the correlated effectiveness of the two ductility mechanisms in high-throughput screening. Among all 63 solutes, 81% of them are located in the first and third quadrants, having the same effects of ductilizing Mg (−,−) or not (+, +). The detailed tables for this figure are referred to Ref. [23].

Machine-Learning Reliably Reproduces the Predictions of the Two Descriptors

Identifying reliable descriptors usually takes time since we need to understand the deformation process, while ML does not, given sufficient data from experiment/simulations. It is of great significance if ML can reproduce the predictions of the two descriptors. This will demonstrate ML is not just an efficient method but also a reliable and accurate one. We have performed DFT screening on 21 hexagonal close-packed elements and tensile tests on 5 Mg alloys guided by DFT [13]. The experiments indeed confirmed the enhanced ductility in these alloys. Using the available 21 DFT-screened elements in several Gaussian Process Classification (GPC) models, we show the ML predictions are highly consistent with the two descriptors [10]. As is shown in Fig. 3, the major area is in pink, implicating the consistency of ML results with the two descriptors.

Reliable ML models with carefully selected ML algorithms are not just mathematical black boxes but also can be contributors to understand the physics in the materials science problems. When the two descriptors are not consistent in predictions, ML as a third method can help pick the more promising descriptor whose prediction is consistent with it. The mechanism whose descriptors are in better agreement with ML is more likely the one active during deformation. Also, we can use ML to discover new descriptors that are able to reproduce the ML results, as descriptors usually do not need computer programming and are simpler than ML in application. We give one example in Ref. [10] and show it is possible to rediscover the descriptor for mechanism I.

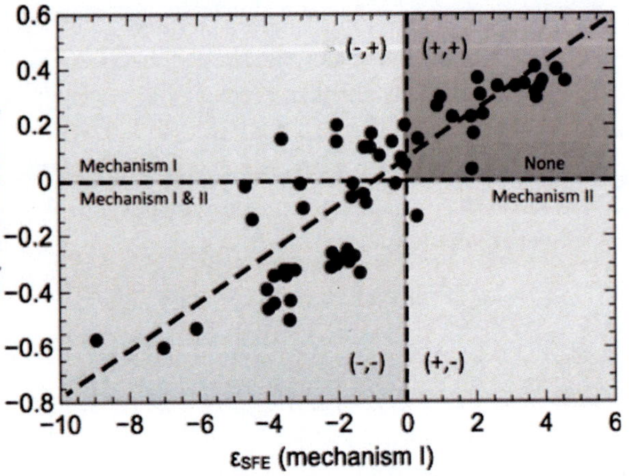

Fig. 2 The correlation between the two mechanisms. The correlation between the two ductility descriptors has a Pearson's r = 0.86. Based on the mathematical signs of the two descriptors, the whole space is divided into four quadrants by the signs of the descriptors. The minus sign indicates a solute contributes to the enhancement of ductility through the descriptor/mechanism, while the plus sign means that there is no such effect. For example, solutes in the third quadrant are expected to improve the intrinsic ductility of Mg through both mechanisms, while those in the first quadrant are expected not to have the effect. The figure is reproduced from Ref. [23]

Design Maps Constructed with United dislocation- and Twin-Based Descriptors

In the previous sections, we already gave high-throughput screening examples of ML and mechanisms-based descriptors. Both descriptors originate from the behavior of non-basal dislocations of various characters. Here we continue to show that descriptors based on multiple defects can be combined for screening with multiple objectives, e.g., high strength and high ductility. More accurately, twin-based descriptors [24] are combined with the dislocation-based descriptor for ductility (mechanism I).

It is well known that solutes can reduce the mobility of twin boundaries through segregation, which strengthens the Mg alloys [26]. The higher concentrations of the solutes, the more significant the strengthening effect. Hence, we take the solute coverages of the profuse twin boundaries in Mg at room temperature as descriptors to screen for Mg alloys with high yield stresses. The descriptors (denoted by DX) are calculated using Langmuir-McLean model with DFT-computed segregation energies as inputs [24]. We do not consider the interactions between twin and dislocations and assume both defects act independently in Mg alloys. The complicate interactions between the defects merit further systematic studies using atomistic simulations [27].

Two profuse twin boundaries, i.e., the {10–11} compression and {10–12} tensile twins are considered. Results are shown in Fig. 4a for compression twin and (b) for tensile

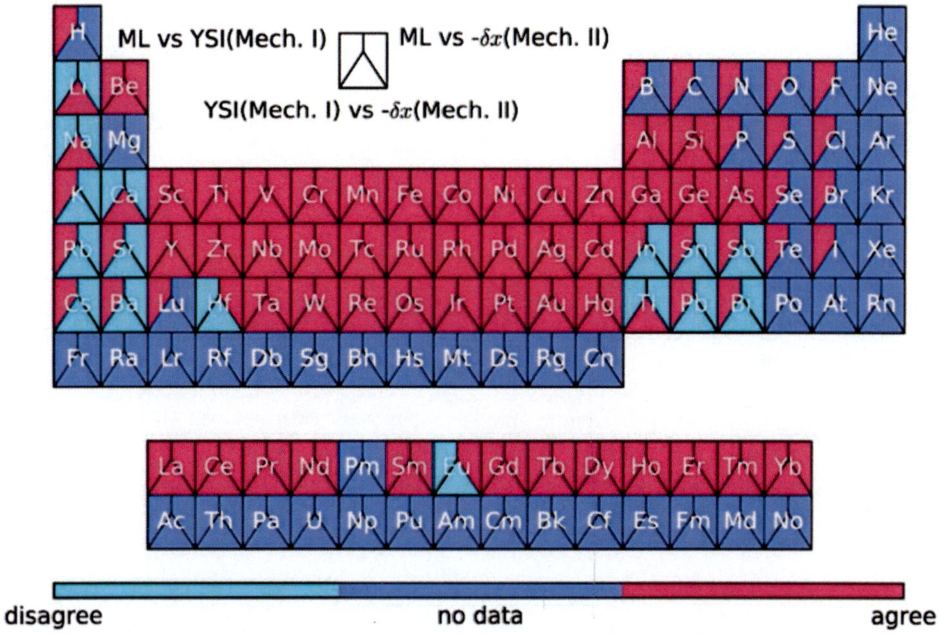

Fig. 3 Machine-learning is able to generally reproduce the predictions of the two mechanism-based descriptors [10]. The comparisons are made between each two of the machine learning (ML) solution, YSI descriptor (Mechanism I) and $-\delta_X$ (Mechanism II). The cyan blocks represent where the two predictions disagree, the blue ones represent no data is available, and the pink ones represent where the two predictions agree. (Color figure online)

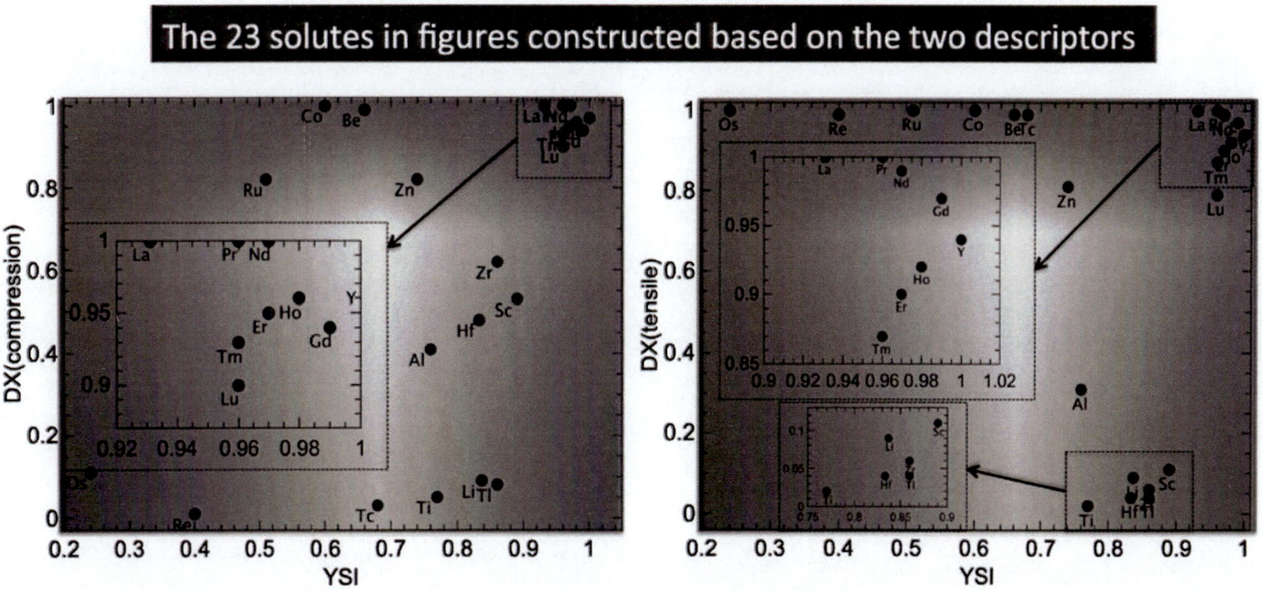

Fig. 4 The design maps constructed with descriptors based on strengthening through twin boundaries and ductility through dislocations in Mg [27, 28]. **a** The design map of the yttrium similarity index (YSI) verse solute coverage-based strengthening descriptor DX (compression twin) for Mg. **b** The design map of the YSI verse solute coverage-based strengthening descriptor DX (tensile twin) for Mg

twin. Green area indicates the promising solutes from the viewpoints of both objectives. The elements simultaneously appear in the compression twin (Fig. 4a) and tensile twin (Fig. 4b) maps are particularly promising as candidates for improving both ductility and strength of Mg. These predictions are consistent with available experimental results. For example, Tb and Ho are clearly confirmed to be able to increase both the ductility and yield stress of Mg [13]. It is expected new Mg alloys can be synthesized with the guidance of the design maps.

Conclusions

We have demonstrated several methods and perspectives for the high-throughput design of novel Mg alloys, where both mechanisms and machine learning play important roles. The relation between different mechanisms, and the relation between mechanisms and machine learning are explored and discussed. Machine-learning models with carefully selected algorithms can help understand physics in materials science problems, in addition to its usual role as a mathematical black toolbox. Using Mg alloys as an excellent example, we also demonstrate it is possible to perform high-throughput screening with multiple objectives, such as high strength and high ductility. The methods and perspectives described here can be extended to guide the design of other novel alloys as well.

References

1. Z. Pei, J. Yin, J.A. Hawk, D.E. Alman, M.C. Gao, Machine-learning informed prediction of high-entropy solid solution formation: Beyond the Hume-Rothery rules, npj Computational Materials 6(1) (2020) 1–8.
2. R. Ouyang, S. Curtarolo, E. Ahmetcik, M. Scheffler, L.M. Ghiringhelli, SISSO: A compressed-sensing method for identifying the best low-dimensional descriptor in an immensity of offered candidates, Physical Review Materials 2(8) (2018) 083802.
3. C.J. Bartel, C. Sutton, B.R. Goldsmith, R. Ouyang, C.B. Musgrave, L.M. Ghiringhelli, M. Scheffler, New tolerance factor to predict the stability of perovskite oxides and halides, Science advances 5(2) (2019) eaav0693.
4. C. Draxl, M. Scheffler, NOMAD: The FAIR concept for big data-driven materials science, Mrs Bulletin 43(9) (2018) 676-682.
5. S. Curtarolo, W. Setyawan, G.L. Hart, M. Jahnatek, R.V. Chepulskii, R.H. Taylor, S. Wang, J. Xue, K. Yang, O. Levy, AFLOW: an automatic framework for high-throughput materials discovery, Computational Materials Science 58 (2012) 218-226.
6. J.E. Saal, S. Kirklin, M. Aykol, B. Meredig, C. Wolverton, Materials design and discovery with high-throughput density functional theory: the open quantum materials database (OQMD), JOM 65(11) (2013) 1501-1509.
7. A. Jain, S.P. Ong, G. Hautier, W. Chen, W.D. Richards, S. Dacek, S. Cholia, D. Gunter, D. Skinner, G. Ceder, Commentary: The Materials Project: A materials genome approach to accelerating materials innovation, Apl Materials 1(1) (2013) 011002.
8. L. Talirz, S. Kumbhar, E. Passaro, A.V. Yakutovich, V. Granata, F. Gargiulo, M. Borelli, M. Uhrin, S.P. Huber, S. Zoupanos, Materials Cloud, a platform for open computational science, arXiv preprint arXiv:2003.12510 (2020).
9. E.A. Holm, In defense of the black box, Science 364(6435) (2019) 26-27.
10. Z. Pei, J. Yin, Machine learning as a contributor to physics: Understanding Mg alloys, Materials & Design 172 (2019) 107759.
11. S. Sandlöbes, M. Friák, S. Zaefferer, A. Dick, S. Yi, D. Letzig, Z. Pei, L.-F. Zhu, J. Neugebauer, D. Raabe, The relation between ductility and stacking fault energies in Mg and Mg–Y alloys, Acta Materialia 60(6-7) (2012) 3011-3021.
12. Z. Pei, L.-F. Zhu, M. Friák, S. Sandlöbes, J. von Pezold, H.W. Sheng, C.P. Race, S. Zaefferer, B. Svendsen, D. Raabe, Ab initio and atomistic study of generalized stacking fault energies in Mg and Mg–Y alloys, New Journal of Physics 15(4) (2013) 043020.
13. S. Sandlöbes, Z. Pei, M. Friák, L.-F. Zhu, F. Wang, S. Zaefferer, D. Raabe, J. Neugebauer, Ductility improvement of Mg alloys by solid solution: Ab initio modeling, synthesis and mechanical properties, Acta Materialia 70 (2014) 92-104.
14. Z. Pei, An overview of modeling the stacking faults in lightweight and high-entropy alloys: Theory and application, Materials Science and Engineering: A 737 (2018) 132-150.
15. S. Sandlöbes, S. Zaefferer, I. Schestakow, S. Yi, R. Gonzalez-Martinez, On the role of non-basal deformation mechanisms for the ductility of Mg and Mg–Y alloys, Acta Materialia 59(2) (2011) 429-439.
16. Z. Pei, D. Ma, M. Friák, B. Svendsen, D. Raabe, J. Neugebauer, From generalized stacking fault energies to dislocation properties: Five-energy-point approach and solid solution effects in magnesium, Physical Review B 92(6) (2015) 064107.
17. M. Yoo, Slip, twinning, and fracture in hexagonal close-packed metals, Metallurgical transactions A 12(3) (1981) 409-418.
18. S.R. Agnew, Ö. Duygulu, Plastic anisotropy and the role of non-basal slip in magnesium alloy AZ31B, International Journal of plasticity 21(6) (2005) 1161-1193.
19. S. Agnew, M. Yoo, C. Tome, Application of texture simulation to understanding mechanical behavior of Mg and solid solution alloys containing Li or Y, Acta Materialia 49(20) (2001) 4277-4289.
20. Z. Pei, M. Friák, S. Sandlöbes, R. Nazarov, B. Svendsen, D. Raabe, J. Neugebauer, Rapid theory-guided prototyping of ductile Mg alloys: from binary to multi-component materials, New Journal of Physics 17(9) (2015) 093009.
21. S. Sandlöbes, M. Friák, S. Korte-Kerzel, Z. Pei, J. Neugebauer, D. Raabe, A rare-earth free magnesium alloy with improved intrinsic ductility, Scientific reports 7(1) (2017) 1-8.
22. Z. Wu, R. Ahmad, B. Yin, S. Sandlöbes, W. Curtin, Mechanistic origin and prediction of enhanced ductility in magnesium alloys, Science 359(6374) (2018) 447-452.
23. Z. Pei, J. Yin, The relation between two ductility mechanisms for Mg alloys revealed by high-throughput simulations, Materials & Design 186 (2020) 108286.
24. Z. Pei, R. Li, J.-F. Nie, J.R. Morris, First-principles study of the solute segregation in twin boundaries in Mg and possible descriptors for mechanical properties, Materials & Design 165 (2019) 107574.
25. J.A. Yasi, L.G. Hector Jr, D.R. Trinkle, First-principles data for solid-solution strengthening of magnesium: From geometry and chemistry to properties, Acta Materialia 58(17) (2010) 5704-5713.
26. J.F. Nie, Y. Zhu, J. Liu, X.-Y. Fang, Periodic segregation of solute atoms in fully coherent twin boundaries, Science 340(6135) (2013) 957-960.
27. Z. Pei, H. Sheng, X. Zhang, R. Li, B. Svendsen, Tunable twin stability and an accurate magnesium interatomic potential for dislocation-twin interactions, Materials & Design 153 (2018) 232-241.
28. Z. Pei, X. Zhang, T. Hickel, M. Friák, S. Sandlöbes, B. Dutta, J. Neugebauer, Atomic structures of twin boundaries in hexagonal close-packed metallic crystals with particular focus on Mg, NPJ Computational Materials 3(1) (2017) 1-7.

Three-Dimensional Atomistic Simulations of $\{10\bar{1}2\}$ Non-cozone Twin–Twin Interaction in Mg—Role of Twin Stability and Mobility

Khanh Dang, Carlos N. Tomé, and Laurent Capolungo

Abstract

Given the ease of activation of tensile twinning on the $\{10\bar{1}2\}$ planes in Mg, multiple $\{10\bar{1}2\}$ twin variants can be activated and interact with each other. The outcomes of these interactions are twin–twin junctions (TTJs) that can serve as initiation sites for microcracks. Here, we investigate the 3D structural characteristic and evolution of the non-cozone $\{10\bar{1}2\}$ twin–twin junctions using atomistic simulations. This comprehensive approach allows us to identify additional twin–twin boundaries (TTBs) such as the TTB_{BP} and TTB_{K2}. They formed due to the interaction between the basal prismatic (BP) and conjugate twin (K_2) interfaces with the coherent twin boundary (CTB). Moreover, the TTJs associated with the $\{\bar{1}2\bar{1}2\}$ TTBs are found to promote the growth of the 3-D twin along the normal and forward direction of the twin during the interaction and hinder the detwinning process when loading is reversed.

Keywords

Twinning • Twin–twin interaction • HCP metals

Given the ease of activation of tensile twinning on the $\{10\bar{1}2\}$ planes, multiple $\{10\bar{1}2\}$ twin variants can be activated and interact with each other. The outcomes of these interactions are twin–twin junctions (TTJs) [1–6] that can serve as nucleation sites for dislocation and as well as initiation sites for microcracks. Crystallographically, these interactions can be categorized into two types: cozone (type I between T1-T4 twin variants) and non-cozone (type II between T1 and T2 or T3 twin variants) depending on whether the twin variants share the same zone axis or not. Depending on the reactions of twinning disconnections (TDs), twin–twin boundaries (TTBs) associated with these junctions can (1) retard the growth of the two interacting twins [7] and (2) serve as barriers for other defects, which cause strain hardening. Due to their important role in deformation mechanism, the structural characteristics of these TTJs has been of great interest. However, most of the current studies limit in 2-D settings, which prevents a complete understanding of the interactions.

Here, we investigate the 3-D structural characteristic and role on the twin growth of the $\{10\bar{1}2\}$ non-cozone interactions using the classical atomistic simulation code LAMMPS [8]. Figure 1a, b show the simulation cell employed in this work, with both a 3-D twin and a flat CTB. This setting represents many interactions observed in experiments where a small twin interact with a much larger twin [2, 9]. Here, non-cozone interactions between variant 2 and 5 with variant 1 are used to represent Type II(a) and (b), respectively, to simplify the analysis. The modified embedded atom method (MEAM) interatomic potential developed by Wu et al. [10], which is a reparameterization of a prior MEAM potential by Kim et al. [11] is used to model Mg systems due to its accurate reproduction of the lattice parameters, defect energies, and cohesive energy. Since the MEAM potential accounts for angular dependence of atomic bonding, this potential is more appropriate for HCP Mg than earlier embedded atom method (EAM) potentials by Sun et al. [12]. Results from atomistic simulations are visualized using the Open Visualization Tool (OVITO) [13].

In this talk, we will present new TTBs that can only be seen in 3-D MD simulations. These newly observed TTBs also explain the discrepancy between 2-D MD simulation and experimental study. Moreover, the TTJ associated with this TTB is found to promote the growth of the 3-D twin in both normal and forward direction of the twin. It Is also found that during cyclic loading, the $\{\bar{1}2\bar{1}2\}$ TTB hinder the

K. Dang (✉) · C. N. Tomé · L. Capolungo
MST-8, Los Alamos National Laboratory,
Los Alamos, NM 87544, USA
e-mail: kqdang@lanl.gov

V. M. Miller et al. (eds.), *Magnesium Technology 2021*, The Minerals, Metals & Materials Series,
https://doi.org/10.1007/978-3-030-65528-0_11

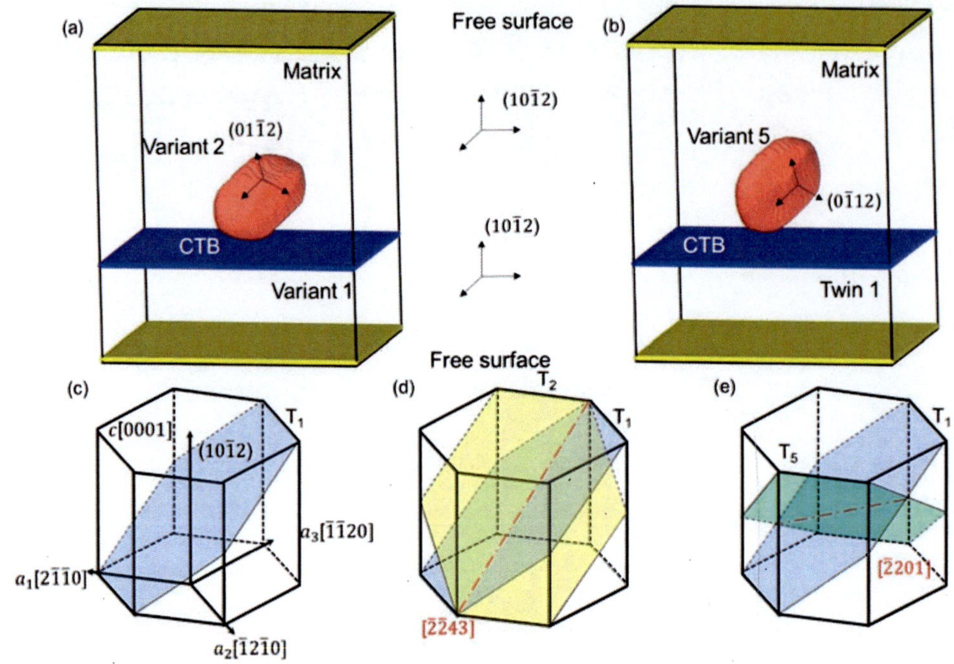

Fig. 1 Simulation setup for **a** type II **a** interaction and **b** type II **b** interaction. **c** Coordinate systems used in this study. **d** Type II **a** interaction with the interaction line along $[\bar{2}\bar{2}43]$. **e** Type II **b** interaction with the interaction line along $[\bar{2}201]$. (Color figure online)

mobility of the twin, which causes cyclic hardening behavior. These results highlight the importance of the $\{\bar{1}2\bar{1}2\}$ TTB to the $\{10\bar{1}2\}$ twins' morphology and mobility after interaction with other non-cozone variants.

References

1. El Kadiri H, Kapil J, Oppedal AL, Hector LG, Agnew SR, Cherkaoui M, Vogel SC (2013) Acta Mater. 61:3549–3563
2. Yu Q, Wang J, Jiang Y, McCabe RJ, Li N, Tomé CN (2014) Acta Mater. 77:28–42
3. Yu Q, Wang J, Jiang Y, McCabe RJ, Tomé CN (2014) Mater. Res. Lett. 2:82–88
4. Gong M, Xu S, Jiang Y, Liu Y, Wang J (2018) Acta Mater. 159:65–76
5. Sun Q, Zhang XY, Ren Y, Tan L, Tu J (2015) Mater. Charact. 109:160–163
6. M. Gong, W. Wu, J. Mater. Res. (2020) 1–13.
7. Morrow BM, McCabe RJ, Cerreta EK, Tomé CN (2014) Metall. Mater. Trans. A. 45:5891–5897
8. Plimpton S (1995) J. Comput. Phys. 117:1–19
9. Chen H, Liu T, Xiang S, Liang Y (2017) J. Alloys Compd. 690:376–380
10. Wu Z, Francis MF, Curtin WA (2015) Model. Simul. Mater. Sci. Eng. 23:015004
11. Kim Y-M, Kim NJ, Lee B-J (2009) Calphad. 33:650–657
12. Sun DY, Mendelev MI, Becker CA, Kudin K, Haxhimali T, Asta M, Hoyt JJ, Karma A, Srolovitz DJ (2006) Phys. Rev. B. 73:024116
13. Stukowski A (2010) Model. Simul. Mater. Sci. Eng. 18:015012

Part III

Mechanical Behavior

Understanding Twinning-Detwinning Behavior of Unalloyed Mg During Low-Cycle Fatigue Using High Energy X-ray Diffraction

Aeriel D. Murphy-Leonard, Darren C. Pagan, Armand Beaudoin, Matthew P. Miller, and John E. Allison

Abstract

It is well understood that twinning during deformation plays an important role in deformation of Mg and its alloys [1–8]. In hexagonal close packed (HCP) Mg alloys, the dominant deformation mode at room temperature is <a> slip on the basal (0001) plane Mg [9, 10]. The other slip systems—prismatic <a> slip, pyramidal <a> slip, and pyramidal <c + a> slip—require much higher stresses to activate during deformation [11]. Mechanical twinning allows for grains to easily deform along their c-axis [12] and has been the focus of significant, active research [e.g., 13–23].

Keywords

Low-cycle fatigue • Synchrotron diffraction • Magnesium

It is well understood that twinning during deformation plays an important role in deformation of Mg and its alloys [1–8]. In hexagonal close packed (HCP) Mg alloys, the dominant deformation mode at room temperature is <a> slip on the basal (0001) plane Mg [9, 10]. The other slip systems—prismatic <a> slip, pyramidal <a> slip, and pyramidal <c + a> slip—require much higher stresses to activate during deformation [11]. Mechanical twinning allows for grains to easily deform along their c-axis [12] and has been the focus of significant, active research [e.g., 13–23].

In unalloyed Mg and Mg alloys, with a c/a ratio less than the ideal value of 1.633, the $\{10\bar{1}2\}\langle 10\bar{1}1\rangle$ extension twinning is the dominant deformation mode, where extension along the c-axis can be accommodated, but not contractions along that same direction [16, 24]. As a result, during mechanical loading the tensile yield strength is significantly higher than the compressive yield strength resulting in a tension–compression asymmetry [25]. Begum et al., found that the tensile yield strength was much higher than the compressive yield strength during low-cycle fatigue (LCF) of an AM30 extruded Mg alloy and related this to twinning that occurs during compression and detwinning that occurs during tension [5]. During compression, twins form causing an 86.3° reorientation of the basal pole [9, 11, 15, 22]. During reversed unloading or tensile loading these twinned regions can undergo detwinning in which twins become narrower and/or disappear [11, 25, 26]. Detwinning causes a reorientation of the c-axis from the twin back to the matrix or parent grain [9, 25–27]. Twins can reappear upon reloading and thus, the twinning-detwinning behavior continues until the end of life [28].

In this study, the twinning detwinning behavior of extruded, polycrystalline unalloyed Mg under cyclic loading conditions was investigated at the Cornell High Energy Synchrotron Source (CHESS) using in-situ high energy X-ray diffraction. Measurements were conducted at three different strain amplitudes. The initial crystallographic texture was such that the c-axis in most grains was normal to the loading direction and therefore, favorable for extension twinning during compressive loading. The experimental results showed that an increase or decrease in the {0002} basal X-ray peak intensity was observed during low-cycle fatigue and these changes are indicative of the occurrence of twinning and

A. D. Murphy-Leonard (✉)
Department of Materials Science and Engineering, The Ohio State University, Columbus, OH, USA
e-mail: aerielm@umich.edu

D. C. Pagan
Department of Materials Science and Engineering, Pennsylvania State University, State College, PA, USA

A. Beaudoin
Cornell High Energy Synchrotron Source, Cornell University, Ithaca, NY, USA

M. P. Miller
Sibley School of Mechanical and Aerospace Engineering, Cornell University, Ithaca, NY, USA

J. E. Allison
Department of Materials Science and Engineering, University of Michigan, Ann Arbor, MI, USA

V. M. Miller et al. (eds.), *Magnesium Technology 2021*, The Minerals, Metals & Materials Series,
https://doi.org/10.1007/978-3-030-65528-0_12

detwinning. During cyclic loading complete twinning-detwinning occurred for the first few hundred cycles where all of the twins formed in compression were removed during tensile loading of the following cycle. Eventually, this phenomenon ceases and residual twins remain in the material throughout each cycle. At strain amplitudes below 0.5%, there was no indication of twinning during compressive loading. The complete article on this study can be found in the *International Journal of Fatigue* [29].

References

1. C. Bettles, M. Gibson, Current wrought magnesium alloys: strengths and weaknesses, J. Metals 57(2005) 46–49.
2. Q. Yu, Y. Jiang, J. Wang, Cyclic deformation and fatigue damage in single-crystal magnesium under fully reversed strain-controlled tension-compression in the [1010] direction, Scr. Mater. 96 (2015) 41–44.
3. Q. Yu, J. Zhang, Y. Jiang, Fatigue damage development in pure polycrystalline magnesium under cyclic tension-compression loading, Mater. Sci. Eng. A 528 (2011) 7816–7826.
4. S. Begum, D.L. Chen, S. Xu, A. Lou, Low cycle fatigue properties of an extruded AZ31 magnesium alloy, Int. J. of Fat. 31 (2009) 726–735.
5. S. Begum, D.L. Chen, S. Xu, A. Lou, Strain-controlled low-cycle fatigue properties of a newly developed extruded magnesium alloy, Metall. Mater. Trans A 39 (2008) 3014–3026.
6. F. Mokdad, D.L. Chen, Strain-controlled low cycle fatigue properties of a rare-earth containing ZEK100 magnesium alloy, Materials & Design 67 (2015) 436–447.
7. F.A. Mirza, D.L. Chen, D.J. Li, X. Q. Zeng, Low cycle fatigue of a rare-earth containing extruded magnesium alloy, Mater. Sci. Eng. A 575 (2013) 65–73.
8. L. Chen, C. Wang, E.I.W. Wu, Z. Liu, G.M. Stoica, L. Wu, P.K. Liaw, Low-Cycle Fatigue Behavior of an As-Extruded AM50 Magnesium Alloy, Metall. Mater. Trans. A 38 (2007), 2235–2241.
9. X.Y. Lou, M. Li, R.K. Boger, S.R. Agnew, R.H. Wagoner. Hardening evolution of AZ31B Mg sheet, Int. J. Plast. 23 (2007) 44-86
10. C.S. Roberts, Magnesium and its alloys, Wiley, New York/London 1960
11. S. Kleiner, S. Uggowitzer, Mechanical anisotropy of extruded Mg-6%Al-1%Zn alloy, Mater. Sci. Eng. A 379 (2004) 258-263
12. M.H. Yoo, J.R. Morris, K.M. Ho, S.R. Agnew, Nonbasal deformation modes of HCP metals and alloys: role of dislocation source and mobility, Metall. Mater. Trans. A 33 (2002) 813.
13. M. Knezevic, A. Levinson, R. Harris, R. Mishra, R. Doherty, S. Kalidindi, Deformation twinning in AZ31: Influence in strain hardening and texture evolution, Acta Mater. 58 (2010) 6230-6242
14. L. Wu, S.R. Agnew, Y. Ren, D.W. Brown, B. Clausen, G.M. Stoica, D.R. Wenk, P.K. Liaw, The effects of texture and extension twinning on the low cycle fatigue behavior of a rolled Mg alloy, AZ31B, Mater. Sci. & Eng. A 527 (2010) 7057-7067.
15. L.Wu, A. Jain, D.W. Brown, G.M. Stoics, S.R. Agnew, B. Clausen, D.E. Fielden, P.K. Liaw, Twinning-detwinning behavior during the strain controlled low-cycle fatigue testing of a wrought magnesium alloy, ZK60A, Acta. Mater. 56 (2008) 688-695
16. S.R. Agnew, O. Duygulu, Plastic anisotropy and the role of non-basal slip in magnesium alloy AZ31B, Int. J. Plast. 21 (2005) 1161-1193.
17. S.R. Agnew, C.N. Tome, D.W. Brown, S.C. Vogel, Study of slip mechanisms in a magnesium alloy by neutron diffraction and modeling, Scr. Mater. 48 (2003) 1003-1008
18. S.B. Yi, H.G. Brokmeier, R. Bolmaro, K.U. Kainer, J. Homeyer, The texture evolutions of Mg alloy, AZ31 under uni-axial loading. Materials Science Forum 495-497 (2005) 1665–1674
19. D.W. Brown, S.R. Agnew, M.A.M. Bourke, T.M. Holden, S.C. Vogel, C.N. Tome, Internal strain and texture evolution during deformation twinning in magnesium, Mater. Sci. Eng. A 399 (2005) 1–12
20. M.A. Gharghouri, G.C. Weatherly, J.D. Embury, J. & Root, Study of the mechanical properties of Mg-7. 7at. % Al by in-situ neutron diffraction, Philos. Mag. 79 (1999) 1671–1695
21. C.H. Caceres, T. Sumitomo, M. Veidt, Pseudoelastic behaviour of cast magnesium AZ91 alloy under cyclic loading – unloading, Acta. Mater. 51 (2003) 6211–6218.
22. D. W. Brown, A. Jain, S.R. Agnew, B. Clausen, Twinning and Detwinning During Cyclic Deformation of Mg Alloy AZ31B, Mater. Sci. Forum 539-543 (2007) 3407–3413.
23. A. Jain, S.R. Agnew, Modeling the temperature dependent effect of twinning on the behavior of magnesium alloy AZ31B sheet Modeling the temperature dependent effect of twinning on the behavior of magnesium alloy AZ31B sheet, Mater. Sci. Eng. A 462 (2007) 29-36
24. Q. Yu, J. Zhang, Y. Jiang, Q. Li, An experimental study on the cyclic deformation and fatigue of extruded ZK60 magnesium alloy, Int. J. Fat. 36 (2012) 47-58
25. Y. N. Wang, J. C. Huang, The role of twinning and untwinning in yielding behavior in hot-extruded Mg-Al-Zn alloy, Acta Mater. 55 (2007) 897–905,
26. J. W. Christian, S. Mahajan, Deformation twinning, Prog. Mater. Sci. 39 (1995) 1-157
27. J. Wang, I. J. Beyerlein, and C.N. Tome, Anatomic and probabilistic perspective on twin nucleation in Mg, Scr. Mater. 63 (2010) 741–746
28. Q. Yu, J. Wang, Y. Jiang, Inverse slip accompanying twinning and detwinning during cyclic loading of magnesium single crystal, J. Mater. 2013 (2013) 1-8
29. A. Murphy-Leonard, D. Pagan, A. Beaudoin, M. Miller, J. Allison, Quantification of Twinning-Detwinning Behavior During Low-Cycle Fatigue of Pure Magnesium Using High Energy X-Ray Diffraction, International Journal of Fatigue 125 (2019) pp 314-323

The Effects of Basal and Prismatic Precipitates on Deformation Twinning in AZ91 Magnesium Alloy

B. Leu, M. Arul Kumar, and Irene J. Beyerlein

Abstract

The advancement of Mg-based applications is motivated by inherently the high specific-strength and low densities of Mg and its alloys. The AZ91 alloy is one such exemplary cast Mg alloy that is being considered due to its relatively high strength, excellent corrosion resistance and castability. However, like most Mg alloys, it is not yet widely used due to poor formability and a complex, anisotropic plastic response. These behaviors have been linked to their propensity to deform by twinning, in addition to slip. Furthermore, the grains in the AZ91 alloy contain β-phase precipitates, which are similar in size as the twin lamellae. In this study, we use an elasto-viscoplastic fast-Fourier-transform (EVP-FFT) crystal plasticity micromechanics model to examine the interactions between $\{10\bar{1}2\}$-type extension twins and two types of precipitates—basal and prismatic precipitates. The calculations focus on conditions that determine whether the twin is blocked by the precipitate or can *cross* the precipitate by nucleating a new twin on the other side of it. The effects of twin thickness, relative to the size of the precipitate, and precipitate types are investigated. The analysis shows that both precipitates are effective at blocking the propagation of twins and limit their ability to thicken. The basal precipitates are, however, substantially easier to cross by the twin than prismatic precipitates. Furthermore, the thicker the twin is when it impinges on the precipitate, the higher the chance it can cross.

Keywords

AZ91 mg-alloy • Precipitates • Twinning

Introduction

The Mg-9wt.%Al-1wt.%Zn-0.2wt.%Mn (AZ91) alloy is one of the most widely used Mg-alloys today in the automotive and aerospace industries due to its relatively high strength, good corrosion resistance, ductility, and castability [1–3]. The enhanced mechanical and chemical properties of these alloys are attributed to the β-phase body-centered-cubic (BCC) $Mg_{17}Al_{12}$ intermetallic phases that emerge during heat treatment [4]. The β-phase precipitates out via both continuous and discontinuous precipitation. Continuous precipitation nucleates and grows within the parent grain, while discontinuous precipitation occurs along the grain boundaries [5]. Thin lath-shaped precipitates, whose habit plane is parallel to basal plane of the matrix, are commonly observed and referred to here as *b-precipitates*. The precipitate forms the Potter orientation relationship with the parent matrix, which is described by $(0001)_{Mg}$ 2° from $(011)_{\beta}$, $[2\bar{1}\bar{1}0]_{Mg}$ // $[1\bar{1}1]_{\beta}$, and $(01\bar{1}1)_{Mg}$ // $(110)_{\beta}$ [4–6]. Another hexagonal prism-shaped rod precipitate is also commonly observed and lies on the prismatic plane of the parent matrix with its long axis of the rod parallel to the c-axis of the matrix crystal, which we denote as *p-precipitates*. These rod precipitates mostly form a Crawley orientation relationship, described by $(0001)_{Mg}$ // $(111)_{\beta}$, $[11\bar{2}0]_{Mg}$ // $[11\bar{2}]_{\beta}$, or $[1\bar{1}00]_{Mg}$ // $[1\bar{1}0]_{\beta}$ [7]. These two precipitates are among the most commonly found in AZ91 alloys and contribute greatly to their mechanical strength [4, 7–10]. Precipitation hardening has been studied as a way to reduce the mechanical asymmetry by differentially strengthening slip along the basal, prismatic, and pyramidal slip systems [8, 10]. Precipitates can block dislocation motion in the matrix and, therefore, strengthen the alloy to

B. Leu (✉) · I. J. Beyerlein
Materials Department, University of California at Santa Barbara, Santa Barbara, CA 93106, USA
e-mail: brandonleu@ucsb.edu

M. Arul Kumar
Materials Science and Technology Division, Los Alamos National Laboratory, Los Alamos, NM 87545, USA

I. J. Beyerlein
Department of Mechanical Engineering, University of California at Santa Barbara, Santa Barbara, CA 93106, USA

V. M. Miller et al. (eds.), *Magnesium Technology 2021*, The Minerals, Metals & Materials Series,
https://doi.org/10.1007/978-3-030-65528-0_13

varying degrees depending on their size, distribution, shape, orientation, and habit plane. Robson et al. described the strengthening effects through an Orowan based equation that calculates the stress required to bow dislocations around the unshearable precipitates [10].

A major limitation of AZ91 alloy stems from its low-symmetry hexagonal close-packed (HCP) crystal structure and thus the limited number of easily activatable slip systems. In AZ91 alloy, the critical resolved shear stress (CRSS) for the activation of basal slip along the $\langle a \rangle$-direction is typically low, while the CRSS for $\langle c+a \rangle$ slip is typically higher (by 2–5 times or more), making accommodation of strain along the $\langle c \rangle$-direction difficult [11]. Thus, deformation twinning, when easier than $\langle c+a \rangle$ slip, can be a prevalent inelastic deformation mode for accommodating strain along the c-axis in AZ91 and other Mg-alloys. A commonly seen twin type is the $\{10\bar{1}2\}$ tensile twin, which can be activated when the c-axis of the crystal is subjected to tension.

Twinning deformation involves a lattice reorientation and localized shear in a subdomain of a grain, which results in the creation of new interfaces and heterogenous distributions of internal stress [12–16]. The formation and growth of twins influence the strength, ductility, and stability of AZ91 alloy, and thus, it is important to understand the influence that precipitates have on twinning [17–22]. It has been shown that precipitates can be effective in blocking slip; however, only a few studies have considered precipitate interactions with deformation twins. These studies mainly focused on the additional strengthening effects precipitates provide by the pinning twin boundaries [8, 10, 23]. While important, there are still many aspects of twin propagation, growth, and multiple twin formation that precipitates can affect.

In this study, we investigate the effects that two commonly occurring types of precipitates (*b- and p-precipitates*) have on twinning. We employ a crystal plasticity fast Fourier transform (CP-FFT) model that is capable of treating multiple phases and discrete twins within crystals. The calculations consider a single twin of varying thicknesses that has arrested on one side of either the b- or p-precipitate in the center of the grain. The model is used to calculate the local internal stresses that develop around twin/precipitate junction to evaluate where in the microstructure further twinning is favorable. We find that when the twin impinges on the b-precipitate, there is a forward stress concentration on the opposite side of the precipitate that is favorable for the nucleation of another twin of the same variant or covariant. Furthermore, the larger/thicker the twin is, the larger the forward stress concentration becomes. However, this is not the case for p-precipitates. P-precipitates are found to be much better at shielding the stress fields at the twin front. Instead, if the twin is near the edge of the p-precipitate, then a strong forward stress concentration appears on the adjacent side of the precipitate that favors the nucleation of a new twin of the same variant or covariant. Similarly, the stress concentration on the adjacent side scales similarly to the stress concentration on the opposite side of the twin in the b-precipitate case.

EVPFFT Model

In order to calculate the local stress fields around twins, we employ a crystal-plasticity fast-Fourier-transform (EVP-FFT) model [24]. In prior work, this model has been used to study the development of local stresses and effective mechanical response of heterogenous polycrystalline materials with spatial variations in crystal orientation, and elastic and plastic properties [25–27]. More recently, this model has been adapted to study deformation twinning in single- and polycrystals of various sizes [27, 28].

The model builds upon continuum mechanics principles of equilibrium, kinematic relationships, and constitutive laws under an infinitesimal strain approximation. The model uses explicit twinning, which is briefly described here. The stress field at every material point x, or voxel, is solved for by using an implicit time discretization of the form:

$$\sigma^{t+\Delta t}(x) = C(x) : \left(\varepsilon^{t+\Delta t}(x) - \varepsilon^{p,t}(x) - \dot{\varepsilon}^{p,t+\Delta t}(x)\Delta t - \varepsilon^{tw,t}(x) - \Delta\varepsilon^{tw,t+\Delta t}(x)\right) \quad (1)$$

In the above equation, $\boldsymbol{\sigma}(x)$ is the Cauchy stress, $\mathbf{C}(\mathbf{x})$ is the elastic stiffness tensor. The term inside the bracket is the elastic strain tensor, which is given by the total strain minus the plastic strain, $\boldsymbol{\varepsilon}^p$, and the twinning transformation strain, $\boldsymbol{\varepsilon}^{tw}$. The plastic strain evolves due to dislocation slip on crystallographic slip systems:

$$\begin{aligned}\dot{\varepsilon}^p &= \sum\nolimits_{s=1}^{N} m^s(x)\dot{\gamma}^s(x) \\ &= \dot{\gamma}_o(x)\sum\nolimits_{s=1}^{N} m^s(x)\left(\frac{|m^s(x) : \sigma(x)|}{\tau_c^s(x)}\right)^n sgn(m^s(x) : \sigma(x))\end{aligned} \quad (2)$$

$$m^s = \frac{1}{2}(b^s \otimes n^s + n^s \otimes b^s) \quad (3)$$

$\tau_c^s(x)$ is the critical resolved shear stress (CRSS) associated with the slip system s, and n is the stress exponent (inverse of the rate-sensitivity exponent). The tensor $\mathbf{m}^s$ is the symmetric part of the Schmid tensor, and $\mathbf{b}^s$ and $\mathbf{n}^s$ are the unit vectors along the slip direction and normal to the glide plane of slip system s, respectively. Twinning in the model is simulated explicitly by reorienting by following the crystallographic twinning relationship and imposing characteristic twining shear, $\mathbf{g}^{tw}$, as an eigenstrain in a predetermined twin region. The twinning transformation strain is incremented over N^{tw} steps on a particular twinning plane in the twinning shear direction.

$$\Delta\varepsilon^{tw} = m^{tw}\frac{g^{tw}}{N^{tw}} \qquad (4)$$

In simulation, $\Delta t = 10^{-4}$s and $N^{tw} = 1000$ is kept sufficiently large to ensure convergence. Note that ε^{tw} everywhere outside of the twin domain is zero.

Results and Discussion

Figure 1 presents the model set up, on which the calculations are based. It consists of a $\{10\bar{1}2\}$ tensile twin (orange) inside a single grain of AZ91 (light blue). The twin is arrested on one side by a thin lath-shaped b-precipitate (red) in Fig. 1a and prism rod-shaped p-precipitate (red) in the Fig. 1b. In both cases, the precipitates are β-phase body-centered-cubic (BCC) $Mg_{17}Al_{12}$ intermetallic. Surrounding the parent matrix is a polycrystalline layer (dark blue) with uniformly distributed crystal orientations that approximate the bulk response of the AZ91 alloy. Periodic boundary conditions are imposed that allows the polycrystalline layer to fully encompass the parent matrix. The simulation cell size is 3 (x) × 400 (y) × 400 (z) voxels. The elastic constants of parent matrix used in the calculation for C_{11}, C_{12}, C_{13}, C_{33}, and C_{44} are 59.75, 23.24, 21.70, 61.70, and 16.39 GPa, respectively [29]. In the Mg matrix, prismatic $\langle a\rangle$, basal $\langle a\rangle$ and pyramidal type-I $\langle c+a\rangle$ slip was allowed with CRSS values of 100, 35, 160 MPa, respectively [11]. The elastic constants of the BCC $Mg_{17}Al_{12}$ precipitate phase used for C_{11}, C_{12}, and C_{44} are 86.8, 29.0, and 20.0 GPa, respectively [30]. The precipitates are assumed to deform elastically, consistent with reports in the literature [5, 10, 30].

The crystallography of the model is made to mirror the common scenario seen experimentally. The orientation of the parent grain is (0°, 90°, 0°), the b-precipitate is (39°, 114°, 63°) and the p-precipitate is (−35°, 134°, 0°), according to the Bunge convention. The b-precipitate and p-precipitate follow the Potter and Crawley orientation relationships, respectively, as described in Sec. 1. The twin variant $(01\bar{1}2)[0\bar{1}11]$ is chosen such that the twin plane normal and twin shear direction are kept in-plane (y–z plane). For simplicity, the precipitate dimensions are held constant. For the b-precipitate, the short edge is 7 voxels and the long edge is 35 voxels. For the p-precipitate, the short edge is 21 voxels and the long edge is 35 voxels. The dimensions of the precipitates are representative of what is commonly observed in the literature [4, 7, 8, 10, 31]. Since there are no intrinsic length scales in the model formulation, the length scales in the model are normalized. Accordingly, the twin thickness, t, is varied such that the ratio between t and L_{opp} ranges from 0.5 to 3 representing the broad range of twin thicknesses that may encounter the precipitates. Here L_{opp} is the short edge length of the precipitates, see Fig. 1, and it remains constant for all the simulations. On the other hand, L_{adj} varies depending on the twin thickness, as it is defined as the distance from the twin boundary to the short edge of the precipitate, see Fig. 1.

The model outputs are the stress fields induced in the parent crystal due to the twin/precipitate interaction. To determine driving forces for the further propagation/growth of the twin, the model calculated full stress tensor projected onto the twinning plane and twinning shear direction of the $\{10\bar{1}2\}$ twin variant studied, denoted as the TRSS. Twinning is a unidirectional mechanism; so, in addition to intensity, the sign of the TRSS is also important. Negative values of TRSS signify a driving stress that opposes twinning.

Figure 2 shows the calculated TRSS fields that develop when twins of varying thickness are arrested at either a b- or p-precipitates. The calculations indicate that heterogenous stress fields develop in the matrix, which are the result of heterogeneities of the twin and precipitate, which differ by: lattice orientation, plasticity in the twin, purely elastic response in the precipitate, and local shearing of the twin domain according to its characteristic twin shear [32]. The TRSS fields are colored such that positive values of TRSS are red and negative values are blue. It is seen in Fig. 2 that intense positive TRSS are generated on the other side of the precipitate, while on the boundaries of the twin, the TRSS is negative. This indicates that the driving force for twinning ahead of the twin is favorable for nucleation while twin growth in the lateral direction is unfavorable in comparison. This result agrees well with experimental observations that precipitate-twin interactions can cause the total number of twins to increase while maintaining the same twin volume fraction [33]. These observations imply that precipitates promote the nucleation of new twins while suppressing their growth, consistent with our model predictions.

The TRSS fields generated by forming a twin inside a single crystal without explicit representation of precipitates were calculated as well; the results are not shown here in the interest of space. When the fields with a precipitate are compared to the TRSS field generated ahead of the twin in a single crystal without a precipitate, the TRSS values at the precipitate/matrix interface opposite of the twin tip are 60% lower, indicating, as expected, that the precipitate acts as an

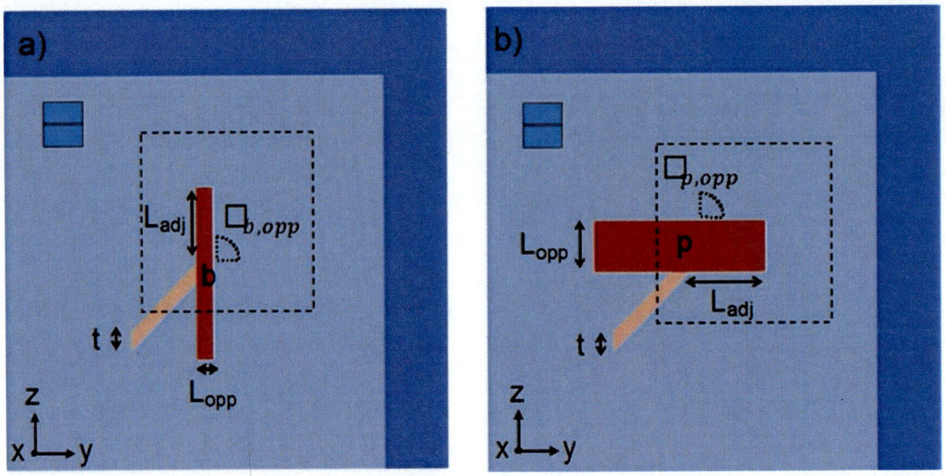

Fig. 1 Schematic representation of EVPFFT unit-cell to simulate $\{10\bar{1}2\}$ tensile twin of varying thicknesses in a single grain of AZ91 alloy with **a** thin lath-shaped b-precipitate and **b** prism rod-shaped p-precipitate. The hexagonal insets represent the orientation of the parent matrix. The dimensions of the precipitates, L, remain constant while the twin thickness, t, is varied. A subsection is outlined for further investigation of the twin-plane resolved shear stresses (TRSS) that develop after forming the twins. (Color figure online)

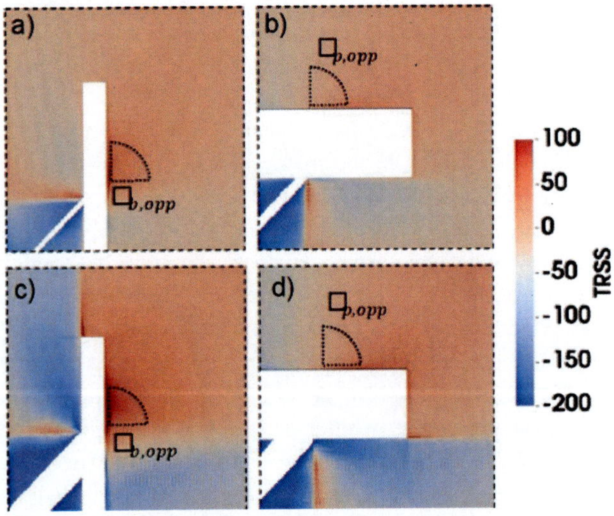

Fig. 2 EVPFFT model predicted Twin-plane Resolved Shear Stress (TRSS) fields (in MPa) that developed from forming the twin. The precipitate and twin regions have been omitted for clarity. The top row shows the TRSS fields that develop after forming the thinnest twin arrested by the **(a)** b-precipitate ($t/L_{opp} = 0.5$) and **(b)** p-precipitate ($t/L_{opp} = 0.35$), respectively. The bottom row shows the TRSS fields that develop after forming the thickest twin arrested by the **(c)** b-precipitate ($t/L_{opp} = 3$) and **(d)** p-precipitate ($t/L_{opp} = 1$), respectively. The model predicts local regions with stress concentrations that are favorable for the nucleation of a new twin. Averages are taken in the outlined regions, ω, that lie on the opposite side of the precipitate where the twin is incident. (Color figure online)

impenetrable hard obstacle and blocks the twin from propagating forward. Without the precipitates, an intense forward stress at the twin tip in the parent Mg crystal develops. This value is sufficiently intense to drive the twin forward, explaining why twins are generally driven to propagate rapidly across its parent crystal [14, 25]. The negative TRSS (blue region) that develops along the twin boundary is a backstress. It has been shown that for a twin terminating in the interior of the grain, a similar backstress is generated due to the reaction of the surrounding Mg crystal to the twin shear [26, 27, 34]. In the case of a twin impinging on the hard-elastic precipitate, the backstresses can be up to 20% more intense.

The stress concentration at the front of the twin tip is shielded by the precipitate and the size of the precipitates influences the amount of shielding. It can be seen from Fig. 2 that for thicker twins, the stress concentrations at the twin tip become stronger and are able to overcome the shielding effects of the precipitates. In some cases, a stress concentration can appear along the opposite edge of the precipitate across from where the twin is incident. In these regions, ω_{opp}, the stress concentrations are high and can favor the nucleation of a new twin. The TRSS for this twin variant is similar to its cozone variant, and both are much higher than those of the other four variants. Thus, the more likely variants to nucleate on the other side of the precipitate on further loading are of the same or cozone variant of the impinging twin.

To assess the propensity for continued twin propagation across the precipitates via nucleation of new twins on the opposite side of precipitates, Fig. 3 shows the average TRSS in the regions, ω_{opp}, plotted with respect to the ratio of the twin thickness, t, to the short edge length of the precipitate, L_{opp}. For b-precipitates (black solid line), it can be seen that with increasing twin thickness, the stress concentration on the opposite side increases and favors the nucleation of additional twins in the ω_{opp} region. This indicates that b-precipitates can be effective at stopping the propagation of thin twins. However, if a thicker twin were to impinge on the same precipitate, another twin of the same (or cozone) variant on the opposite side may nucleate. The twins would appear to *cross* the precipitate, a scenario that is consistent with experimental observations of twin interactions and basal precipitates [31]. In contrast, for p-precipitates (red solid line), it can be seen that increasing twin thickness has little effect on the stress within the region ω_{opp} and the stress is nearly zero. Due to their shape, size, and crystallography, p-precipitates act as more effective shields against twin crossing, which is also consistent with experimental observations that prismatic precipitates are effective at blocking twins and often lead to lower twin volume fractions than basal precipitates [8, 31, 35].

In the cases studied thus far, the twin impinges on the precipitates at the center of the precipitates and L_{adj} is large compared to t and L_{opp}. Another situation worth considering concerns the case in which the twin thickness is comparable or larger than L_{adj}. Figure 4 shows the TRSS field for this case in which the twin thickness is broad. With increasing twin thickness, the calculations show the highest TRSS region, and thus, the most likely nucleation site for a new twin, is not on the opposite side of the precipitate, but rather on the side of the precipitate, in the region ω_{adj}, marked in Fig. 4c. The change in the TRSS stress concentration on the adjacent ω_{adj} region is plotted in Fig. 3 with respect to the ratio of the twin thickness, t, over the adjacent length, L_{adj}. It suggests that when the twin is wide enough such that its boundary impinges near the edge of p-precipitates, nucleation of new twin is more likely to occur on the side of the precipitate. If such a propagation path were to happen, it could appear as if the twin grows around or *engulfs* the precipitate, unlike the basal precipitates.

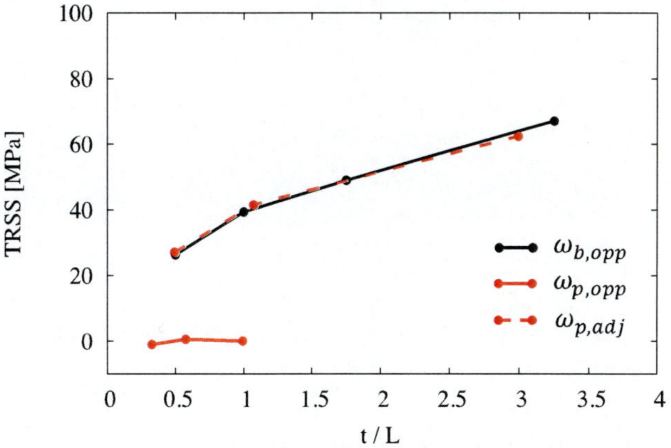

Fig. 3 The effect twin thickness on the forward stress for the nucleation of new twins on the other side of the precipitate. TRSS averaged in the region ω for various twin thicknesses. Solid lines indicate values on the side opposite the twin tip while dashed lines indicate values on the side adjacent to the twin tip. The black line represents the basal precipitate case while the red lines represent values from the prismatic precipitate case. For basal precipitates, increases in twin thickness increases the forward stress. For prismatic precipitates, increases in twin thickness does not increase forward stress on the opposite side. However, if the twin is near the edge of the prismatic precipitate, increases in twin thickness can increase the forward stress on the adjacent side. (Color figure online)

These predictions can help to explain some reported observations in which twins may partially engulf some precipitates [10, 35]. In AZ91, $\{10\bar{1}2\}$ tensile twins partially engulfing both basal and prismatic precipitates have been reported [10]. In other studies on AZ91 and Mg-9wt.%Al binary alloys, twin crossing of only the basal precipitates is mainly observed in [8, 31]. In general, prismatic precipitates have been found superior in blocking twin propagation against crossing, via nucleation on the other side of the precipitate. However, if the precipitate length is too small or if the twin impinges near the edge of the precipitate, then nucleation of new twins on the adjacent side becomes favorable. This may lead to partial engulfment of the precipitate by the twin. Basal precipitates, on the other hand, are less effective at blocking twin crossing, which may result in a network structure of twins connected by precipitates.

Summary

In this work, we present a multi-phase crystal plasticity elastic-viscoplastic fast Fourier transform method to study the interaction between a discrete twin lamella and precipitates within an AZ91 crystal. The analysis focuses on the localization of stress around the precipitate when a twin has impinged on one of its boundaries. The intensity of the stress concentration and its suitability for forming a new twin on the other side of the precipitate are assessed to determine if the precipitate blocks or permits continued twin propagation. We show that the thicker the twin is, relative to the precipitate, when it meets the precipitate, the more likely it will nucleate a new twin on the other side, thereby appearing to *cross* the precipitate. For the same geometry, we find that prismatic precipitates are more effective at blocking propagation of twins than basal precipitates. To overcome prismatic precipitates, the twin is more likely to grow contiguously around them, as if to *engulf* them. For simplicity, we studied the interaction of a single twin with one precipitate, in which the twin is finer than the half length of the precipitate and intersects it in the precipitate center. Worthy of further study are much thicker twins or twins that impinge on one end of the precipitate.

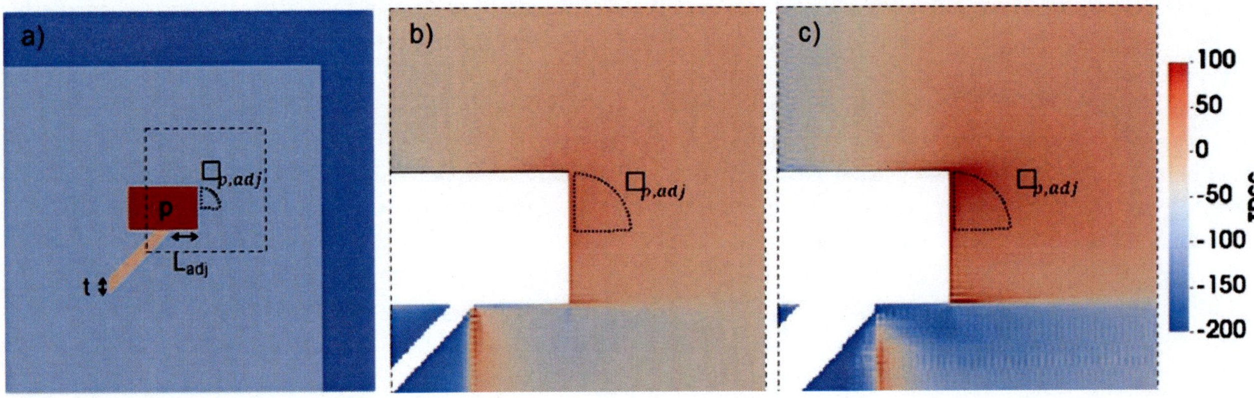

Fig. 4 EVPFFT model predicted Twin-plane Resolved Shear Stress (TRSS) fields that developed from forming the twin. Only the TRSS fields resulting from the (**b**) thinnest twin and (**c**) thickest twin are shown for brevity. The p-precipitate and twin regions have been omitted for clarity. The model predicts that thicker twins can cause local stress concentrations on the adjacent edge of the precipitate that are favorable for the nucleation of a new twin of the same variant or cozone variant. Averages are taken in the outlined regions, ω_{adj}, that lie on the adjacent side of the precipitate where the twin is incident. (Color figure online)

Acknowledgements B. L. was supported by the Department of Defense (DoD) through the National Defense Science & Engineering Graduate Fellowship (NDSEG) Program. M. A. K. acknowledges the financial support from the U.S. Department of Energy, Office of Basic Energy Sciences (OBES) FWP-06SCPE401. I. J. B. acknowledges financial support from the National Science Foundation (NSF CMMI-1728224).

Conflict of Interest The authors declare no conflict of interest.

References

1. L. Čížek, M. Greger, L. Pawlica, L. A. Dobrzański, and T. Tański, "Study of selected properties of magnesium alloy AZ91 after heat treatment and forming," *J. Mater. Process. Technol.*, vol. 157–158, no. SPEC. ISS., pp. 466–471, 2004, https://doi.org/10.1016/j.jmatprotec.2004.07.149.
2. M. C. Zhao, M. Liu, G. Song, and A. Atrens, "Influence of the β-phase morphology on the corrosion of the Mg alloy AZ91," *Corros. Sci.*, vol. 50, no. 7, pp. 1939–1953, 2008, https://doi.org/10.1016/j.corsci.2008.04.010.
3. Y. Z. Lit, Q. D. Wang, W. J. Ding, X. Q. Zeng, and Y. P. Zhu, "Fracture behavior of AZ91 magnesium alloy," *Mater. Lett.*, vol. 44, no. 5, pp. 265–268, 2000, https://doi.org/10.1016/S0167-577X(00)00041-0.
4. J. P. Zhou *et al.*, "In situ observation of ageing process and new morphologies of continuous precipitates in AZ91 magnesium alloy," *Mater. Lett.*, vol. 61, no. 25, pp. 4707–4710, 2007, https://doi.org/10.1016/j.matlet.2007.03.013.
5. M. X. Zhang and P. M. Kelly, "Crystallography of Mg17Al12 precipitates in AZ91D alloy," *Scr. Mater.*, vol. 48, no. 5, pp. 647–652, 2003, https://doi.org/10.1016/S1359-6462(02)00555-9.
6. D. Duly, M. C. Cheynet, and Y. Brechet, "Morphology and chemical nanoanalysis of discontinuous precipitation in MgAl alloys-I. Regular growth," *Acta Metall. Mater.*, vol. 42, no. 11, pp. 3843–3854, 1994, https://doi.org/10.1016/0956-7151(94)90450-2.
7. A. F. Crawley and B. Lagowski, "Effect of Two-Step Aging on the Precipitate Structure in Magnesium Alloy Az91.," *Met. Trans*, vol. 5, no. 4, pp. 949–951, 1974, https://doi.org/10.1007/BF02643153.
8. J. D. Robson, N. Stanford, and M. R. Barnett, "Effect of precipitate shape and habit on mechanical asymmetry in magnesium alloys," *Metall. Mater. Trans. A Phys. Metall. Mater. Sci.*, vol. 44, no. 7, pp. 2984–2995, 2013, https://doi.org/10.1007/s11661-012-1466-0.
9. Y. Cui *et al.*, "Impact of solute elements on detwinning in magnesium and its alloys," *Int. J. Plast.*, vol. 91, pp. 134–159, 2017, https://doi.org/10.1016/j.ijplas.2016.09.014.
10. J. D. Robson, N. Stanford, and M. R. Barnett, "Effect of precipitate shape on slip and twinning in magnesium alloys," *Acta Mater.*, vol. 59, no. 5, pp. 1945–1956, 2011, https://doi.org/10.1016/j.actamat.2010.11.060.
11. J. J. Bhattacharyya, S. R. Kada, M. R. Barnett, and S. R. Agnew, "Crystal plasticity and in-situ diffraction-based determination of the dislocation strengthening and load-sharing effects of precipitates in Mg alloy, AZ91," *Materialia*, vol. 6, March, p. 100308, 2019, https://doi.org/10.1016/j.mtla.2019.100308.
12. M. H. Yoo and J. K. Lee, "Deformation twinning in h.c.p. metals and alloys," *Philos. Mag. A Phys. Condens. Matter, Struct. Defects Mech. Prop.*, vol. 63, no. 5, pp. 987–1000, 1991, https://doi.org/10.1080/01418619108213931.
13. P. G. Patridge, "The crystallog raphy and defor- mation modes of hexagonal close-packed metals," *Metall. Rev.*, vol. 12, pp. 169–194, 1967, https://doi.org/10.1179/mtlr.1967.12.1.169.
14. M. Arul Kumar, A. K. Kanjarla, S. R. Niezgoda, R. A. Lebensohn, and C. N. Tomé, "Numerical study of the stress state of a deformation twin in magnesium," *Acta Mater.*, vol. 84, pp. 349–358, 2015, https://doi.org/10.1016/j.actamat.2014.10.048.
15. I. J. Beyerlein, L. Capolungo, P. E. Marshall, R. J. McCabe, and C. N. Tome, "Statistical analyses of deformation twinning in magnesium," *Philos. Mag.*, vol. 90, no. 16, pp. 2161–2190, 2010, https://doi.org/10.1080/14786431003630835.
16. M. R. Barnett, N. Stanford, A. Ghaderi, and F. Siska, "Plastic relaxation of the internal stress induced by twinning," *Acta Mater.*, vol. 61, no. 20, pp. 7859–7867, 2013, https://doi.org/10.1016/j.actamat.2013.09.024.
17. M. R. Barnett, M. D. Nave, and A. Ghaderi, "Yield point elongation due to twinning in a magnesium alloy," *Acta Mater.*, vol. 60, no. 4, pp. 1433–1443, 2012, https://doi.org/10.1016/j.actamat.2011.11.022.
18. A. A. Salem, S. R. Kalidindi, R. D. Doherty, and S. L. Semiatin, "Strain hardening due to deformation twinning in α-titanium: Mechanisms," *Metall. Mater. Trans. A Phys. Metall. Mater. Sci.*,

vol. 37, no. 1, pp. 259–268, 2006, https://doi.org/10.1007/s11661-006-0171-2.
19. S. R. Kalidindi, A. A. Salem, and R. D. Doherty, "Role of deformation twinning on strain hardening in cubic and hexagonal polycrystalline metals," *Adv. Eng. Mater.*, vol. 5, no. 4, pp. 229–232, 2003, https://doi.org/10.1002/adem.200300320.
20. M. Lentz, M. Risse, N. Schaefer, W. Reimers, and I. J. Beyerlein, "Strength and ductility with {10 11}-{1012} double twinning in a magnesium alloy," *Nat. Commun.*, vol. 7, pp. 1–7, 2016, https://doi.org/10.1038/ncomms11068.
21. Q. Ma, B. Li, E. B. Marin, and S. J. Horstemeyer, "Twinning-induced dynamic recrystallization in a magnesium alloy extruded at 450 °c," *Scr. Mater.*, vol. 65, no. 9, pp. 823–826, 2011, https://doi.org/10.1016/j.scriptamat.2011.07.046.
22. B. A. Simkin, B. C. Ng, M. A. Crimp, and T. R. Bieler, "Crack opening due to deformation twin shear at grain boundaries in near-γ TiAl," *Intermetallics*, vol. 15, no. 1, pp. 55–60, 2007, https://doi.org/10.1016/j.intermet.2006.03.005.
23. C. Liu *et al.*, "On the interaction of precipitates and tensile twins in magnesium alloys," *Acta Mater.*, vol. 178, pp. 146–162, 2019, https://doi.org/10.1016/j.actamat.2019.07.046.
24. R. A. Lebensohn, A. K. Kanjarla, and P. Eisenlohr, "An elasto-viscoplastic formulation based on fast Fourier transforms for the prediction of micromechanical fields in polycrystalline materials," *Int. J. Plast.*, vol. 32–33, pp. 59–69, 2012, https://doi.org/10.1016/j.ijplas.2011.12.005.
25. M. Arul Kumar, I. J. Beyerlein, and C. N. Tomé, "Effect of local stress fields on twin characteristics in HCP metals," *Acta Mater.*, vol. 116, pp. 143–154, 2016, https://doi.org/10.1016/j.actamat.2016.06.042.
26. M. Arul Kumar, I. J. Beyerlein, R. A. Lebensohn, and C. N. Tomé, "Modeling the effect of neighboring grains on twin growth in HCP polycrystals," *Model. Simul. Mater. Sci. Eng.*, vol. 25, no. 6, 2017, https://doi.org/10.1088/1361-651X/aa7bbb.
27. M. Arul Kumar, I. J. Beyerlein, and C. N. Tomé, "Grain size constraints on twin expansion in hexagonal close packed crystals," *J. Appl. Phys.*, vol. 120, no. 15, 2016, https://doi.org/10.1063/1.4965719.
28. M. Arul Kumar, B. Leu, P. Rottmann, and I. J. Beyerlein, "Characterization of Staggered Twin Formation in HCP Magnesium," in *Magnesium Technology 2019*, pp. 207–213, https://doi.org/10.1007/978-3-030-05789-3_31.
29. G. Simmons, *Single Crystal Elastic Constants and Calculated Aggregate Progress*, vol. 34. 1965.
30. N. Wang, W. Y. Yu, B. Y. Tang, L. M. Peng, and W. J. Ding, "Structural and mechanical properties of Mg17Al12 and Mg24Y5 from first-principles calculations," *J. Phys. D. Appl. Phys.*, vol. 41, no. 19, 2008, https://doi.org/10.1088/0022-3727/41/19/195408.
31. X. Ma, Q. Jiao, L. J. Kecskes, J. A. El-Awady, and T. P. Weihs, "Effect of basal precipitates on extension twinning and pyramidal slip: A micro-mechanical and electron microscopy study of a Mg–Al binary alloy," *Acta Mater.*, vol. 189, pp. 35–46, 2020, https://doi.org/10.1016/j.actamat.2020.02.037.
32. Y. Zhu and X. Wu, "Perspective on hetero deformation induced (HDI) hardening and backstress," *Mater. Res. Lett.*, vol. 7, no. 10, pp. 393–398, 2019.
33. J. D. Robson, N. Stanford, and M. R. Barnett, "Effect of particles in promoting twin nucleation in a Mg-5 wt.% Zn alloy," *Scr. Mater.*, vol. 63, no. 8, pp. 823–826, 2010, https://doi.org/10.1016/j.scriptamat.2010.06.026.
34. M. Arul Kumar, I. J. Beyerlein, R. J. McCabe, and C. N. Tomé, "Grain neighbour effects on twin transmission in hexagonal close-packed materials," *Nat. Commun.*, vol. 7, no. May, p. 13826, 2016, https://doi.org/10.1038/ncomms13826.
35. N. Stanford and M. R. Barnett, "Effect of particles on the formation of deformation twins in a magnesium-based alloy," *Mater. Sci. Eng. A*, vol. 516, no. 1–2, pp. 226–234, 2009, https://doi.org/10.1016/j.msea.2009.04.001.

Check for updates

On the Role of Crystallographic Anisotropy and Texture in Damage Tolerance of Magnesium and Its Alloys

Shahmeer Baweja, Padmeya P. Indurkar, and Shailendra P. Joshi

Abstract

The remarkable crystallographic plastic anisotropy of magnesium and its alloys reflects in its polycrystal response via texture. While texture-strength linkages have been studied, the role of textural variability on damage remains elusive. The challenge is to obtain relevant metrics that relate the net plastic anisotropy to macroscopic modes of damage. A possible approach is to adopt mechanistic descriptions of the damage. Motivated by the recent experimental and theoretical works in this direction, here we appeal to the Hill yield function to characterize the net plastic anisotropy of polycrystalline magnesium via the Hill plastic anisotropy tensor $\mathbb{h}$. Metrics based on the components of $\mathbb{h}$ offer a way to predict damage as a possible damage predictor. Using the results from our recent extensive three-dimensional crystal plasticity simulations for a wide range of textures, we map the net plastic anisotropy on to the coefficients of $\mathbb{h}$, separately for the tensile and compressive responses. Metrics based on these coefficients serve as indicators for the propensity of textured polycrystals to damage by: (i) porosity evolution, or (ii) shear instability. An attempt is made to understand the potential roles textural variability and crystallographic plastic anisotropy play in damage under different loading conditions.

Keywords

HCP materials • Damage tolerance • Textural variability • Plastic anisotropy • Crystal plasticity

S. Baweja · S. P. Joshi (✉)
Department of Mechanical Engineering, University of Houston, Houston, TX 77204-4006, USA
e-mail: shailendra@uh.edu

P. P. Indurkar
Department of Mechanical Engineering, National University of Singapore, Singapore, 117575, Singapore

Introduction

It is recognized that ductile failure is a multi-scale phenomenon [1, 2]. Crystallography defines deformation mechanisms at the atomic scale, which interact with microstructural length-scales defined by the size and distribution of microscopic defects (e.g. second-phase particles) to trigger damage through void nucleation and growth. Coarser length-scales appear with inter-flaw interactions that ultimately coalesce to form mesoscopic damage zones. Interaction of meso-scale structures with specimen/component scale forms the final feature of macroscopic failure. Coupling between these length-scales is often complicated by the anisotropic nature of plasticity but is of particular significance in hexagonal close-packed (HCP) materials that deform by protean slip and twinning mechanisms.

The remarkable crystallographic plastic anisotropy, tension-compression asymmetry, and strong texture effects exhibited by many HCP materials are often referred to as origins of damage intolerance. Their plastic anisotropy and strength asymmetry are a result of intricate slip and twinning mechanisms [3–5]. Such materials also tend to exhibit more complex damage evolution [6–8]. Huez et al. [9] reported void evolution at twins, twin-twin intersections, and twin-particle intersections in an α-Ti alloy under different triaxiality levels with unexpectedly rapid growth rates. Crépin [6] and Caré [10] observed hexagonal shaped tubular dimples in Zr alloys, attributed to the prismatic slip. More recent investigations on single [7] and alloyed polycrystals [11, 12] of Mg, as well as on Ti alloys [13] serve as potent markers for the need to assess the role of slip and twinning on anisotropic ductile damage.

Recent experiments on two dilute solute-strengthened Mg alloys (AZ31B and WE43) with different levels of plastic anisotropy reveal the following salient characteristics [14, 15]: (i) a non-monotonic dependence of fracture strain on the stress state, (ii) a strong effect of the stress triaxiality on the

V. M. Miller et al. (eds.), *Magnesium Technology 2021*, The Minerals, Metals & Materials Series,
https://doi.org/10.1007/978-3-030-65528-0_14

forensics of fracture ranging from quasi-brittle (cleavage-like) to ductile (dimpled) features on fracture surfaces, and (iii) a dramatic dependence of the tensile ductility on the level of plastic anisotropy. Quasi-brittle fracture with the linkage of twin induced microcracks in WE43 versus more ductile fracture AZ31B via void linkage alludes to the role of alloying and texture-induced plastic anisotropies on ultimate fracture [12]. Even for the same material (AZ31B), subjected to similar loading conditions, contrasting observations on the nature of damage mechanisms viz. micro-void sheeting [16] versus micro-void growth and coalescence [11], present an intriguing dichotomy. This is apparent even in single crystals where synergistic correlation between $\{10\bar{1}2\}$ extension twinning and toughness [7] conflicts with the low toughness and ductility associated with twinning [4, 5].

Another critical aspect that remains unresolved in our understanding of damage in HCP materials is the role of plastically soft slip and twinning mechanisms in shear failure. This is important for several reasons: first, shear failure may set limits to ductility that are more severe than by mere damage accumulation (void growth to coalescence) [17]. Second, experiments indicate that damage in the form of flattened micro-voids (at twins or second-phase particles) remains relatively flat as damage evolves, in contrast with continuum porous metal plasticity models, which predict significant blunting of such initial penny-shaped voids [18]. Uniaxial tensile experiments using flat tapered tensile specimens of an Mg rare-earth alloy (ZEK100) with AZ31B alloy [12] reveal that while the ZEK100 exhibits a quasi-brittle (flat fracture surface) failure compared to the more ductile (dimpled) characteristics in AZ31B, the final failure occurs via macroscopic shear localization in both alloys.

These experimental observations hint at more complex interactions between plastic anisotropy and failure mechanisms than currently appreciated. On the sole basis of experiments, it is unclear how, and to what extent, matrix plastic anisotropy affects the rates and states of damage accumulation in structural components. Computational investigations of boundary value problems based on damage-free polycrystal plasticity modeling [19] as well as those based on homogenization-based porous plasticity [17] shed some light in this regard. For failure induced by internal damage accumulation via porosity evolution of Hill-type anisotropic materials, a so-called AED (anisotropy effect on ductility) index has been proposed [20, 21]. On the other hand, Benzerga et al. [17] indicate that shear failure is plausible even in geometries and under tensile boundary conditions that are stiff against the formation of shear bands, and that the net plastic anisotropy may be the main driving factor for shear fracture. In particular, the role of shear anisotropy ratios may be important, although a robust index (akin to the AED index) characterizing their role in shear fracture remains elusive.

The foregoing concepts of the AED index and, to some extent, the shear anisotropy ratios, have been shown to corroborate well with broader experimental observations on materials exhibiting Hill-type plastic anisotropy [17, 20]. On that backdrop, Mg and its alloys cannot be well described by a Hill-type yield function owing to the tension-compression asymmetry as a result of the polar nature of twinning. Nevertheless, the recent analysis of Mg alloys based on these ideas [21] suggests it may still provide a useful basis for a theory-based design paradigm of damage-tolerant HCP materials. Hence, it serves as a motivation for the present analysis.

In this work, we view the role of *net* plastic anisotropy that embeds the intrinsic (crystallographic) and textural effects on the potential macroscopic damage modes in HCP materials with a focus on Mg alloys. In particular, we attempt to address the following questions in the context of Mg and its alloys:

1. How do crystallographic and textural effects influence the AED index and shear anisotropy ratios under tensile and compressive loading? An allied question is: to what extent do the differences in these metrics computed from tensile versus compressive loading qualitatively affect the broader conclusions?
2. Are there combinations of crystallographic and texture-induced plastic anisotropy (hereafter referred to as the net plastic anisotropy) that satisfy minimum requirements for a material to be potentially damage-tolerant?

To address these questions, we rely on the dataset from our recent three-dimensional crystal plasticity investigation on polycrystal statistical volume elements (SVEs) performed for a range of synthetic textures that mimic experimental rolling textures at two different levels of crystallographic plastic anisotropies [22]. The explicit resolution of the SVE combined with crystallographic kinetics of plasticity provides detailed insight into the evolution of microstructure-property linkages including plastic anisotropy and tension-compression asymmetry [22].

Computational Setup

A detailed account of the three-dimensional crystal plasticity modeling and simulation of the polycrystalline SVE is provided in Ref. [22]. It considers eleven textures and two crystallographic plastic anisotropies (one representing an Mg alloy and the other representing pure Mg). For each texture, we perform simulations under uniaxial stress condition along six directions: (i) three along material principal axes (L, T, S), and (ii) three off-axes (LT, LS, TS), characterizing the full macroscopic plastic anisotropy for each microstructure. Moreover, to characterize the tension-compression asymmetry, monotonic uniaxial compressive and tensile loading states are applied. Thus, for each crystallographic plastic anisotropy, we perform 132 simulations.

The present work builds upon those detailed calculations with a focus on characterizing the role of net plastic anisotropy in the potential damage response of Mg and its alloys. In what follows, we provide a brief background of the key microstructural descriptors that are adopted in the discussion of the results.

Texture Descriptor

In the polycrystal plasticity simulations, we create three-dimensional Voronoi tessellations of a cubic domain in NEPER [10], giving in an SVE with 300 grains ($N_{grains} = 300$) that are discretized into a fine finite element mesh. The discretized SVE is imported into ABAQUS/STANDARD® for subsequent mechanical analysis. To describe a texture of an SVE, individual grains are characterized by distinct Euler angle sets in the Bunge representation: $[E] = \left[\overline{\varphi}_1 \pm \varphi_1^{\sigma}, \overline{\Phi} \pm \Phi^{\sigma}, \overline{\varphi}_2 \pm \varphi_2^{\sigma}\right]$ where $[\overline{\varphi}_1, \overline{\Phi}, \overline{\varphi}_2]$ denote the mean values and $\varphi_1^{\sigma}, \Phi^{\sigma}, \varphi_2^{\sigma}$ their respective standard deviations. Using a normal distribution with $\overline{\varphi}_1 = \overline{\Phi} = \overline{\varphi}_2 = 0$, each texture is then described by the maximum standard deviation $[E^{\sigma}] = [\varphi_1^{\sigma}, \Phi^{\sigma}, \varphi_2^{\sigma}]$.

Under a uniaxial loading controlled via applied nominal strain rate $\left(\dot{\varepsilon}_{app}\right)$, the macroscopic stress state in the SVE is $\boldsymbol{\Sigma} = \Sigma_{yy}\left(\boldsymbol{e}_y \otimes \boldsymbol{e}_y\right)$, where $\boldsymbol{e}_y$ is the unit vector in the y-direction (loading axis). The corresponding macroscopic logarithmic strain state is: $\mathbf{E} = \mathrm{E}_{xx}(\boldsymbol{e}_x \otimes \boldsymbol{e}_x) + \mathrm{E}_{yy}\left(\boldsymbol{e}_y \otimes \boldsymbol{e}_y\right) + \mathrm{E}_{zz}(\boldsymbol{e}_z \otimes \boldsymbol{e}_z)$where $\mathrm{E}_{xx} = \ln(L_x/L_0)$, $\mathrm{E}_{yy} = \ln\left(L_y/L_0\right)$, and $\mathrm{E}_{zz} = \ln(L_z/L_0)$. The von-Mises equivalent stress is: $\Sigma_{eq} = \sqrt{(3/2)\boldsymbol{\Sigma}' : \boldsymbol{\Sigma}'} = |\Sigma_{yy}|$ and the corresponding effective strain is: $\mathrm{E}_{eq} = \sqrt{(2/3)\mathbf{E}' : \mathbf{E}'}$ where $\boldsymbol{\Sigma}'$ are the deviatoric stresses and $\mathbf{E}'$ are the corresponding deviatoric strains. Finally, the macroscopic strain ratios (also known as the Lankford ratios) are given by: $R_L = \mathrm{E}_{TT}/\mathrm{E}_{SS}; R_T = E_{LL}/E_{SS}; R_{LT} = E_{LT^{\perp}}/E_{SS}; R_{LS} = E_{LS^{\perp}}/E_{TT}; R_{TS} = E_{LL}/E_{TS^{\perp}}$ [see [22] for further details].

HCP Crystal Plasticity

The finite strain crystal plasticity model comprises of eighteen slip systems (3 basal, 3 prismatic, 6 pyramidal $\langle a \rangle$, and 6 pyramidal $\langle c + a \rangle$) and twelve twin systems (6 extension twinning and 6 contraction twinning) with rate-dependent viscoplastic flow rules that embed the physics of plasticity kinetics and incorporate twinning-induced lattice reorientation [23]. Two sets of material parameters are considered (Table 2); one set of parameters is representative of an Mg alloy (AZ31B) and those in the brackets are for 99.97% pure Mg [22].

Micromechanical Theory of Anisotropic Damage

As noted earlier, the hypothesis underlying this work is that both, intrinsic plastic anisotropy and texture-induced effect play a role in the damage tolerance of Mg alloys. For a pristine (*damage-free*) material, these contributions emerge in the form of the net plastic anisotropy at the macroscopic scale, which may be characterized by one of the several anisotropic yield functions [24]. In this work, we choose a Hill-type representation of the net plastic anisotropy. Admittedly, such a representation may be deemed too simplistic for materials such as Mg that exhibit a complex plastic anisotropy and perhaps even inadequate as it does not cater for the tension-compression asymmetry [13]. Notwithstanding this caveat, the choice of this yield model is driven by two considerations. First, it is a simple representation of the plastic anisotropy that offers a physical basis to the anisotropy coefficients in the yield function. Second, the yield function has been extended to incorporate damage by porosity evolution [25]. These features make the theory attractive in gaining insights by correlating the coefficients of Hill anisotropy tensor to the two primary mechanisms of macroscopic ductile fracture: (i) internal damage accumulation by void growth, and (ii) mechanical instability driven by shear bands. Indeed, the trends based on this model have been shown to broadly corroborate with the experimentally observed ductility trends in some Mg alloys [21, 26]. Below, we briefly elucidate the main ingredients of the Hill plastic anisotropy and its relation to damage. For a Hill-type material, the yield condition is:

$$\sigma_{eq}^2 = \frac{3}{2}\mathbf{s} : \mathbb{h} : \mathbf{s} = \tilde{\sigma}^2$$

where σ_{eq} is the equivalent stress of the Hill yield criterion,$\mathbf{s}$ the stress deviator, $\mathbb{h}$ the fourth-order Hill anisotropy tensor, and $\tilde{\sigma}$ the flow stress. The flow stress is arbitrarily chosen along a direction, which serves as the reference direction. Referring to Fig. 1, with loading axes aligned with the principal directions of plastic orthotropy, the above equation reduces to:

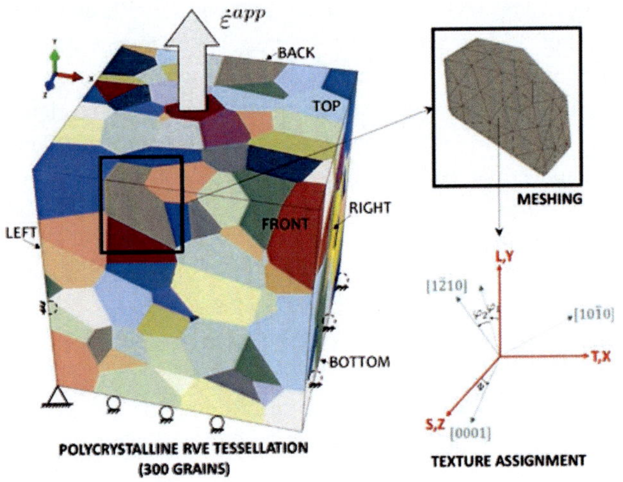

Fig. 1 Polycrystal SVE. The material directions are denoted by L, T, and S. Uniaxial loading is applied along the global y-axis. The particular case shown here is L-tension. (Color figure online)

$$\sigma_{eq}^2 = \frac{3}{2}\left(h_L s_{11}^2 + h_T s_{22}^2 + h_S s_{33}^2 + 2h_{LT} s_{12}^2 + 2h_{LS} s_{13}^2 + 2h_{TS} s_{23}^2\right) = \bar{\sigma}^2 \quad (1)$$

the anisotropy coefficients h_i are related to those used by Hill [27]. In the present work, a strain-based approach is used in the calculation of these coefficients. Given the principal directions of plastic orthotropy L-T-S, the strain relationships between h_i and the strain ratios are as follows:

$$\frac{h_T}{h_L} = 1 - \frac{3(R_L R_T - 1)}{R_L R_T - 2R_L - 2}; \frac{h_s}{h_L} = 1 - \frac{3R_L(R_T - 1)}{R_L R_T - 2R_L - 2}; \frac{h_{LT}}{h_L} = -\frac{1}{2}\frac{(2R_{LT}+1)(R_L R_T + 1)}{R_L R_T - 2R_L - 2}$$

$$\frac{h_{LS}}{h_L} = -\frac{1}{2}\frac{(2R_{LS}+1)(R_L+1)}{R_L R_T - 2R_L - 2}; \frac{h_{TS}}{h_L} = -\frac{1}{2}\frac{(2R_{TS}+1)(R_L+1)}{R_L R_T - 2R_L - 2}$$

These anisotropy coefficients enter in the micromechanics of damage via porosity evolution as follows [20]:

$$\frac{\dot{f}}{\dot{\varepsilon}_{eq} f} \approx \frac{3}{h}\sinh\left(\frac{3}{h}\mathcal{T}\right) \quad (2)$$

where f is the current void volume fraction, $\dot{\varepsilon}_{eq}$ is the effective plastic strain rate, $\mathcal{T}$ is the stress triaxiality ratio, and h is a scalar invariant of $\mathbb{h}$, given by:

$$h = 2\left[\frac{2}{5}\frac{h_L + h_T + h_S}{h_L h_T + h_T h_S + h_L h_S} + \frac{1}{5}\left(\frac{1}{h_{LT}} + \frac{1}{h_{LS}} + \frac{1}{h_{TS}}\right)\right]^{\frac{1}{2}} \quad (3)$$

Equation (2) shows that the rate of growth of porosity $\dot{f}$ in a plastically anisotropic material is a function of the *degree of plastic anisotropy*, characterized by h. As such, it is referred to as the Anisotropy Effect on Ductility (AED) index [21]. In the limiting case of plastic isotropy, all $h_i = 1$, which gives $h = 2$.

Another canonical mode of failure is the propensity to failure by mechanical instability via shear bands, referred to as shear failure. As expounded in Ref. [28], shear failure is distinct from failure in shear; while the former may occur under remote loading states devoid of a shear component, the latter occurs under dominant shear loading conditions. While there is no robust index (even for a Hill-type plastic anisotropy) yet that describes the propensity of a plastically anisotropic material to macroscopic shear failure, the role of plastic anisotropy in shear failure may to zeroth order, be viewed from the vantage point of the ratios of shear anisotropy coefficients [17]. In orthotropic plasticity, the coefficients of interest are h_{LS} and h_{TS}. We discuss this further in the next section.

Results and Discussion

In the following, we present the evolution and textural dependencies of three quantities computed from the expressions in the preceding section using the data from crystal plasticity simulations [22]. These quantities are: the AED index h (Eq. 3), and two shear anisotropy ratios $\widehat{h}_{LST} = h_{LS}/h_T$ and $\widehat{h}_{TSL} = h_{TS}/h_L$ (cf. Eq. 1). In what follows, we focus on the results for an Mg alloy (Table 2), and later comment on the role of crystallographic anisotropy by examining the trends for pure Mg (Section "Role of Crystallographic Anisotropy").

AED Index

Figure 2 shows the evolution of the AED index with equivalent strain E_{eq} calculated from tensile loading along the L, T, S, LT, LS, and TS directions as detailed in Ref. [22]. The different curves are for different textures ranging from A through K (cf. Table 1).

Beyond the initial transient, all curves show a systematic increase with a tendency toward steady-state values at large strains ($E_{eq} \sim 0.2$). The dashed line demarcates an isotropic plastic response. There is a dependence of h on texture. This is clearer from Fig. 3, which captures steady-state values of the AED index for each texture. For tensile loading, the steady-state values of the AED index are in the range $1.9 \lesssim h \lesssim 2.5$. Recall that for a plastically isotropic material

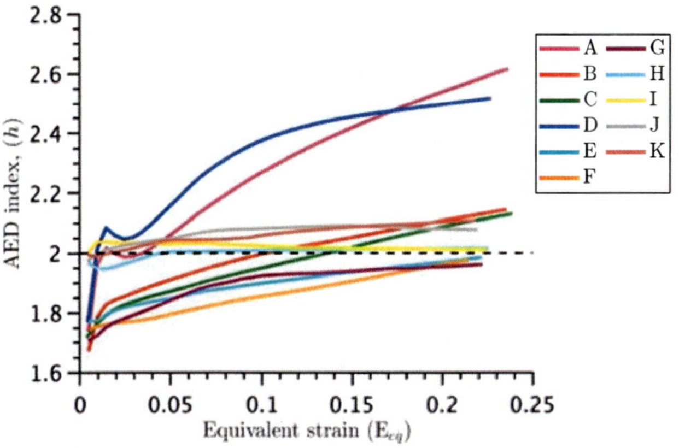

Fig. 2 Evolution of AED index with strain for different textures under tensile loading (Mg alloy). The dashed line indicates the limiting case of isotropic plasticity. (Color figure online)

$h = 2$. The AED index exhibits a non-monotonic correlation with the peak intensity of the (0001) pole figure. With reference to Table 1, textures with the highest and the lowest (0001) peak intensities show $h > 2$ with stronger textures showing higher values. In comparison, textures with intermediate (0001) peak intensities exhibit tempered values of $h \lesssim 2$. While the corresponding correlation with $(10\bar{1}0)$ peak intensities is not as clean, the trends are roughly similar. Broadly, it appears then that that stronger textures are likely to be more resistant to void growth than their weaker counterparts and isotropic materials.

A similar analysis using compression datasets along the same six directions allows computing AED indices, whose steady-state values are also shown in Fig. 3. For a Hill-type material (no tension-compression asymmetry), the AED index calculated from the compression data should be identical to its tensile counterpart. In such a scenario, the choice of compressive tests is driven by practical considerations; they are more cost-effective and consume less material per test. For most Mg alloys, the tension-compression asymmetry precludes the use of the AED index computed from the compression dataset. As seen from Fig. 3, the *compressive* AED indices lie below the isotropic limit whereas the *tensile* AED indices indicate a more sensitivity to textural variations. For instance, given a fixed intrinsic plastic anisotropy, textures E, F, and G exhibit a higher propensity to void growth ($h_{tensile} < 2$) compared to a plastically isotropic material. On the other hand, textures A and D indicate a higher resistance to damage by porosity evolution than the other textures. Such an assessment would not be possible from the compressive AED data if they were to be used as indicators, as they would suggest that all textures being highly susceptible to porosity growth. Hence, one conclusion we make is that for materials such as Mg alloys, it is not advisable to use compression data to characterize this particular mode of damage.

Table 1 Texture cases considered in this work [22]

Angles↓/Cases→	A	B	C	D	E	F	G	H	I	J	K
φ_1^σ	20°	10°	20°	30°	30°	30°	30°	30°	30°	45°	45°
Φ^σ	10°	15°	20°	10°	20°	20°	20°	30°	45°	60°	75°
φ_2^σ	0°	0°	0°	0°	0°	10°	20°	30°	30°	45°	45°
Peak intensity [0001]	31	24	19	30	18	20	18	13	8	6	4
Peak intensity $[10\bar{1}0]$	6	10	6	5	5	5	4	4	3	3	2

Table 2 Material properties for alloyed Mg (AZ31B) alloy and pure Mg (in brackets)

Mechanisms	τ_0(MPa)	h_0(MPa)	τ_s(MPa)		$\tau_0^i/\tau_0^{pris.\langle a\rangle sl.}$
Basal $\langle a\rangle$ slip	10 (0.5)	50 (20)	–		0.18 (0.02)
Prismatic $\langle a\rangle$ slip	55 (25)	1500	110 (85)		1
Pyramidal $\langle a\rangle$ slip	55 (25)	1500	110 (85)		1
Pyramidal $\langle c+a\rangle$ slip	60 (40)	3000	170 (150)		1.09 (1.6)
Extension twinning	τ_0(MPa)	h_{et}(MPa)	τ_{s_et}(MPa)	h_{et_sl}(MPa)	0.27 (0.14)
	15 (3.5)	120 (100)	30 (20)	100	
Contraction twinning	τ_0(MPa)	H_{ct}(MPa)	H_{ct_sl}(MPa)	b	1.55(2.2)
	85 (55)	6000	15	0.05	

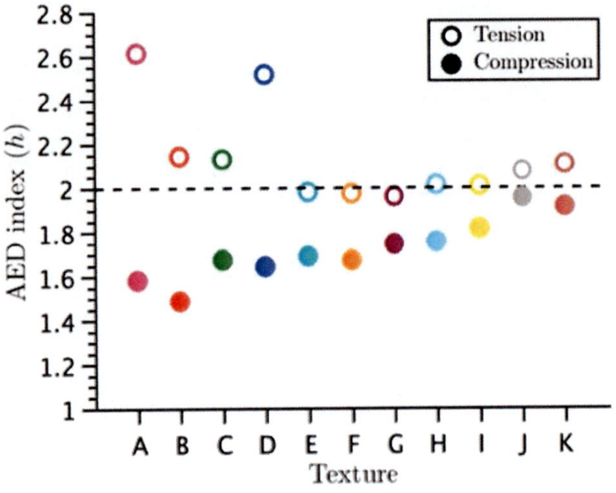

Fig. 3 Steady-state values of the AED index computed for tensile and compressive loading (Mg alloy). The dashed line indicates the limiting case of isotropic plasticity. (Color figure online)

Figure 4 compares the trends of the tensile and compressive AED indices for three textures whose textural intensities are comparable with the experiments of Kondori and Benzerga [14]. As seen, the computed trends are consistent with the particular experimental data.

Shear Anisotropy Ratios

While the compressive AED index is not of much consequence in the context of damage prediction by porosity evolution, the shear anisotropy ratios $(\widehat{h})$ do have relevance in both tension and compression. As noted before, $\widehat{h} \leq 1$ denotes a shear resistant material while $\widehat{h} > 1$ indicates a material that is weak against shear failure. On that backdrop, Fig. 5 reveals some interesting features. Note that the vertical axis is plotted on the logarithmic scale because of the

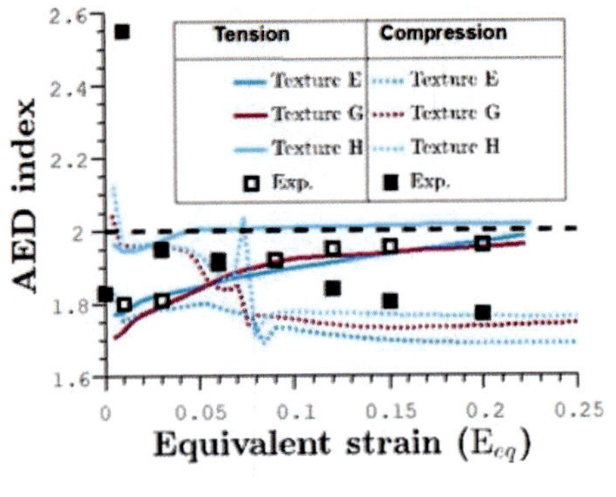

Fig. 4 Comparison of the computed AED index with experimental results [14] for an Mg alloy. The dashed line indicates the limiting case of isotropic plasticity. (Color figure online)

large range of $\widehat{h}$ values particularly in compression. For the given level of crystallographic anisotropy, the propensity of the material to shear instability seems to depend on both, the texture and loading. For a given texture, susceptibility to instability in compression is generally higher than in tension.

Under tensile loading, stronger textures appear to be more susceptible to shear bands compared to the weaker ones. In general, the $\widehat{h}_{LST}$ correlates with $(10\bar{1}0)$ peak intensity whereas $\widehat{h}_{TSL}$ correlates with (0001) peak intensity in that the higher the peak intensities the higher the corresponding $\widehat{h}$ (i.e. less shear resistant). There appear to be three broad families: (i) weak in shear along both planes comprising textures A and B, (ii) weak in shear along one plane (TS) but shear resistant in the other (LS), comprising textures C, D, and F, and (iii) shear resistant along both planes, comprising textures E, and G-K.

In compression, the situation is more straightforward in that all textures exhibit the propensity to shear bands, although the weaker the texture the lesser its susceptibility to instability compared to the stronger counterparts. Here, both $\widehat{h}_{LST}$ and $\widehat{h}_{TSL}$ appear to correlate better with the between (0001) peak intensity compared to the $(10\bar{1}0)$ peak intensity.

Failure Map

With the foregoing results, we propose a failure map in the $\widehat{h} - h$ space, Fig. 6. The limiting cases for an isotropic material $(\widehat{h} = 1, h = 2)$ split the $\widehat{h} - h$ space into four quadrants. Figure 6a, collates the shear anisotropy data from both tension and compression along both planes and plots it against the tensile AED indices for the same set of textures. The green region is the goldilocks zone; textures whose all four datapoints lie in this zone should be ideal choices in terms of their crystallographic plastic anisotropy and texture. It turns out that for the level of crystallographic anisotropy representative of AZ31B, none of the eleven textures considered here come across as ideal candidates. Several candidates exhibit desirable characteristics insofar as the resistant to porosity evolution is considered but they are expected to generally exhibit poor characteristics against shear failure particularly in compression (northeast). These include: A, B, C, D, H, and I. A few candidates, which occupy the northwest region are altogether undesirable. These comprise: textures E, F, and G. A limited set of textures seem to be somewhat desirable as they occupy the space in the vicinity of the goldilocks zone, e.g. textures J and K.

Notably, the role of compressive loading in shear instability appears crucial. This suggests that the performance of these materials against shear instability in compression is a

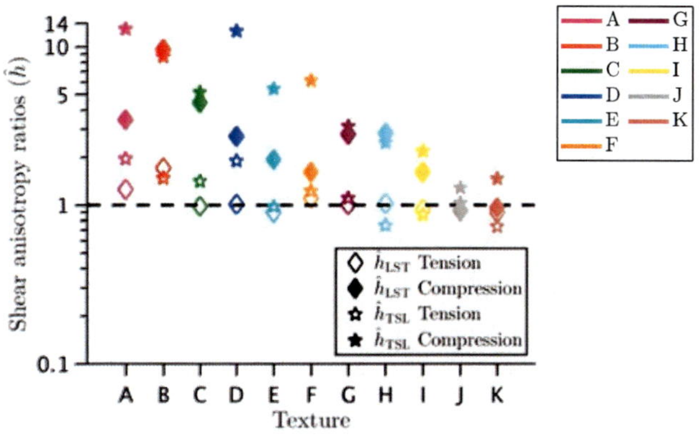

Fig. 5 Texture-dependent steady-state values of shear anisotropy ratios under tensile and compressive loading (Mg alloy). (Color figure online)

factor that may need a deeper investigation. If one ignores the compressive values of $\widehat{h}$, the situation is more forgiving as several textures serve as potential candidates as seen from Fig. 6b.

Role of Crystallographic Anisotropy

We briefly comment on the role of crystallographic plastic anisotropy in the failure landscape. Figure 7 shows the failure map for pure Mg, whose intrinsic crystallographic plastic anisotropy is significantly larger compared to the Mg alloy considered here (cf. Table 2). The figure includes a limited dataset in that only the tensile values of $\widehat{h}$ are plotted and with fewer texture cases (due to much higher computational costs). As seen from the figure, at this level of crystallographic plastic anisotropy there seem to be more possibilities of a damage-tolerant material design.

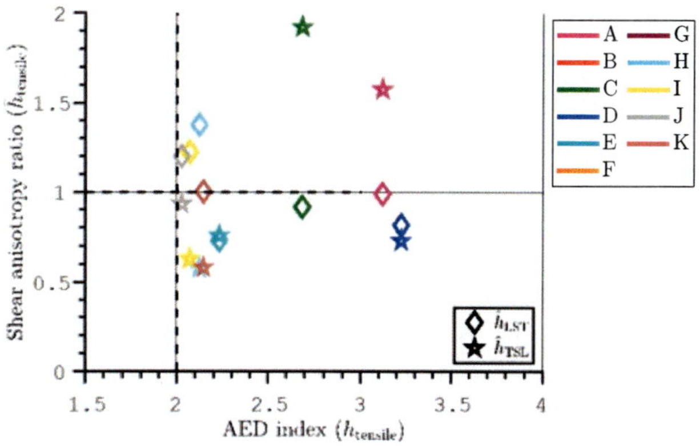

Fig. 7 Failure map for pure Mg. (Color figure online)

Textures D, E, K exhibit the characteristics of a material that is both, shear resistant and tolerant to internal damage by porosity growth. Even the remaining textures (A, C, H, I, and J) show resistance to porosity growth that is better than an isotropic material and their propensity to shear instability is relatively tempered.

Summary

A predictive understanding of the interacting effects of crystallographic and textural-induced plastic anisotropy in the ductile response of Mg and its alloys is an outstanding challenge. In this work, we present an elementary exploration towards mapping the net plastic anisotropy (derived from crystallographic and texture effects) to the macroscopic damage tolerance against two main agents of ductile failure, porosity growth and shear bands. Motivated by recent proofs of concept, we adopt a simple, yet elegant micromechanical

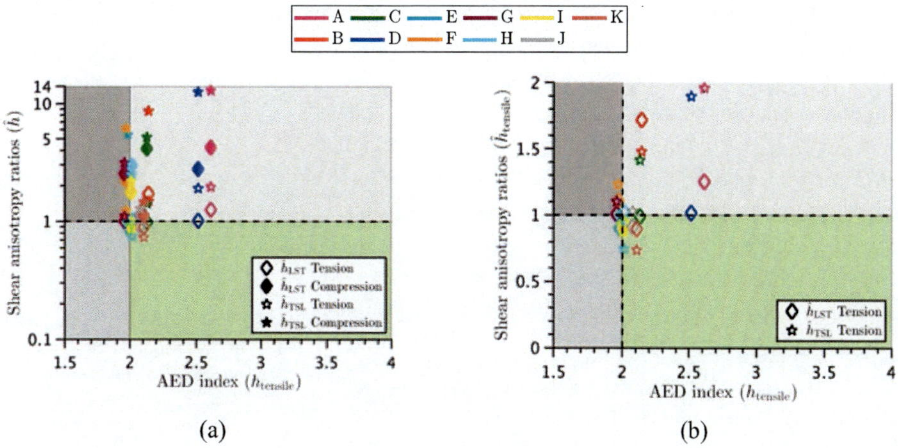

Fig. 6 Proposed failure map characterized by metrics describing propensity to shear instability $\left(\widehat{h}\right)$ and damage by porosity evolution (h). Panel (**a**) includes shear anisotropy ratios for both, tension and compression. Panel (**b**) shows the shear anisotropy ratios computed from the tensile dataset only. (Color figure online)

model of anisotropic plasticity based on the Hill yield criterion, which delivers the AED index that indicates the effect of plastic anisotropy to porosity growth, and offers insights into potency of shear instability by way of shear anisotropy ratios. Using detailed analysis of the AED indices and shear anisotropy ratios in both tension and compression, we present preliminary failure maps for two levels of crystallographic plastic anisotropies over a wide range of textures. With regard to the questions posed in Introduction, the main observations are as follows:

1. For a material resembling AZ31B Mg, the AED index shows a non-monotonic correlation with textural strengths. The trend suggests that the strongest and the weakest textures may provide a better resistance to porosity driven ductile damage than intermediate textures.
2. On the other hand, for the same material the trends of shear anisotropy ratios suggest that, for tensile loading states, weaker textures may be somewhat more shear resistant compared to stronger and intermediate textures. For compressive loading states, the situation is rather challenging in that most textures may show propensity to shear failure, with the situation being much worse for the strongest textures.
3. For a material with a high crystallographic plastic anisotropy (e.g. pure Mg), the scenario is more forgiving with a wider range of textures showing resistance to both modes of failure.

The analysis presented here needs to be served with caution. While the model has been shown to corroborate with some experimental observations on the anisotropy-ductility linkages in some Mg alloys, it discounts intricacies associated with the crystallographic aspects of deformation mechanisms in Mg that can play a role in the fundamental micromechanics of ductile damage [29]. As such, while the qualitative trends may provide some guidance, a rigorous quantitative assessment may be an overreach. Notwithstanding these caveats, the present work perhaps serves as a preliminary concept towards in silico design of damage-tolerant Mg microstructures guided by the micromechanics of HCP crystal plasticity and coupled into sophisticated homogenization-based mechanics of anisotropic ductile failure.

Acknowledgements SPJ and SB acknowledge financial support from the National Science Foundation, USA under Grant Number CMMI-1932976. PPI is grateful for the support via NUS Research Scholarship. The authors acknowledge the use of the Opuntia Cluster and the advanced support from the Research Computing Data Core at the University of Houston to carry out the research presented here.

References

1. Van der Giessen, E., & Needleman, A. (2002). Micromechanics simulations of fracture. *Annual Review of Materials Research, 32* (1), 141–162.
2. Besson, J. (2010). Continuum models of ductile fracture: a review. *International Journal of Damage Mechanics, 19*(1), 3–52.
3. Christian, J. W., & Mahajan, S. (1995). Deformation twinning. *Progress in Materials Science, 39*(1-2), 1–157.
4. Barnett, M. R. (2007). Twinning and the ductility of magnesium alloys: Part I: "Tension" twins. *Materials Science and Engineering: A, 464*(1-2), 1–7.
5. Barnett, M. R. (2007). Twinning and the ductility of magnesium alloys: Part II: "Contraction" twins. *Materials Science and Engineering: A, 464*(1–2), 8–16.
6. Crépin, J., Bretheau, T., & Caldemaison, D. (1996). Cavity growth and rupture of β-treated zirconium: a crystallographic model. *Acta Materialia, 44*(12), 4927–4935.
7. Kaushik, V., Narasimhan, R., & Mishra, R. K. (2014). Experimental study of fracture behavior of magnesium single crystals. *Materials Science and Engineering: A, 590*, 174–185.
8. Nemcko, M. J., Li, J., & Wilkinson, D. S. (2016). Effects of void band orientation and crystallographic anisotropy on void growth and coalescence. *Journal of the Mechanics and Physics of Solids, 95*, 270–283.
9. Huez, J., Helbert, A. L., Feaugas, X., Guillot, I., & Clavel, M. (1998). Damage process in commercially pure α-titanium alloy without (Ti40) and with (Ti40-H) hydrides. *Metallurgical and Materials Transactions A, 29*(6), 1615–1628.
10. Caré, S., & Zaoui, A. (1996). Cavitation at triple nodes in α-zirconium polycrystals. *Acta Materialia, 44*(4), 1323–1336.
11. Prasad, N. S., Kumar, N. N., Narasimhan, R., & Suwas, S. (2015). Fracture behavior of magnesium alloys–role of tensile twinning. *Acta Materialia, 94*, 281–293.
12. Ray, A. K., & Wilkinson, D. S. (2016). The effect of microstructure on damage and fracture in AZ31B and ZEK100 magnesium alloys. *Materials Science and Engineering: A, 658*, 33–41.
13. Revil-Baudard, B., Cazacu, O., Flater, P., Chandola, N., & Alves, J. L. (2016). Unusual plastic deformation and damage features in titanium: Experimental tests and constitutive modeling. *Journal of the Mechanics and Physics of Solids, 88*, 100–122.
14. Kondori, B., & Benzerga, A. A. (2014). Effect of stress triaxiality on the flow and fracture of Mg alloy AZ31. *Metallurgical and Materials Transactions A, 45*(8), 3292–3307.
15. Kondori, B., & Benzerga, A. A. (2015). On the notch ductility of a magnesium-rare earth alloy. *Materials Science and Engineering: A, 647*, 74–83.
16. Steglich, D., & Morgeneyer, T. F. (2013). Failure of magnesium sheets under monotonic loading: 3d examination of fracture mode and mechanisms. *International Journal of Fracture, 183*(1), 105–112.
17. Benzerga, A. A., Thomas, N., & Herrington, J. S. (2019). Plastic flow anisotropy drives shear fracture. *Scientific reports, 9*(1), 1–9.
18. Kondori, B., Morgeneyer, T. F., Helfen, L., & Benzerga, A. A. (2018). Void growth and coalescence in a magnesium alloy studied by synchrotron radiation laminography. *Acta Materialia, 155*, 80–94.
19. Selvarajou, B., Joshi, S. P., & Benzerga, A. A. (2017). Three dimensional simulations of texture and triaxiality effects on the plasticity of magnesium alloys. *Acta Materialia, 127*, 54–72.
20. Benzerga, A. A., & Besson, J. (2001). Plastic potentials for anisotropic porous solids. *European Journal of Mechanics-A/Solids, 20*(3), 397–434.

21. Basu, S., Dogan, E., Kondori, B., Karaman, I., & Benzerga, A. A. (2017). Towards designing anisotropy for ductility enhancement: A theory-driven investigation in Mg-alloys. *Acta Materialia, 131*, 349–362.
22. Indurkar, P. P., Baweja, S., Perez, R., & Joshi, S. P. (2020). Predicting textural variability effects in the anisotropic plasticity and stability of hexagonal metals: Application to magnesium and its alloys. *International Journal of Plasticity*, 102762.
23. Zhang, J., & Joshi, S. P. (2012). Phenomenological crystal plasticity modeling and detailed micromechanical investigations of pure magnesium. *Journal of the Mechanics and Physics of Solids, 60*(5), 945–972.
24. Indurkar, PP (2020). On the multiscale mechanics of deformation, stability and damage in magnesium. Ph.D. Thesis, National University of Singapore.
25. Benzerga, A. A., Besson, J., & Pineau, A. (2004). Anisotropic ductile fracture: Part II: theory. *Acta Materialia, 52*(15), 4639–4650.
26. Benzerga, A. A. (2019). A Theory for Designing Ductile Materials with Anisotropy. In *Magnesium Technology 2019* (pp. 359–362). Springer, Cham.
27. Hill, R. (1948). A theory of the yielding and plastic flow of anisotropic metals. *Proceedings of the Royal Society of London. Series A. Mathematical and Physical Sciences, 193*(1033), 281–297.
28. Pineau, A., Benzerga, A. A., & Pardoen, T. (2016). Failure of metals I: Brittle and ductile fracture. *Acta Materialia, 107*, 424–483.
29. Selvarajou, B., Joshi, S. P., & Benzerga, A. A. (2019). Void growth and coalescence in hexagonal close packed crystals. *Journal of the Mechanics and Physics of Solids, 125*, 198–224.

Eliminating Yield Asymmetry and Enhancing Ductility in Mg Alloys by Shear Assisted Processing and Extrusion

Dalong Zhang, Jens Darsell, Nicole Overman, Darrell R. Herling, and Vineet V. Joshi

Abstract

Solid phase processing techniques such as friction stir welding, Shear assisted processing and extrusion (ShAPE)/friction extrusion and cold spray have been successfully demonstrated as promising thermomechanical methods to produce metallic materials with enhanced performance. In this study, AZ series with and without silicon, ZK60 Mg alloys in as-received forms (as-cast or as-extruded) were processed using Shear Assisted Processing and Extrusion (ShAPE). Microstructural characterization was performed using EBSD and TEM and revealed that as compared to the feedstock materials/billets, friction extruded Mg alloys had more uniform microstructure, equiaxed grains, finer and homogeneously distributed precipitates and chemical homogeneity. It was also observed that basal planes were not oriented parallel to extrusion axis. As a result, rod products exhibited significantly reduced (in some cases eliminated) yield asymmetry and achieved enhanced ductility, which were uncommon or difficult to attain using conventional processing techniques. In addition, modified texture likely suppressed deformation twinning under compressive deformation.

Keywords

Shear assisted processing and extrusion • ShAPE • Magnesium alloys • Yield asymmetry

Introduction

Due to the common basal texture and easily activated basal slip and twinning [1–4], wrought Mg alloys produced by extrusion or rolling often suffer from tension-compression yield asymmetry (YA) [5–7] and strength differential (SD) [8]. YA is quantified as the ratio between compressive yield strength (CYS) and tensile yield strength (TYS), whereas SD is the qualitative difference between the tensile flow curve and the compressive one. The presence of YA and SD significantly limits the application of Mg alloys because useful parts or products often need to withstand both tensile and compressive loads [9, 10].

In the past two decades, various approaches have been proposed to alleviate YA in wrought Mg alloys. One effective approach is adding rare earth (RE) alloying elements such as Ce, La, Y to weaken the basal texture [5, 11, 12], leading to similar yield behavior in both tension and compression. However, RE element addition inevitably makes the Mg alloys expensive. Another approach is microstructural refinement by severe plastic deformation (SPD) based techniques, including ball milling [13], friction stir processing (FSP) [14], and equal-channel angular extrusion (ECAP) [15]. The resultant small grain size often impedes twinning activity under compression load [13], making the CYS comparable to TYS. SPD techniques, especially ECAP, can also have effect of texture modification which may contribute to reducing YA [15]. However, the major challenge with SPD based techniques is they can only produce lab-scale research samples, with no cost-effective pathway towards scale-up manufacturing.

Moreover, most of above-mentioned approaches could at best reduce YA, whereas SD remains pronounced. Rare exceptions exist when using microstructural refinement in conjunction with adding RE elements, as the case for an ultrafine grained Mg-Y alloy [2]. However, the scale-up manufacturing of such RE-containing alloys is rather cost-prohibitive. In recent years, a cost-effective advanced

D. Zhang · J. Darsell (✉) · N. Overman ·
D. R. Herling · V. V. Joshi
Pacific Northwest National Laboratory, Richland, WA, USA
e-mail: jens.darsell@pnnl.gov

V. M. Miller et al. (eds.), *Magnesium Technology 2021*, The Minerals, Metals & Materials Series,
https://doi.org/10.1007/978-3-030-65528-0_15

manufacturing method, namely shear assisted processing and extrusion (i.e. ShAPE), has been developed at Pacific Northwest National Laboratory (PNNL). With ShAPE, tubes of 50.8 mm diameter were successfully extruded for ZK60 Mg alloy, with refined grain size (4–5 μm) and modified texture [16]. It was hypothesized that the modified texture, namely a basal texture that was tilted ~20° away from the extrusion axis, would lead to reduced YA and SD, as well as improved corrosion resistance.

In this work several different non-rare earth containing alloys of magnesium were extruded using the ShAPE process. The primary goal was to attain uniform microstructure, equiaxed grains, finer and homogeneously distributed precipitates and chemical homogeneity along with desired texture to eliminate yield asymmetry and strength differential in these alloys. It is to be noted that the work presented here is comprehensive and additional mechanical and microstructural characterization has been performed and will be available in subsequent publications.

Methods

Commercially extruded AZ31B, as-cast AZ31, and cast ZK60 alloys were purchased from MetalMart International. Inc (Commerce, CA, USA). Canmet (Hamilton, ON, Canada) cast AZ31-2Si. Mg-3Si alloy was cast at Brunel University (London, UK). Sourced magnesium alloys were first machined into billets measuring approximately 31.7 mm diameter by 18 mm thick. The magnesium alloy extruded billets were extruded into 5 mm diameter rods using ShAPE process; a diagram of the apparatus is shown in Fig. 1. The billet was loaded into a cylindrical container measuring 31.8 mm inside diameter and 20 mm tall that is held stationary. A rotating die is brought into contact with the material and an axial force is applied. The combination of rotation and applied force causes heating due to friction between the tool and material interface, and from plastic deformation of the material. An extrusion is formed as the softened material is pushed out the die orifice due to the axial pressure and motion of material flow from scrolls. A 4-scroll die was used for this work as shown in Fig. 1. The samples discussed in this paper were all extruded at the same process parameters of 7.62 mm/min and 150 RPM and an extrusion ratio of 40.

ShAPE processing was performed with a custom-designed die and material container assembly that was loaded into and actuated using a Friction Stir Welding (FSW) machine designed and built by Bond Technologies (previously named Transformation Technologies Incorporated). This FSW machine is unique in that it is one of the stiffest gantry machines in North America. It is a servo-operated machine capable of an axial force up to 130 kN and uses a nominally rated 30 kW spindle that can operate up to 2000 RPM. In addition, it is fully instrumented to record forces in the x, y, and z directions as well as torque, power, and temperature. In this work, the force in the z-direction corresponds to the axial forge force and will be the only force discussed further. Die face temperature was recorded by inserting a type-K thermocouple approximately 1–2 mm from the working face of the die. Torque is calculated from direct measurements of current drawn by the spindle motor at a given operating voltage and calibrated with an external dynamometer. Power is then calculated based on the torque and rotational velocity.

After processing, the microstructures of both the extruded rod and the billet (puck) were analyzed. Portions of the extrudate (middle of rod for this work) and puck were mounted in epoxy, cross-sectioned using a diamond saw, and polished in steps to a final polish using 0.05 μm colloidal silica. Care was taken in timely coating and transporting polished samples to microscope chambers to avoid surface oxidation. For optical microscopy (OM), specimens were briefly etched with dilute acid solution to reveal grain boundaries and slip lines. A JEOL 7600F Field Emission Scanning Electron Microscope (FE-SEM) was used to perform microstructural evaluations. Electron backscatter diffraction (EBSD) mapping was performed using an accelerating voltage of 20 keV and working distance of approximately 22 mm. Indexing was accomplished using a magnesium hexagonal crystal structure, Laue group 9, space group 194, and unit cell parameters a = 3.209 Å, b = 3.209 Å, c = 5.211 Å, α = 90°, β = 90°, γ = 120°. Low magnification maps were generated using a 1 μm step size while high magnification mapping was performed using more refined step sizes of 150–300 nm. Scanning transmission electron microscopy (STEM) observations were conducted on a cold field-emission JEOL ARM200CF microscope operated at 200 kV, equipped with a hexapole type Cs-corrector (CESCOR, CEOS). The instrument is configured with a JEOL Centurio energy-dispersive X-ray spectroscopy (EDS) detector. TEM specimens were prepared using a FEI (now Thermo Fisher Scientific) Quanta 3D focus ion beam (FIB)/scanning electron microscope (SEM). FIB lamellae were extracted using a 30 keV Ga beam for the initial lift-out and thinning, then thinned to approximately 100 ~ 200 nm using a 5 keV beam followed by a 2 keV step.

The mechanical testing and samples were prepared as per ASTM E8 standard with a 3 mm gauge diameter. Tensile and compression specimens were machined with the gauge length along the extrusion direction for extruded rods. Tension and compression tests were carried out at room temperature with quasi-static strain rate of 10^{-3}/s.

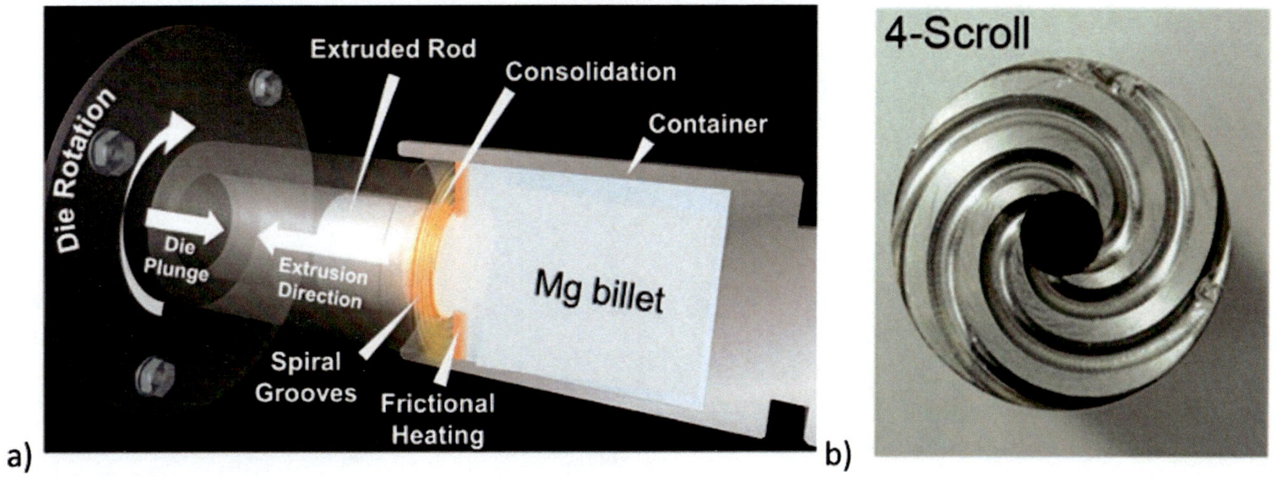

Fig. 1 A diagram of the ShAPE process and 4-scroll die. (Color figure online)

Results and Discussion

Figure 2 shows the approximately 300 mm long 5 mm diameter rods extruded from a cast Mg-3Si (AS31), AZ31, AZ31-2Si (AZ312), and ZK60 at the conditions previously mentioned. A portion of the remaining cast billet can be seen below the rod in each case. Process data corresponding to ShAPE extrusion of previously extruded AZ31, and cast alloys AZ31, Mg-3Si, AZ312, and ZK60 are shown in Fig. 3. There are slightly different behaviors for extrusion force, die temperature, and power that will not be fully elaborated in this paper. In general, the extrusion force (Fig. 3a) exhibits two behaviors. There is an initial break-through force that ranges from 56 kN for Mg-3Si up to 84 kN for a previously extruded AZ31. We postulate that this is related to the pressure and temperature needed to cause each material to flow. The force is then reduced to a steady state that comes on earlier and is lower in value ($\sim$20 kN) for cast AZ31 and ZK60 while AZ31-2Si is slow to reduce and is still 33 kN at the end of the extrusion. We would expect that the high temperature strength of each material will affect this but also note that each material may also have differences in ability to flow in the die that may not always be strength related. The die temperature for this set of samples is shown in Fig. 3b. The range of die temperatures used to extrude the different alloys occurs in two groups. The extruded AZ31, Mg-3Si, and ZK60 alloys reach approximately 400 °C, while AZ31 and AZ31-2Si reach 475 and 500 °C, respectively. Spindle power used to turn the die during extrusion is shown in Fig. 3c. There is an initially increase in power up to 4–5 kW of power during the high rate of temperature increase and high force. Then the power decreases closer to a steady state between 2 and 3 kW. Mg-3Si appears to experience the least amount of power application while AZ31-2Si experiences higher power levels. A spike in power near the 64 s mark is due to a change in spindle rotational speed that occurs at the end of the extrusion and start of die extraction.

The ability of ShAPE process to modify microstructures are illustrated in the following micrographs. Here we show preliminary microstructural data for some of the samples but not all the extruded samples. ShAPE's ability to alter grain size can be seen by comparing Figs. 4 and 5. Figure 4 shows a large-area EBSD map of the as-extruded AZ31B, together with the {0001} pole figure. The extrusion direction (ED) was vertical as represented by the blue arrow. This is a typical bimodal grain structure with large grains on the order of 1 mm surrounded by smaller grains on the scale of 20 μm. The extruded sample has typical fiber basal texture with most of the grains' {0001} poles approximately normal

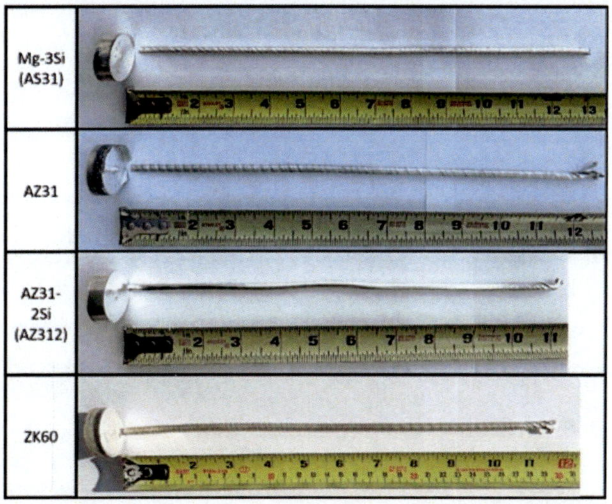

Fig. 2 Images of rods produced by ShAPE from cast Mg-3Si (AS31), AZ31, AZ31-2Si (AZ312), and ZK60. (Color figure online)

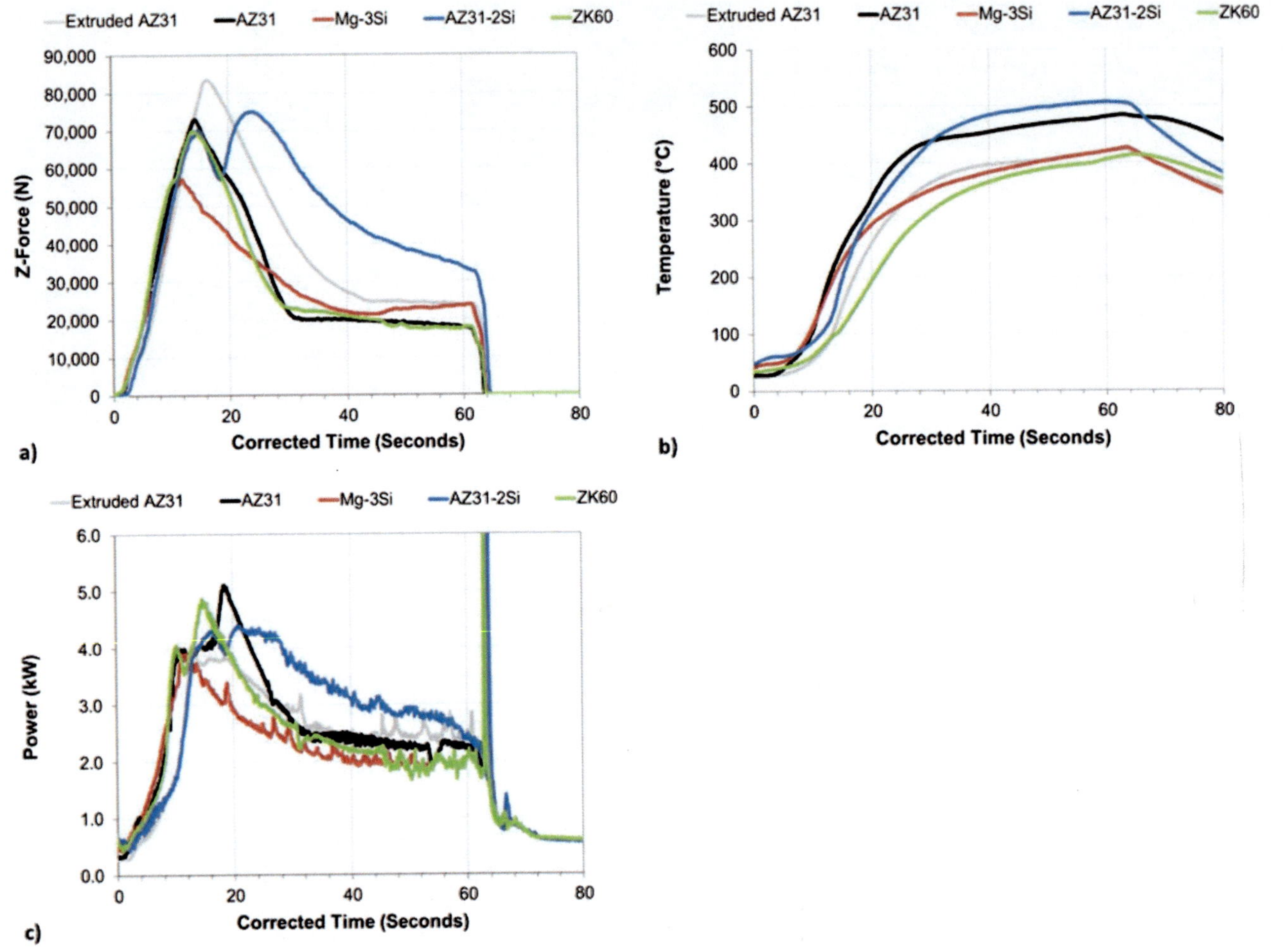

Fig. 3 Process data with corrected time during ShAPE extrusion of extruded AZ31 and cast alloys AZ31, Mg-3Si, AZ312, and ZK60 into rods at 150RPM and 7.62 mm/min. Time was corrected to show the first buildup of force due to die contacting material as t = 0 s. (Color figure online)

to the ED. After ShAPE extrusion, the AZ31B alloy (ShAPE AZ31B hereafter) had equiaxed grains with an average size of 27.9 μm, as revealed by OM (Fig. 5a) and EBSD map (Fig. 5b). The ShAPE extrusion direction (SED for short) was also vertical as represented by the blue arrow. Since the IPF color code was based on SED, it is obvious in Fig. 5b that the grains in ShAPE AZ31B were oriented more randomly as opposed to the preferably oriented blue/green grains in as-extruded AZ31B (Fig. 4). The corresponding {0001} pole figure in Fig. 5c agrees with such assessment, in that only a weak texture was present, with most of grains having {0001} pole tilted away from SED by 30–60°, hence a "tilted" basal texture.

In addition to modifications in texture and grain size in ShAPE AZ31B, the size and distribution second phase particles, mostly Al_8Mn_5 [17, 18], also changed, as exemplified by the back scattered electron (BSE) micrographs of as-extruded AZ31B (Fig. 6a) and ShAPE AZ31B (Fig. 6b), respectively. The former had typical "stringers" of particles along the ED with sizes ranging between 0.5 μm and 3.5 μm; whereas the latter had uniformly distributed particles with sizes mostly less than 2 μm. The refining and uniform distribution of precipitates and/or particles by ShAPE have also been demonstrated in ZK60 [16] and AZ91 [19] Mg alloys, as well as an Al alloy with very high alloying contents [20]. Note that ShAPE has been able to retain Al_8Mn_5 precipitates in the 10–20 nm range as observed in the transmission electron micrographs and EDS plots of Fig. 8. Another example of ShAPE's ability to refine and distribute second phases is shown in Fig. 7, where coarse Mg_2Si phase in as-cast Mg-3Si was refined after ShAPE.

The mechanical behavior was investigated and compared for both conventionally extruded AZ31B, ShAPE processed extruded AZ31B, cast AZ31B, and ShAPE processed cast AZ31B. True stress-strain curves for both tension and compression were plotted in Fig. 9. For as-extruded AZ31B, the commonly observed YA and SD were present, with CYS being much lower than TYS; whereas the compressive flow stress rapidly increased after yielding, with ultimate

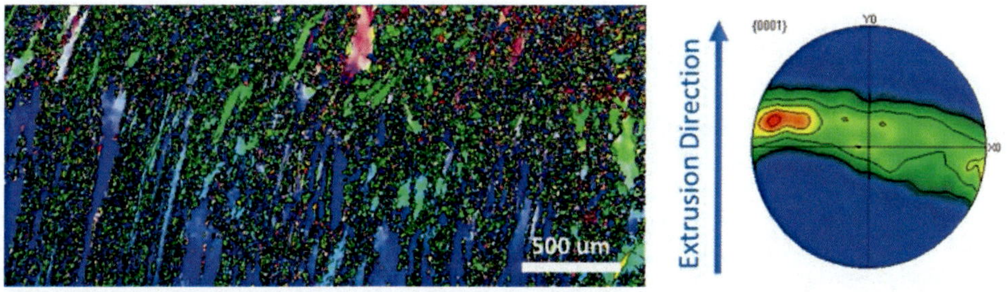

Fig. 4 EBSD map and {0001} pole figure of as-extruded AZ31B shows a bimodal grain size with large grains in the range of 1–2 mm. The extrusion direction (ED) was vertical, as represented by the blue arrow. The IPF color code was based on ED. (Color figure online)

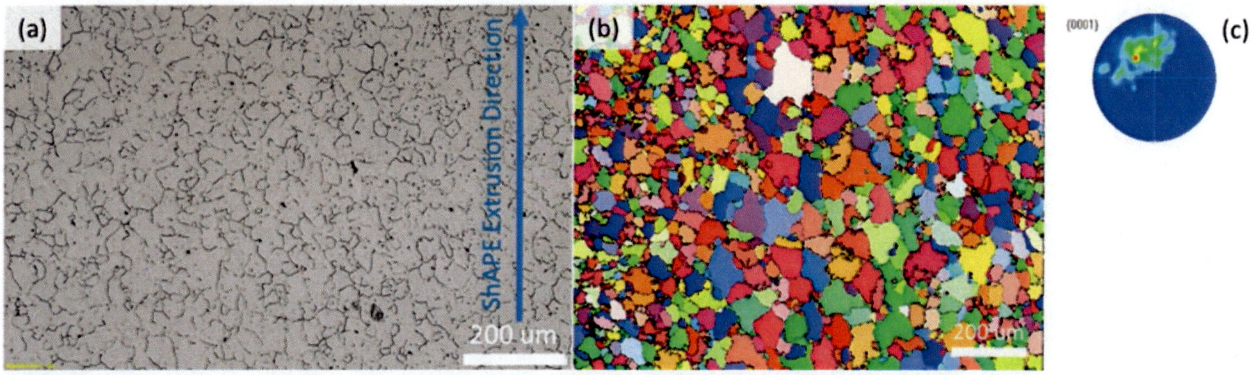

Fig. 5 **a** Optical micrograph of ShAPE AZ31B near center of rod, with vertical ShAPE extrusion direction (SED). **b** EBSD map and shows most grains are grouped near 28 μm and **c** corresponding {0001} pole figure. (Color figure online)

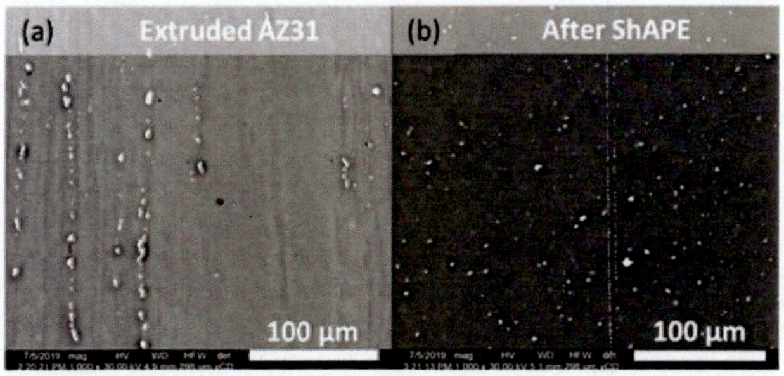

Fig. 6 Backscatter scanning electron microscopy of conventionally extruded and ShAPE extruded AZ31

compressive strength (UCS) significantly higher than ultimate tensile strength (UTS). In contrast, for ShAPE AZ31B, the tensile curve was well aligned with the compressive curve, until the compressive strain level extended far beyond that of tensile strain.

The quantitative comparisons on yield strength, ultimate strength, strain at failure, and uniform elongation (for tensile tests only) were summarized in Table 1 for the extruded, cast, and ShAPE AZ31B samples. One can notice CYS/TYS for ShAPE extruded AZ31B was 1.05, whereas that for as-extruded AZ31B was 0.66. Similarly, UCS/TYS (not shown on table) was 0.99 for the former, and 1.13 for the latter. The table shows that the ShAPE process improved CYS/TYS to closer to 1 for the cast case as well. The decrease in the compressive and tensile yield strength is associated with basal plane orientation [7], which in the case of ShAPE extruded sample was approximately 45° to the extrusion axis. It must be noted that the ShAPE process significantly improved the ductility, with uniform elongation almost doubled, from 7.2 to 13.9% and the strength differential typically observed in the magnesium alloys was completely eliminated. A similar analysis was performed for ShAPE processed Mg-3Si and ZK60 and the results can be found in the lower half of Table 1. Here it is found that CYS/TYS values for ShAPE extruded ZK60 ranged from 1.03 to 1.30, suggesting possible benefit to this alloy. The values for ShAPE processed Mg-3Si ranged from 1.35 to 1.9 indicating that ShAPE processing may not be as beneficial to reducing anisotropy for this alloy. Further investigations will be performed to elucidate this.

The improved ductility in ShAPE AZ31B was also supported by macroscopic OM observations on fractured specimens in Fig. 10. For tensile specimens, as-extruded AZ31B exhibited highly localized necking (Fig. 10a), whereas ShAPE AZ31B had more gradual, "spread-out" necking (Fig. 10e), indicating more uniform elongation prior to fracturing. As for compression specimens, it was obvious

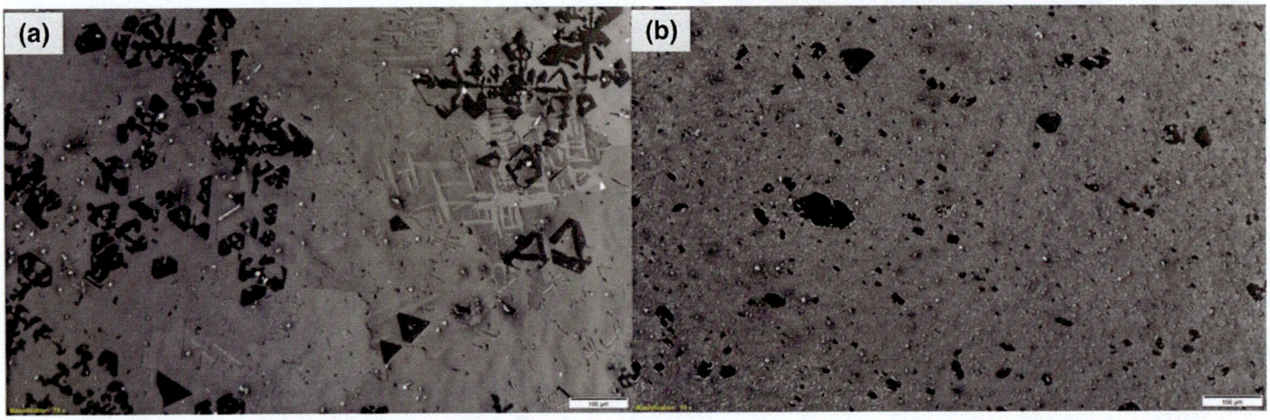

Fig. 7 Backscatter scanning electron microscopy of (**a**) as-cast AZ31-2Si shows a 100 μm "Chinese script" Mg_2Si precipitates that are refined to equiaxed smaller particles with ShAPE extrusion (**b**)

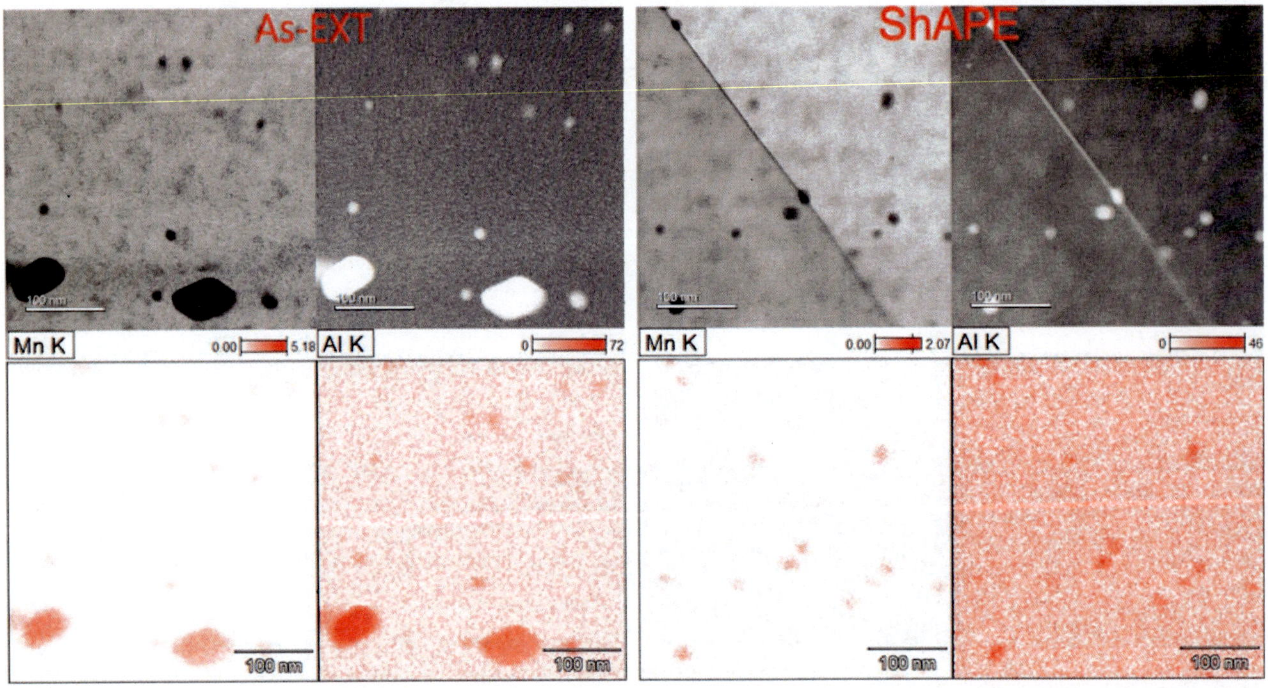

Fig. 8 Transmission electron microscopy of conventionally extruded AZ31B and ShAPE extruded AZ31B. Scale bars are all 100 nm. (Color figure online)

Fig. 9 Tensile and compressive true stress-strain curves for **a** as-extruded AZ31B versus ShAPE and **b** as-cast versus ShAPE. (Color figure online)

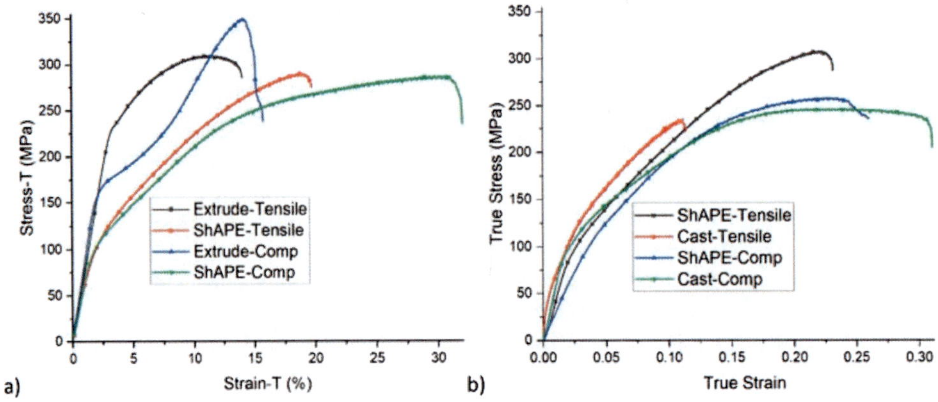

Table 1 Comparisons of mechanical properties of as-extruded AZ31B versus ShAPE of extruded AZ31, and as-cast AZ31 versus ShAPE of cast AZ31. ShAPE drives the CYS/TYS ratio to close to 1 in both cases. Also included are preliminary results of only ShAPE of cast ZK60 and ShAPE of cast Mg-3Si. Preliminary data suggests ShAPE of cast ZK60 results in CYS/TYS ratio of 1.03 to 1.30. ShAPE of cast Mg-3Si results in CYS/TYS ratios of 1.35 and 1.90, suggesting that ShAPE is less successful in driving the ratio to 1 in this alloy

Material	Sample ID	Yield strength (Mpa)	Ultimate strength (Mpa)	Ductility (%)	CYS/TYS
Extruded AZ31	Extrude-Tensile	216.3	309.6	10.40	0.66
	Extrude-Comp	142.2	349.5	10.70	
	Ext-ShAPE-Tensile	82.8	289.3	14.80	1.05
	Ext-ShAPE-Comp	87.1	286.7	27.20	
Cast AZ31	Cast-Tensile	33	235.3	10.10	1.94
	Cast-Comp	64	246.1	26.70	
	Cast-ShAPE-Tensile	71.7	308.7	16.40	1.03
	Cast-ShAPE-Comp	73.7	257.9	15.60	
ZK 60	Cast-ShAPE Tensile #1	169	364	23	1.07
	Cast-ShAPE Comp #2	181	NA	NA	
	Cast-ShAPE Tensile #6	149	340	18	1.30
	Cast-ShAPE Comp #7	194	NA	NA	
	Cast-ShAPE Tensile #11	158	347	21	1.03
	Cast-ShAPE Comp #12	162	NA	NA	
Mg-3Si	Cast-ShAPE Tensile #6	52	174	13	1.35
	Cast-ShAPE Comp #7	70	NA	NA	
	Cast-ShAPE Tensile #11	52	182	13	1.90
	Cast-ShAPE Comp #12	99	NA	NA	

that ShAPE AZ31B (Fig. 10f) went through more shape change than as-extruded AZ31B (Fig. 10b).

OM observation at higher magnification, shown in Fig. 11, for fractured specimens of ShAPE-Tensile (Fig. 11a) and ShAPE-Comp (Fig. 11b) revealed the presence of parallel slip lines within the boundaries of many grains. These parallel slip lines were particularly common in the ShAPE-Comp specimen (Fig. 11b) as the strain at failure was 27.2%, almost twice as much as that for ShAPE-Tensile specimen (Fig. 11a). However, in some selected areas near the fracture surface, as outlined by the yellow dashed box in Fig. 11a, such parallel slip lines can also be seen in ShAPE-Tensile. Considering the titled basal texture in ShAPE AZ31B and the fact that basal slip only occurs on parallel basal planes, these parallel slip lines were most likely the result of dominant basal slip [21, 22] under compressive and tensile deformation, respectively. Dominant basal slip would be a likely reason for low CYS and TYS in ShAPE AZ31B as compared with as-extruded AZ31B. Further study by high-resolution EBSD is ongoing in order to confirm the

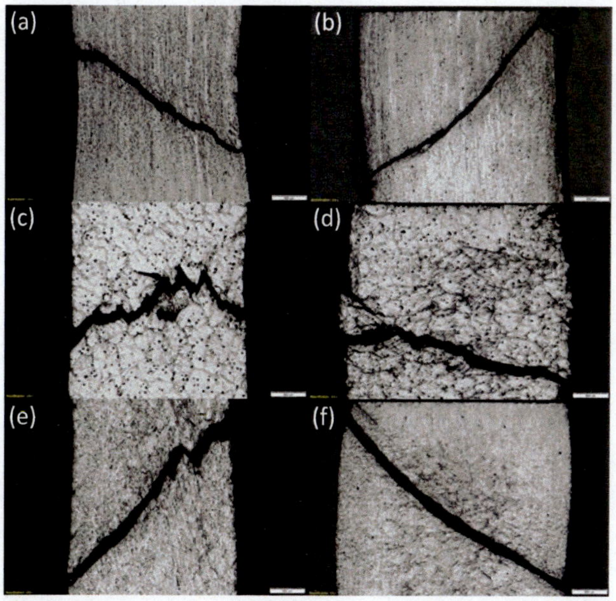

Fig. 10 Macroscopic OM images of fractured specimens of AZ31B **a** Extrude-Tensile, **b** Extrude-Comp, **c** Cast-Tensile, **d** Cast-Comp, **e** ShAPE-Tensile, **f** ShAPE-Comp

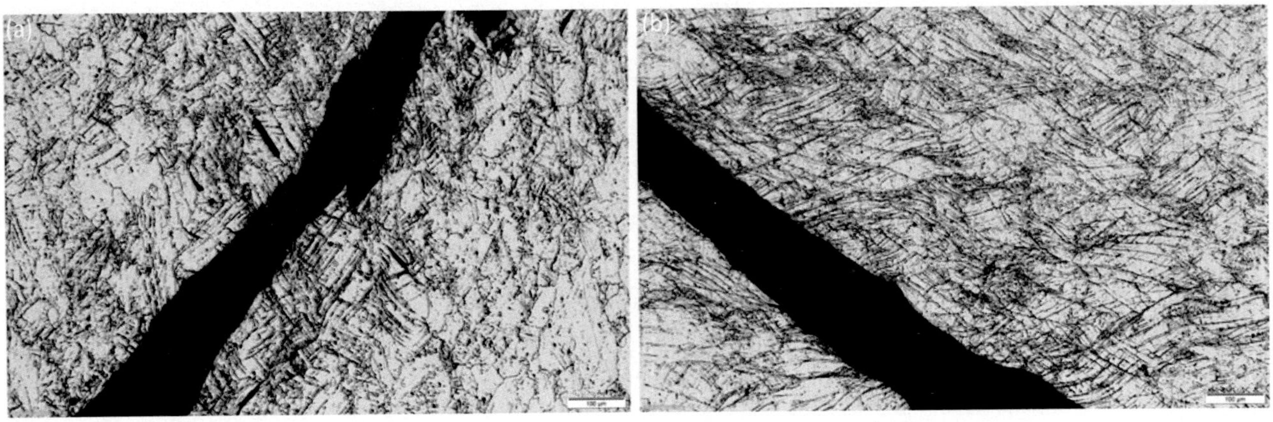

Fig. 11 Deformation microstructure near fracture surface of **a** ShAPE-Tensile, **b** ShAPE-Comp

dominant basal slip activity, as well as the possibility of coordinated basal slip across neighboring grains that might have largely contributed to the enhanced ductility in ShAPE AZ31B.

Conclusions

This work has shown that by using optimized ShAPE process AZ series with and without silicon, ZK60 Mg alloys in as-received forms (as-cast or as-extruded) generated more uniform microstructure, equiaxed grains, finer and homogeneously distributed precipitates and chemical homogeneity. The critical factors necessary to remove the yield anisotropy and minimizing the strength differential in magnesium alloys was achieved using the ShAPE process. The ShAPE processes were responsible for tailoring the microstructure and are readily scalable to produce large parts relevant to automotive applications.

Acknowledgements The authors would like to thank the U. S. Department of Energy, Vehicle Technologies Office for the financial support provided for this project. Pacific Northwest National Laboratory is operated by Battelle Memorial Institute for the United States Department of Energy under Contract DE-AC06-76RLO1830.

References

1. M.R. Barnett, Twinning and the ductility of magnesium alloys: Part I: "Tension" twins, Materials Science and Engineering: A 464 (1–2) (2007) 1–7.
2. D. Zhang, H. Wen, M.A. Kumar, F. Chen, L. Zhang, I.J. Beyerlein, J.M. Schoenung, S. Mahajan, E.J. Lavernia, Yield symmetry and reduced strength differential in Mg-2.5Y alloy, Acta Mater. 120 (2016) 75–85.
3. D. Zhang, L. Jiang, B. Zheng, J.M. Schoenung, S. Mahajan, E. J. Lavernia, I.J. Beyerlein, J.M. Schoenung, E.J. Lavernia, Deformation Twinning (Update), Reference Module in Materials Science and Materials Engineering, Elsevier 2016.
4. M.H. Yoo, Slip, Twinning, and Fracture in Hexagonal Close-Packed Metals, Metall. Mater. Trans. A-Phys. Metall. Mater. Sci. 12(3) (1981) 409–418.
5. J. Bohlen, S. Yi, D. Letzig, K.U. Kainer, Effect of rare earth elements on the microstructure and texture development in magnesium–manganese alloys during extrusion, Materials Science and Engineering: A 527(26) (2010) 7092–7098.
6. D.L. Yin, J.T. Wang, J.Q. Liu, X. Zhao, On tension–compression yield asymmetry in an extruded Mg–3Al–1Zn alloy, J. Alloy. Compd. 478(1–2) (2009) 789–795.
7. Li D., V.V. Joshi, C.A. Lavender, M.A. Khaleel, and S. Ahzi. 2013. "Yield asymmetry design of magnesium alloys by integrated computational materials engineering." Computational Materials Science 79.
8. Joshi V.V., S. Jana, D. Li, H. Garmestani, E.A. Nyberg, and C.A. Lavender. 2014. "High Shear Deformation to Produce High Strength and Energy Absorption in Mg Alloys." In Magnesium Technology 2014, edited by M Alderman, et al, 83–88. Hoboken, New Jersey: John Wiley & Sons, Inc.
9. J. Hirsch, T. Al-Samman, Superior light metals by texture engineering: Optimized aluminum and magnesium alloys for automotive applications, Acta Materialia 61(3) (2013) 818–843.
10. Y. Chino, M. Kado, M. Mabuchi, Enhancement of tensile ductility and stretch formability of magnesium by addition of 0.2 wt% (0.035 at%)Ce, Mater. Sci. Eng. A-Struct. Mater. Prop. Microstruct. Process. 494(1–2) (2008) 343–349.
11. J. Bohlen, M.R. Nürnberg, J.W. Senn, D. Letzig, S.R. Agnew, The texture and anisotropy of magnesium–zinc–rare earth alloy sheets, Acta Mater. 55(6) (2007) 2101–2112.
12. T. Al-Samman, X. Li, Sheet texture modification in magnesium-based alloys by selective rare earth alloying, Materials Science and Engineering: A 528(10–11) (2011) 3809–3822.
13. H.J. Choi, Y. Kim, J.H. Shin, D.H. Bae, Deformation behavior of magnesium in the grain size spectrum from nano- to micrometer, Materials Science and Engineering: A 527(6) (2010) 1565–1570.
14. S.K. Panigrahi, K. Kumar, N. Kumar, W. Yuan, R.S. Mishra, R. DeLorme, B. Davis, R.A. Howell, K. Cho, Transition of deformation behavior in an ultrafine grained magnesium alloy, Materials Science and Engineering: A 549 (2012) 123–127.
15. I.J. Beyerlein, L.S. Toth, Texture evolution in equal-channel angular extrusion, Prog. Mater. Sci. 54(4) (2009) 427–510.
16. S. Whalen, N. Overman, V. Joshi, T. Varga, D. Graff, C. Lavender, Magnesium alloy ZK60 tubing made by Shear Assisted Processing and Extrusion (ShAPE), Materials Science and Engineering: A 755 (2019) 278–288.

17. T. Voisin, N.M. Krywopusk, F. Mompiou, T.P. Weihs, Precipitation strengthening in nanostructured AZ31B magnesium thin films characterized by nano-indentation, STEM/EDS, HRTEM, and in situ TEM tensile testing, Acta Materialia 138 (2017) 174–184.
18. N. Stanford, D. Atwell, The Effect of Mn-rich Precipitates on the Strength of AZ31 Extrudates, Metallurgical and Materials Transactions A 44(10) (2013) 4830–4843.
19. N.R. Overman, S.A. Whalen, M.E. Bowden, M.J. Olszta, K. Kruska, T. Clark, E.L. Stevens, J.T. Darsell, V.V. Joshi, X. Jiang, K.F. Mattlin, S.N. Mathaudhu, Homogenization and texture development in rapidly solidified AZ91E consolidated by Shear Assisted Processing and Extrusion (ShAPE), Materials Science and Engineering: A 701 (2017) 56–68.
20. S. Whalen, M. Olszta, C. Roach, J. Darsell, D. Graff, M. Reza-E-Rabby, T. Roosendaal, W. Daye, T. Pelletiers, S. Mathaudhu, N. Overman, High ductility aluminum alloy made from powder by friction extrusion, Materialia 6 (2019) 100260.
21. C.M. Cepeda-Jiménez, J.M. Molina-Aldareguia, M.T. Pérez--Prado, Effect of grain size on slip activity in pure magnesium polycrystals, Acta Mater. 84(0) (2015) 443–456.
22. C.M. Cepeda-Jiménez, J.M. Molina-Aldareguia, F. Carreño, M.T. Pérez-Prado, Prominent role of basal slip during high-temperature deformation of pure Mg polycrystals, Acta Mater. 85(0) (2015) 1–13.

Numerical Study of Multiaxial Loading Behavior of Mg Alloy AZ31 Extruded Bar

Xiaodan Zhang, Qin Yu, Huamiao Wang, and Peidong Wu

Abstract

During plastic forming of magnesium alloy parts, multiaxial deformation state with shear and normal strain components is frequently observed. Therefore, torsional, torsion-tension coupling, and torsion-compression coupling behaviors of AZ31 magnesium alloy extruded bars are numerically investigated. The elastic visco-plastic self-consistent model with the twinning and detwinning scheme, in conjunction with a torsion specific finite element method, is employed. Stress-strain response, twin volume fraction, relative activities of deformation mechanisms, and deformation textures are obtained and used to interpret the multiaxial deformation behaviors of Mg alloys.

Keywords

Torsion-tension • Torsion-compression • Magnesium alloy • Twinning • Crystal plasticity

X. Zhang · H. Wang (✉)
School of Mechanical Engineering, Shanghai Jiao Tong University, Shanghai, 200240, China
e-mail: wanghm02@sjtu.edu.cn

Q. Yu
Department of Mechanical Engineering, University of Nevada, Reno, NV 89557, USA

Present Address:
Q. Yu
Materials Sciences Division, Lawrence Berkeley National Laboratory, Berkeley, CA 94720, USA

P. Wu
Department of Mechanical Engineering, McMaster University, Hamilton, ON L8S4L7, Canada

Introduction

Magnesium (Mg) alloys gained significant attention as light structural materials for automotive and aerospace applications [1, 2]. However, Mg alloys have exhibited strongly anisotropic behaviors and poor formability to limit the application at ambient temperature, which are ascribed to their hexagonal close-packed (HCP) crystallographic structure, and limited operative deformation modes [3]. To enhance the mechanical properties and widen the applications, understanding the essential deformation mechanism of Mg alloy is worth endless effort. Tremendous investigations under various deformation states have been conducted, e.g., uniaxial/cyclic tension and compression, loading path change, simple shear [3–6].

In contrast to the plenty of tension/compression studies, limited works have been devoted to study the torsional behaviors of Mg alloys, which are attributed to the inhomogeneity of the stress and strain on the cross-sectional plane of the torsional specimens. Similarly, torsion-tension/compression coupled deformations are heterogeneous along the radial direction. But this inhomogeneity is easier to be tracked than an arbitrary deformation state. Taking this feature into consideration, it is possible to apply the efficient SC crystal plasticity model to analyze the torsion-tension/compression coupling behaviors. In this paper, we studied the deformation behaviors of the Mg alloy AZ31 bar under fixed-end torsion, free-end torsion, and coupled torsion-tension/compression by combining the torsion specific finite element method [7] and the EVPSC-TDT model [5, 6].

Numerical Procedure

The EVPSC-TDT model with the Affine self-consistent scheme, which gave the best performance in series of assessments [4–11], was employed in conjunction with the torsion specific finite element approach (denoted as the

V. M. Miller et al. (eds.), *Magnesium Technology 2021*, The Minerals, Metals & Materials Series,
https://doi.org/10.1007/978-3-030-65528-0_16

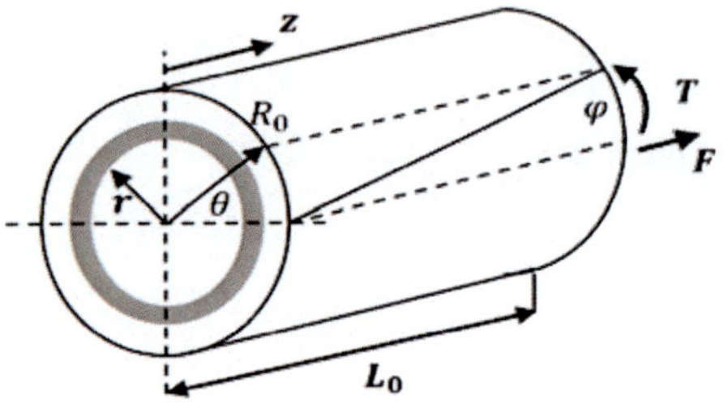

Fig. 1 Schematic of finite element for torsion

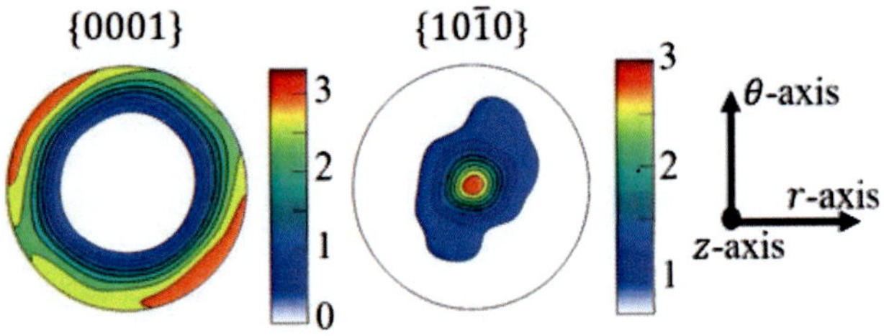

Fig. 2 The initial texture of extruded AZ31 Mg alloy solid rods in term of {0001} and {10$\bar{1}$0} pole figures. (Color figure online)

TFE-EVPSC-TDT model, see Fig. 1) to study the torsion related behaviors of Mg alloys. A uniform cylindrical solid rod with initial radius R_0 and initial length L_0, subject to a twist φ due to an applied torque T and an axial strain ε due to axial force F, is considered (Fig. 1) in TFE. The numerical results of the solid rods based on the TFE-EVPSC-TDT model are sufficiently accurate by using 5 elements [12]. The torsion related results of a solid cylindrical rod are presented in terms of the following quantities:

$$\Gamma = \frac{R_0}{L_0}\varphi; \bar{\tau} = \frac{3T}{2\pi R_0^3}; \bar{\varepsilon} = ln\left(\frac{L}{L_0}\right); \bar{\sigma} = \frac{F}{\pi R_0^2}$$

where Γ represents the shear strain at the outer surface of the cylindrical rod. $\bar{\tau}$, $\bar{\varepsilon}$, and $\bar{\sigma}$ are the average shear stress, axial strain, and axial stress, respectively.

Results and Discussion

The behavior of an AZ31B extruded rod loaded under various loading paths, including fixed-end torsion, free-end torsion, coupled torsion-tension, and coupled torsion-compression, are simulated by the TFE-EVPSC-TDT. The plastic deformation of Mg alloy AZ31B at room temperature is assumed to be composed of basal $\langle a \rangle$ slip, prismatic $\langle a \rangle$ slip, pyramidal $\langle c+a \rangle$ slip, and extension twinning [13]. In the simulation, the reference shear strain rate and rate sensitivity of all slip and twin systems are the same, $\dot{\gamma}_0 = 10^{-3}s^{-1}$, $m = 0.05$. At room temperature, the elastic constants of magnesium single crystal are $C_{11} = 58$, $C_{12} = 25$, $C_{13} = 20.8$, $C_{33} = 61.2$, $C_{44} = 16.6$ GPa [14]. The extruded AZ31 rod texture (Fig. 2) and the hardening parameters (Table 1) are chosen as the same as those used previously [15].

The free-/fixed-end torsion are simulated with a twist rate of $\dot{\varphi} = 1.36 \times 10^{-3}s^{-1}$, whereas torsion-tension/compression are simulated with a twist rate of $\dot{\varphi} = 1.36 \times 10^{-3}s^{-1}$ and axial strain rate $\dot{\bar{\varepsilon}} = \pm 1 \times 10^{-4}s^{-1}$. The simulated response of the stress and strain under the four loading paths are summarized in Fig. 3, where the stress-strain responses all depend strongly on the coupling loading paths. The curves of shear stress-shear strain under free-end and fixed-end torsion are similar because the active deformation modes are almost equivalent under these two loading paths (Fig. 3a, b). The axial strain is contractive and decreases almost linearly with twisting under free-end torsion, while the axial stress increases under fixed-end torsion (Fig. 3b), which is the so-called Swift effect [13]. As shown in Fig. 3c, the curves of shear stress-shear strain and axial stress-axial strain under coupled torsion-tension are similar to fixed-end torsion, while under coupled torsion-compression have the characteristic "S" shapes due to the dominated twinning (Fig. 4d). Relatively, prismatic slip is as active as the basal slip under coupled torsion-tension, while it is relatively less active under free-/fixed-end torsion and coupled torsion-compression.

Figure 5 presents the predicted deformation textures in terms of {0001} and {10$\bar{1}$0} pole figures under four loading paths at shear strain $\Gamma = 0.3$. All textures are evolving towards the typical shear texture, which is consistent with both available experiments and simulations [16–18]. Close observation reveals that the developed textures under coupled torsion-compression are quite different from those under free-/fixed-end torsion and torsion-tension. This is due to the high activity of extension twin under coupled torsion-compression.

Table 1 Values of the hardening parameters associated with the EVPSC-TDT model

Mode	$\tau_0(MPa)$	$\tau_1(MPa)$	$h_0(MPa)$	$h_1(MPa)$	$h^{\alpha\beta}$	A_1	A_2
Basal	23	1	0	0	1		
Prismatic	88	55	400	0	1		
Pyramidal	100	90	2800	0	1		
Extension twin	30	0	0	0	1	0.70	0.75

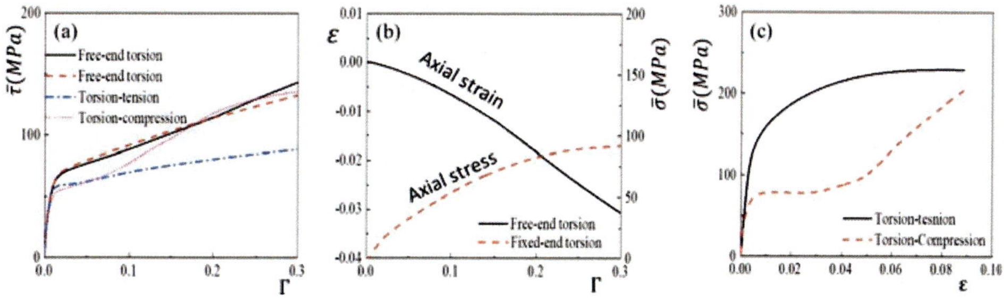

Fig. 3 **(a)** The shear stress-shear strain curves of the rod under free-/fixed-end torsion and coupled torsion-tension/compression; **(b)** the axial strain/stress-shear strain curves under free-/fixed-end torsion; and **(c)** the axial stress-strain curves of the rod under coupled torsion-tension/compression. (Color figure online)

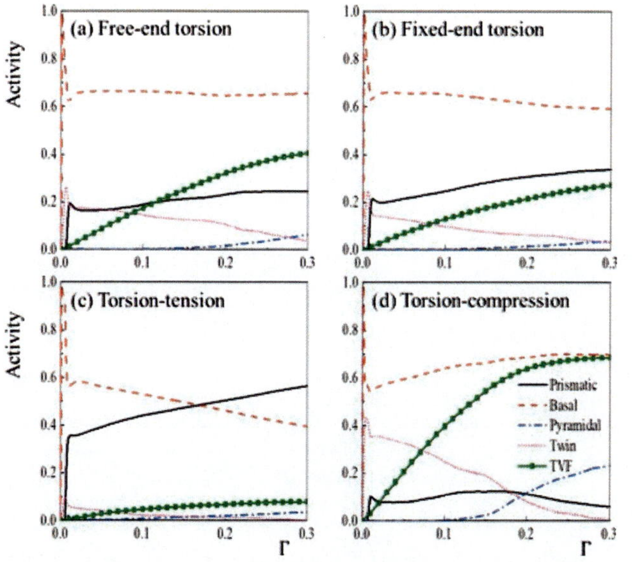

Fig. 4 The relative activities of various deformation mechanisms and the twin volume fraction (TVF) under **(a)** free-end torsion, **(b)** fixed-end torsion, **(c)** coupled torsion-tension, and **(d)** coupled torsion-compression at the outer surface of the rod. (Color figure online)

The distributions of the twin volume fraction (TVF) under different loadings are presented in Fig. 6. The TVFs increase along the radius from the center to the outer surface under fixed-end torsion at difference shear strain. Similarly, under free-end torsion, TVFs increase along the radius at The TVFs increase along the radius from the center to the outer surface under fixed-end torsion at difference shear strain. Similarly, under free-end torsion, TVFs increase along the radius at, but distribute almost uniformly at $\Gamma = 0.1, 0.2$, but distribute almost uniformly at $\Gamma = 0.3$. The TVF remains low under coupled torsion-tension because twinning is very active (Fig. 4c). Interestingly, under coupled torsion-compression, the TVF increases with the radius at $\Gamma = 0.1$, remains nearly constant at $\Gamma = 0.2$, and decreases at $\Gamma = 0.3$. This exceptional distribution of the TVF within the rod under coupled torsion-compression is ascribed to the difference in deformation mechanisms (Fig. 7). It is demonstrated that tuning a variety of gradient twin structures is feasible through torsion. It is particularly significant as it provides an effective and low-cost processing method to adjust the mechanical properties of Mg alloys.

Fig. 5 The developed textures under **(a)** free-end torsion, **(b)** fixed-end torsion, **(c)** coupled torsion-tension, and **(d)** coupled torsion-tension at $\Gamma = 0.3$ in terms of $\{0001\}$ and $\{10\bar{1}0\}$ pole figures in the $\theta - z$ shear plane. (Color figure online)

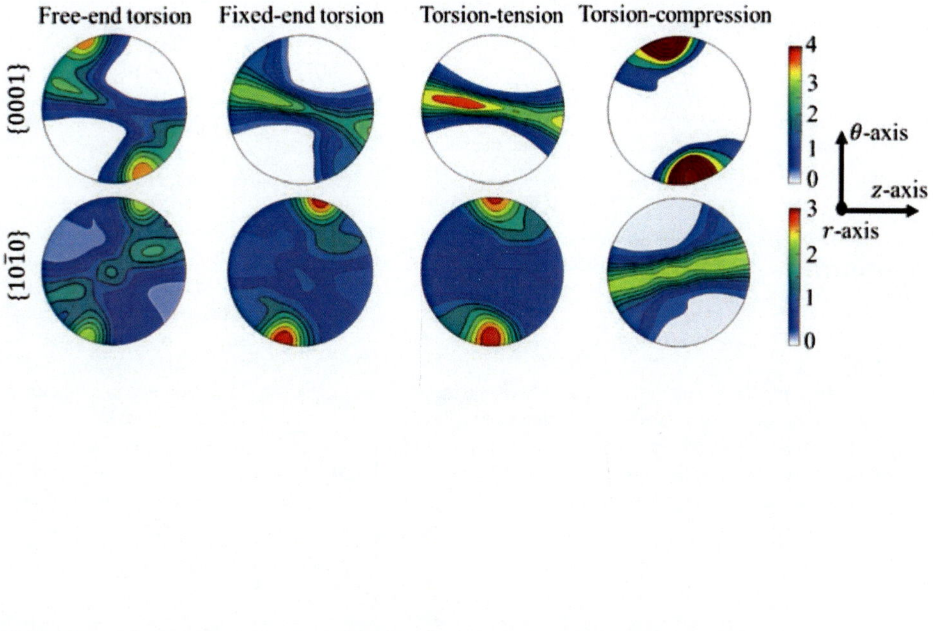

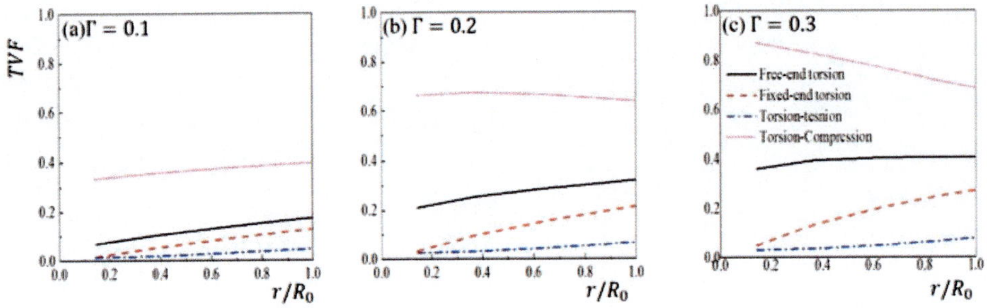

Fig. 6 Distribution of the twin volume fraction under four loading paths at shear strains of **(a)** Γ = 0.1, **(b)** Γ = 0.2, and **(c)** Γ = 0.3. (Color figure online)

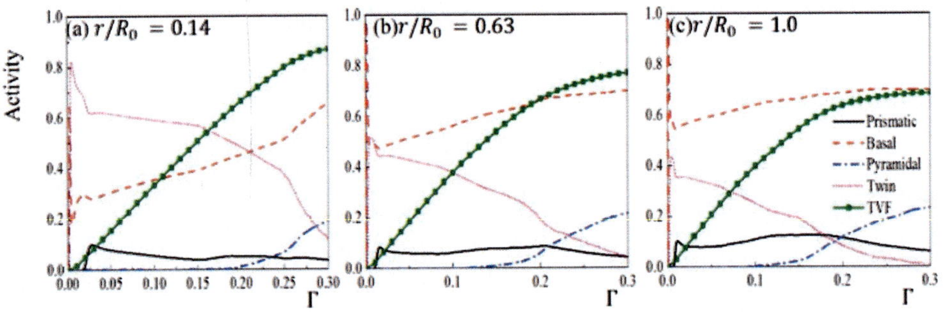

Fig. 7 The relative activities of various deformation mechanisms and the TVF under coupled torsion-compression at Γ = 0.3 inside the rod at **(a)** r = 0.14R$_0$, **(b)** r = 0.63R$_0$, and **(c)** r = R$_0$. (Color figure online)

Conclusions

The torsional, torsion-tension, and torsion-compression coupling behaviors of extruded AZ31 Mg alloy rod are investigated numerically based on the TFE-EVPSC-TDT model. First of all, the stress-strain responses are very sensitive to the loading paths because of the high anisotropy of Mg alloys and the different activity of the extension twin. Also due to its strong anisotropy, Mg alloy has a significant Swift effect. Due to twinning, the axial stress-strain curve exhibits an "S" shape under coupled torsion-compression. Under four loading paths, all develop textures exhibit the characteristic features of a developed shear texture.

Acknowledgements HW was supported by the National Natural Science Foundation of China (No. 51975365) and the Shanghai Pujiang Program (18PJ1405000).

References

1. T.M. Pollock, "Weight loss with magnesium alloys," Science, 328 (2001), 986–987.
2. M.R. Barnett, "Twinning and the ductility of magnesium alloys," Materials Science and Engineering A, 464 (2207), 1–7.
3. S. R. Agnew, O. Duygulu, "Plastic anisotropy and the role of non-basal slip in magnesium alloy AZ31B," International Journal of Plasticity, 21(2005), 1161–1193.
4. H. Wang, P.D. Wu, C.N. Tomé, Y. Huang, "A finite strain elastic-viscoplastic self-consistent model for polycrystalline materials," Journal of the Mechanics and Physics of Solids, 58 (2010), 594–612.
5. H.Wang, P.D. Wu, J. Wang, C.N. Tomé, "A physics-based crystal plasticity model for hexagonal close packed crystals including both twinning and detwinning mechanisms." International Journal of Plasticity 49 (2013), 36–52.
6. H. Wang, P.D. Wu, C.N. Tomé, J. Wang, "A constitutive model of twinning and detwinning for HCP polycrystals," Materials Science and Engineering A, 555 (2012), 93–98.
7. P.D. Wu, E. Van der Giessen, "Analysis of elastic-plastic torsion of circular bars at large strains," Archive of Applied Mechanics, 61 (1991), 89–103.
8. H. Wang, Y. Wu, P.D. Wu, K.W. Neale, "Numerical analysis of large strain simple shear and fixed-end torsion of HCP polycrystals," CMC-Comput Mater Continua, 19 (2010), 255–284.
9. H. Wang, B. Raeisinia, P.D. Wu, S.R. Agnew, C.N. Tomé, "Evaluation of self-consistent polycrystal plasticity models for magnesium alloy AZ31B sheet," International Journal of Solids and Structures, 47 (2010),2905–2917.
10. H. Qiao, P.D. Wu, H. Wang, M.A. Gharghouri, M.R. Daymond, "Evaluation of elastic-viscoplastic self-consistent polycrystal plasticity models for zirconium alloys," International Journal of Solids and Structures, 71 (2015), 308–322.
11. C. Ma, H. Wang, T. Hama, X.Q. Guo, X.B. Mao, J. Wang, "Twinning and detwinning behaviors of commercially pure titanium sheets," International Journal of Plasticity, 121 (2019), 261–279.
12. P.D. Wu, H. Wang, K.W. Neale, "On the large strain torsion of HCP polycrystals," International Journal of Applied Mechanics, 4 (2012), 1250024
13. P. Majors, E. Krempl, "Comments on induced anisotropy, the Swift effect, and finite deformation inelasticity," Mechanics Research Communications, 21 (1994), 465–472.

14. S. Harren, T.C. Lowe, R.J. Asaro, A. Needleman, "Analysis of large-strain shear in rate-dependent Face-Centred Cubic polycrystals: Correlation of micro- and macromechanics," Philosophical Transactions of the Royal Society B Biological Sciences, 328 (1989), 443–500.
15. X.D. Zhang, S.M. Li, X.Q. Guo, H. Wang, Q. Y, P.D. Wu, "Effects of texture and twinning on the torsional behavior of magnesium alloy solid rod: a crystal plasticity approach in comparison with uniaxial tension/compression," International Journal of Mechanical Sciences, 2020, 106062.
16. B. Beausir, L.S. Toth, F. Qods, K.W. Neale, "Texture and mechanical behavior of magnesium during free-end torsion," Journal of Engineering Materials and Technology, 131 (2009), 011108.
17. X.Q. Guo, W. Wu, P.D. Wu, H. Qiao, K. An, P.K. Liaw, "On the swift effect and twinning in a rolled magnesium alloy under free-end torsion," Scripta Materialia, 69 (2013), 319–322.
18. J.Zhang, Q. Yu, Y. Jiang, Q. Li, "An experimental study of cyclic deformation of extruded AZ61A magnesium alloy," International Journal of Plasticity, 27 (2011), 768–787.

Part IV

Alloying and Processing/Primary Production

Absorbable Wire Radiopacity: Influence of Composition and Size on X-ray Visibility

Adam J. Griebel, Aubrey L. Ehle, and Jeremy E. Schaffer

Abstract

Imaging systems employing X-rays (such a 2D projectional radiography, computed tomography, and fluoroscopy) are widely used in medicine to aid in surgical planning, intervention, and follow up. Visualization of medical devices using these techniques relies on differences in x-ray absorption between the medical device and the surrounding tissue. The amount of absorption of a given device is largely a function of its cross-sectional dimensions, density, and atomic properties. Consequently, imaging of relatively small devices, like stents, made of a low-density material, like magnesium, can be very challenging. The aim of this study is to quantify the relative radiopacities of key absorbable metal alloy systems, determine the influence of magnesium alloy composition on radiopacity, and estimate the diameters at which wires are no longer visible under typical clinical conditions. This is accomplished by producing wires from five different magnesium alloys, a zinc alloy, and an iron alloy, in sizes ranging from 0.2 to 1.0 mm diameter, and imaging on a clinical x-ray system.

Keywords

Absorbable • Magnesium • Mg • Iron • Fe • Zinc • Zn • Wire • Radiopacity • Imaging • X-ray • Fluoroscopy

A. J. Griebel (✉) · J. E. Schaffer
Research & Development, Fort Wayne Metals Research Products Corp., Fort Wayne, IN, USA
e-mail: adam_griebel@fwmetals.com

A. L. Ehle
Medical Imaging and Radiologic Sciences, Indiana University School of Medicine, Indiana University Fort Wayne, Fort Wayne, IN, USA

Introduction

The field of radiology has become an indispensable component of modern medical care, allowing for safe and effective diagnosis, surgical planning, intervention, and follow up. There are a wide variety of available imaging modalities (e.g. x-ray, magnetic resonance, ultrasound, positron emission tomography), and each have their own advantages and disadvantages. The field of interventional radiology relies heavily on an x-ray-based modality, fluoroscopy, which allows for real-time imaging during a surgical procedure and is especially useful for intraluminal implant placement [1]. Often contrast agents are used to increase the visibility of the system in question [2], and device visibility requires sufficient x-ray absorbance and scattering (radiopacity) compared to the surrounding tissues.

Radiopacity of a given device will depend on a host of factors, including thickness and density of the surrounding tissue, the imaging parameters, the geometrical thickness of the device, and the composition of the material used to fabricate the device. Absorption of a material relates to the material density, the atomic number, and the atomic mass, and can vary with the x-ray energy employed [3].

Medical devices which benefit from radiopaque materials are generally those with a relatively small cross-section and which rely on live fluoroscopic x-ray imaging for accurate placement. Intraarterial stents, for example, require fluoroscopic guidance to locate the target vessel lesion and ensure the stent has been fully deployed. Stent struts can be as large as 0.2 or 0.25 mm for peripheral devices [4], around 0.1 mm for coronary devices [5, 6], and as small as 0.025 mm for neurovascular devices [7]. Radiopacity is also an important feature for accurate placement of gastrointestinal stents. Larger devices like ligation clips and k-wires have larger wire cross-section (0.8–2.0 mm) and generally do not require fluoroscopy to guide placement, so radiopacity is less critical though it can still aid in post-surgery follow up examination.

V. M. Miller et al. (eds.), *Magnesium Technology 2021*, The Minerals, Metals & Materials Series,
https://doi.org/10.1007/978-3-030-65528-0_17

Production and evaluation of wires of absorbable magnesium, zinc, and iron alloys is well documented, with magnesium receiving the most attention. Many promising magnesium alloys contain rare earth elements which generally serve to increase strength, reduce texture, and reduce corrosion rate. Examples of Mg-rare earth alloys include WE43 and Resoloy® [8–10]. Rare earth elements in other metal systems are known to increase radiopacity [3]. Other Mg alloys contain no rare earth elements, and instead rely on small amounts of alloying elements like Al, Zn, Mn, Ca, and Li for strengthening and corrosion control [11]. Examples of these Mg alloys include ZX10, LZ21, and AZ31. Many zinc-based alloys are being investigated for suitability as an absorbable metal. Zinc alloyed with 7 weight percent silver is one such alloy that has shown encouraging results [12]. Much work has been done in the iron-manganese system, as manganese is an essential micronutrient and its presence in iron can help stabilize the austenite phase and produce anti-ferromagnetic behavior [13]. The high strength and stiffness of iron-based materials makes them attractive for stents, though the corrosion rate still needs to be addressed.

While significant effort has gone into characterizing the microstructural, mechanical, and degradation properties of these materials, there is comparatively little data in the literature on their radiopacity, an important material property when designing an implantable device. Some articles report the X-ray and MRI imaging characteristic of magnesium alloy screws, often comparing to conventional titanium ones [14, 15], but nothing is reported for fine wires.

The aim of this study is to fill this hole in the literature by quantifying the relative radiopacities of key absorbable metal alloy wire systems, determine the influence of magnesium alloy composition on radiopacity, and estimate the diameters at which wires are no longer visible under typical clinical conditions.

Materials and Methods

Alloys investigated in this study include rare earth-free magnesium alloys LZ21, ZX10, and AZ31, rare earth-containing magnesium alloys WE43 and Resoloy®, Zn7Ag, and Fe35Mn. Nominal compositions of the alloys and their measured densities are shown in Table 1.

Wires of each alloy were produced via standard cold-drawing practices to diameters of 1.0, 0.75, 0.5, 0.3, and 0.2 mm and cut into samples of approximately 50 mm in length.

To assess relative radiopacity of the wires, the samples of each diameter were aligned on aluminum plates of 12.5- and 25-mm thicknesses as a body mimic and imaged on a clinical GE Precision 600FP system with a Canon DREX-KL80 X-ray generator (Fig. 1). Each set of wire diameters was imaged at 50 cm SID with parameters of 60 kVp/5 mAs and 100 kVp/0.7 mAs on the 12.5 mm plates, and 60kVp/20mAs, and 100 kVp/2.5 mAs on the 25 mm plates.

To quantify radiopacity of each sample, the pixel intensity difference was calculated per ASTM F-640 [16]. In short, the signal intensity of each wire was measured, and the difference relative to its immediate surrounding background was calculated in ImageJ and Excel.

Results

Typical images generated through this method are shown in Fig. 2. The method proved vary capable of distinguishing between alloys. Complete results of the analysis are shown in Fig. 3, where average pixel difference of both 100 kVp tests and both 60 kVp are shown in the top and bottom plots, respectively. In general, the pixel difference decreased with decreasing wire diameter, as expected. In all cases, the Zn7Ag had the highest pixel difference, followed by Fe35Mn. All magnesium alloys had substantially lower pixel differences than the Zn or Fe alloys. Figure 4 shows only the magnesium alloys to reduce the y-axis scale and facilitate comparison. Here, Resoloy generally had the highest radiopacity, followed by WE43. The LZ21, ZX10, and AZ31 radiopacities were largely equivalent.

Zn7Ag and Fe35Mn were clearly distinguishable at all sizes and imaging conditions. Resoloy and WE43 were very faint at 0.3 and 0.2 mm (pixel differences of 1.0–7.5). LZ21, ZX10, and AZ31 were not distinguishable at 0.2 mm (pixel differences of 0.0).

Table 1 Alloy major composition and density

Alloy	Nominal composition (wt%)	Density (g/cc)
LZ21	Mg-2Li-1Zn	1.68
ZX10	Mg-1Zn-0.3Ca	1.74
AZ31	Mg-3Al-1Zn-0.2Mn	1.77
WE43	Mg-4Y-3Nd-0.4Zr	1.81
Resoloy	Mg-10Dy-1Nd-1Zn-0.2Zr	1.91
Zn7Ag	Zn-7Ag	7.29
Fe35Mn	Fe-35Mn	7.69

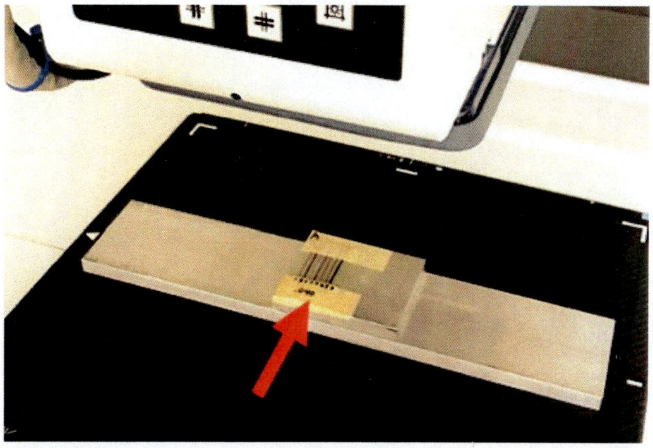

Fig. 1 1 mm wire samples with 25 mm Al plate body mimic on the imaging system. (Color figure online)

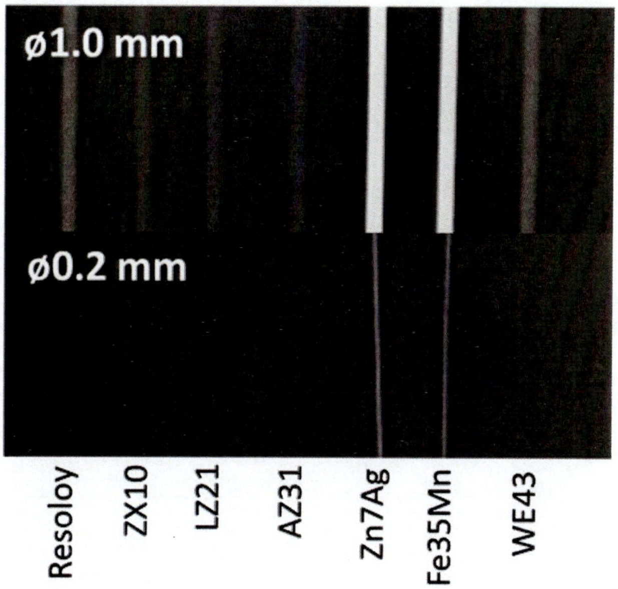

Fig. 2 1 mm (top) and 0.2 mm (bottom) wire samples on 12.5 mm of Al plate imaged at 60 kVp and 5 mAs

Fig. 3 Pixel intensity differences for each alloy and diameter at 100 kVp (top) and 60 kVp (bottom). (Color figure online)

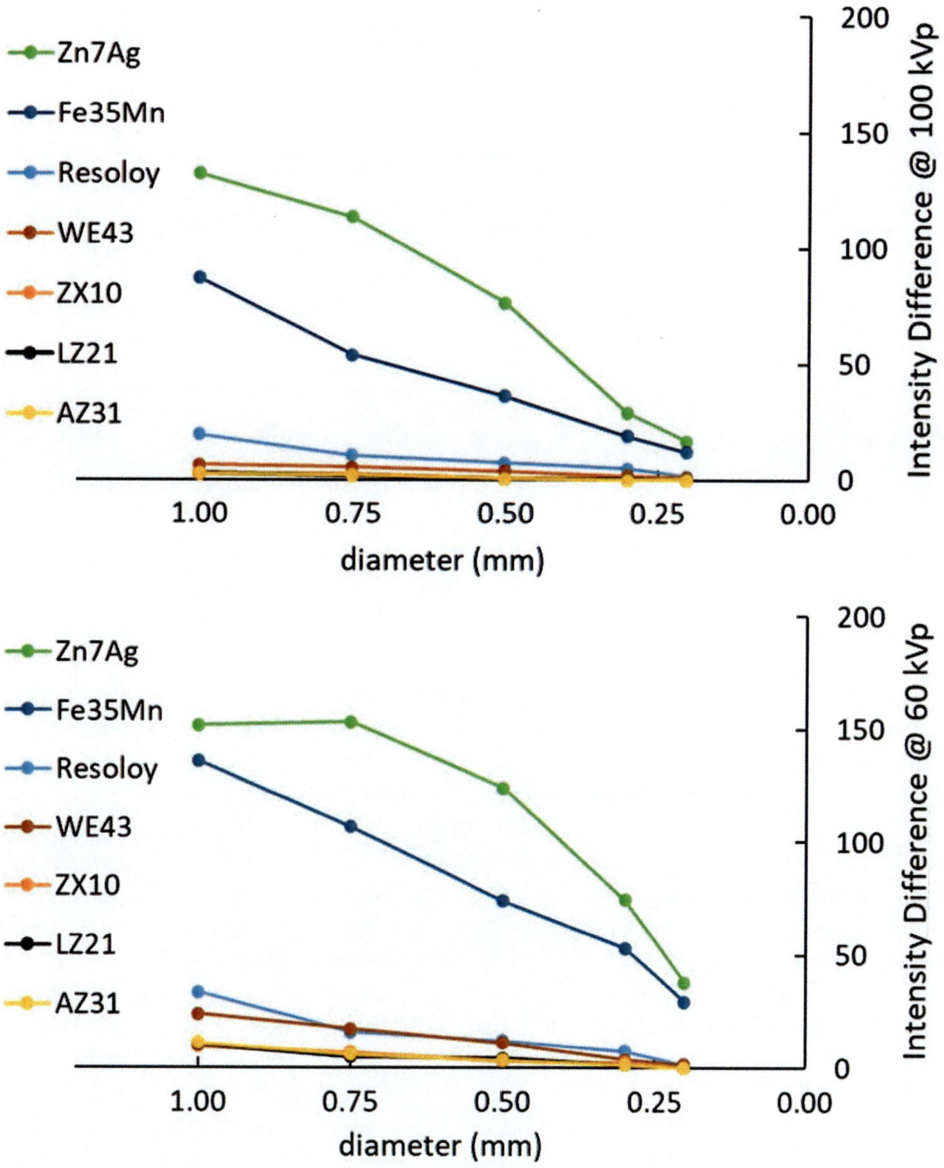

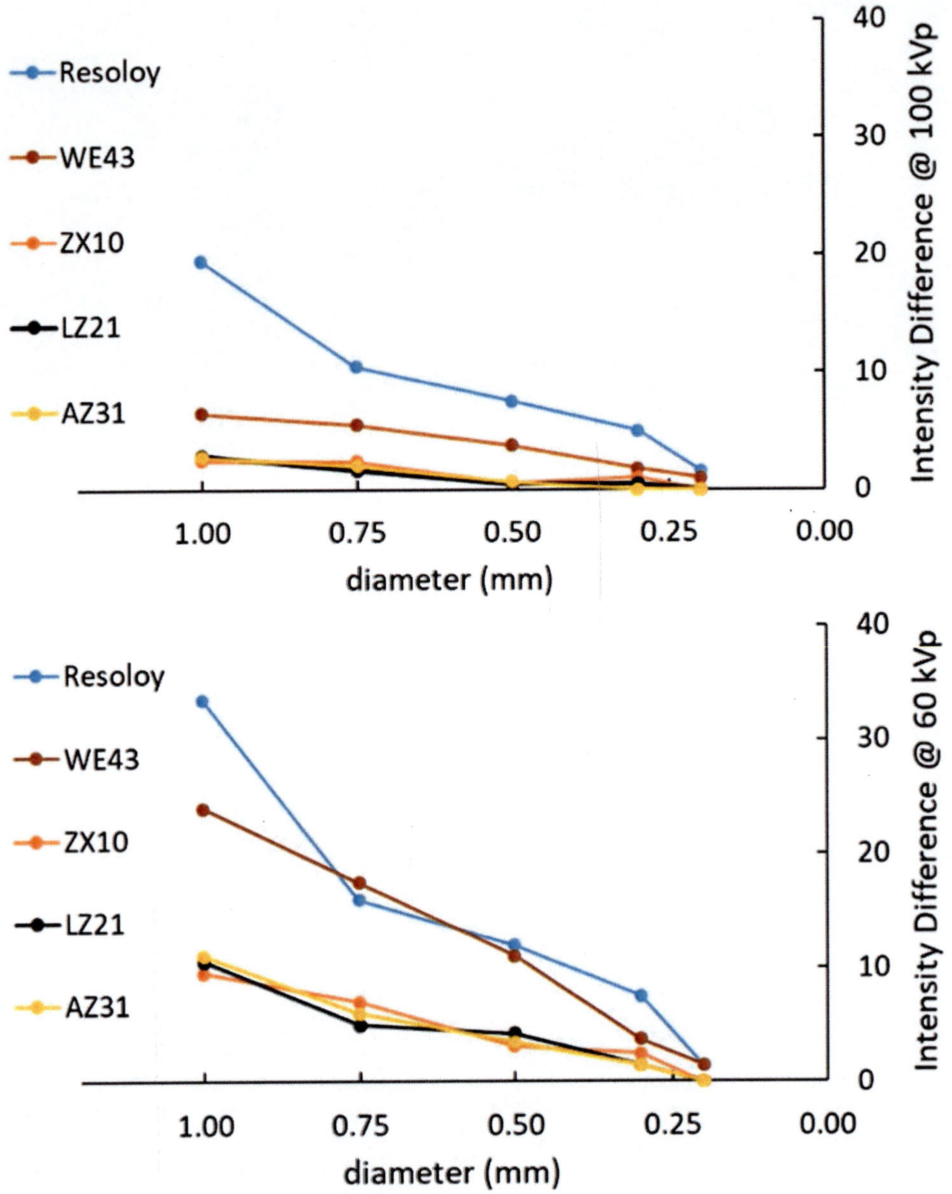

Fig. 4 Pixel intensity differences for each magnesium alloy and diameter at 100 kVp (top) and 60 kVp (bottom). (Color figure online)

Discussion

This work is perhaps the first to quantify the relative radiopacities of absorbable metals. It confirms that which was already known to the field: magnesium alloy radiopacity is low relative to other metals. Interestingly, this study has demonstrated two previously unreported relationships as well: radiopacity variance within magnesium alloys, and relative radiopacities between zinc- and iron-based alloys.

The marked increase in the radiopacity of magnesium alloys which have a significant rare earth metal alloying content (WE43, Resoloy) as shown in Fig. 4 is striking. Resoloy, for example, has between 3 and 10 times the radiopacity over baseline of the LZ21 alloy, depending on the imaging conditions. While this indicates magnesium-rare earth alloys will likely be more visible than magnesium alloys without rare earths, in a practical sense it may have limited utility as most thin magnesium devices like stents which require radiologic guidance during surgery will likely still need radiopaque additions (e.g. tantalum or platinum markers), as even the Resoloy and WE43 were very poorly visible at 0.2 mm while many stents could have struts as thin as 0.1 or 0.125 mm.

Also interesting is the higher radiopacity of the Zn7Ag alloy over the Fe35Mn alloy despite the higher density of the Fe35Mn, pointing to the importance of atomic number and atomic weight in addition to density.

Conclusion

This study provides a quantitative analysis of relative radiopacities of relevant absorbable metals across the three metal classes of interest. It confirms the low radiopacity of magnesium in relation to zinc and iron and shows the influence of magnesium alloying elements on radiopacity. Data generated here will aid in material selection and absorbable device design.

References

1. J. Rosch, F. Keller and J. Kaufman, "The Birth, Early Years, and Future of Interventional Radiology," *Journal of Vascular Interventional Radiology,* vol. 14, pp. 841–853, 2003.
2. H. Chen, M. Rogalski and J. Anker, "Advances in functional X-ray imaging techniques and contrast agents," *Physical Chemistry Chemical Physics,* no. 39, 2012
3. S. A. M. Tofail, J. M. Carlson, A. A. Gandhi, J. Butler, P. Tiernan and L. O'Donoghue, "Nickel-titanium-rare earth alloy and method of processing the alloy". USA Patent 8,440,031 B2, 2013
4. X.-m. Li and e. al, "Evaluation of peripheral artery stent with 64-slice multi-detector row CT angiography: prospective comparison with digital subtraction angiography," *European Journal of Radiology,* vol. 75, no. 1, pp. 98–103, 2010
5. A. Kastrati, J. Mehilli and et al, "Intracoronary Stenting and Angiographic Results," *Circulation,* vol. 103, pp. 2816–2821, 2001.
6. R. Noad, C. Hanratty and S. Walsh, "Clinical Impact of Stent Design," *Interventional Cardiology Review,* vol. 9, no. 2, pp. 89–93, 2014.
7. G. Rajah, S. Narayanan and L. Rangel-Castilla, "Update on flow diverters for the endovascular," *Neurosurgical Focus,* vol. 42, no. 6, 2017.
8. P. Maier, A. Steinacker, B. Clausius and N. Hort, "Influence of Solution Heat Treatment on the Microstructure Hardness and Stress Corrosion Behavior of Extruded Resoloy(R)," *JOM,* pp. 1870–1879, 2020
9. A. Griebel, "An in vitro and in vivo characterization of fine WE43B magnesium wire with varied thermomechanical processing conditions," *Journal of Biomedical Materials Research Part B: Applied Biomaterials,* 2017
10. M. Stekker, N. Hort, F. Feyerabend, E. Hoffmann, M. Hoffmann and R. Horres, "Resorbable stents which contain a magnesium alloy". USA Patent 9522219 B2, 2016
11. Y. Zheng, X. Gu and F. Witte, "Biodegradable Metals," *Materials Science and Engineering R,* vol. 77, pp. 1–34, 2015
12. M. Sikora-Jasinska, "Fabrication, mechanical properties and in vitro degradation behavior of newly developed Zn-Ag alloys for degradable implant applications," *Materials Science and Engineering: C,* pp. 1170–1181, 2017
13. J. Schaffer, "Cold drawn bioabsorbable ferrous and ferrous composite wires: An evaluation of in vitro vascular cytocompatibility," *Acta Biomaterialia,* pp. 8574–8584, 2013
14. Sonnow, "Biodegradable magnesium Herbert screw – image quality and artifacts with radiography, CT and MRI," *BMC Medical Imaging,* 2017
15. J. Espiritu, J. Seitz and M. Meier, "The performance of Mg-based implants in MRI: a systematic review," in *Updates in Bioabsorbable Metals*, Virtual, 2020
16. ASTM International, *F640: Standard Test Methods for Determining Radiopacity for Medical Use,* 2012

Magnesium and Magnesium Alloy Powder Processing Towards the Development of Near Shape Structural Materials

Steven C. Johnson and Dylan G. Goncalves

Abstract

Near shape forming of magnesium (Mg) alloys offers a significant opportunity for structural material lightweight. In this work, conventional press and sinter near shape processing has been applied to commercial AZ91D and pure Mg powders. Results indicate these powders are reasonably compressible achieving green densities of 88 to 98% ρ_{Th} with limited cracking. Sintering of these compacted powders is challenging and requires both solid state and transient liquid phase mass transport apparently due to the inherent powder particle surface oxide layer. Resultant press and sintered materials are characterized for density, porosity distribution, hardness, microstructure, and phase development. Throughout this effort, specific attention is paid to the affect oxygen, as an adsorbed contaminant and a surface thin film, presents as an impediment to solid and transient liquid state densification. Results of this work are intended to advance near shape processing of Mg alloy and Mg powders for potential structural applications.

Keywords

Gas atomized Mg and AZ91D powder • Press and sinter processing • Solid state sintering • Supersolidus liquid phase sintering • Compaction stress • Compaction lubricants • Sintering temperature • Sintering atmosphere • Thermal analysis • Surface oxide layer • Materials characterization

Introduction and Technical Motivation

Fabrication of light alloy materials using near or net shape processes offers a significant opportunity for greater application, more efficient fabrication, and specifically the structural lightweighting of components made from these light alloy materials. In example, Aluminum (Al) alloys with a density of approximately 2.8 gm/cm^3 are commercially relatively abundant, can be fabricated to a desired shape using a variety of methods, and are presently commercially available in powder form for near shape processing by powder consolidation. [1, 2] Alternatively Magnesium (Mg) alloys with a density of approximately 1.8 gm/cm^3 are commercially available but not abundant, are usually processed to shape by casting or wrought processes, and are not commercially available in powders for near shape processing by powder consolidation. As two viable alloy systems applicable for structural lightweighting, the fact that Mg alloys exhibit approximately 36% lower density than Al alloys makes Mg alloys highly desirable for near shape processing investigations.

As stated, there are presently no commercially available powdered Mg alloys intended for near shape powder consolidation. Some work has been performed investigating the viability of Mg and Mg alloys near shape processed by powder consolidation. As example, Liu et al. has consolidated pure Mg powder using spark plasma sintering (SPS). In this work pure Mg powder was consolidated up to 99.67% theoretical density when sintered at 570 °C. It was found that during SPS, by using a pulsed electric current the Mg powder particle surface MgO layer could be melted or even vaporized thus facilitating interparticle diffusion and sintering [3]. Salehi et al. used a binderless three-dimensional printing technique to consolidate powder of a Mg-5.9Zn-0.13Zr alloy. In this work, the green printed shapes were sintered above the alloy solidus temperature to facilitate both solid and liquid state sintering. Results indicate an increase in density from approximately 56%

S. C. Johnson (✉) · D. G. Goncalves
Central Connecticut State University, New Britain, CT 06050, USA
e-mail: scjohnson@ccsu.edu

V. M. Miller et al. (eds.), *Magnesium Technology 2021*, The Minerals, Metals & Materials Series,
https://doi.org/10.1007/978-3-030-65528-0_18

theoretical density in the green condition to a maximum of 71% after sintering primarily due to supersolidus liquid phase sintering [4]. Additionally, Burke and colleagues have performed near shape powder consolidation of pure Mg and Mg elemental powder blends. Results of this work shows the effect of the powder particle surface MgO layer in impeding solid state diffusion, the significance of increased sintering temperatures to facilitate liquid phase sintering in overcoming this surface MgO layer, and the potential of using non-inert sintering atmospheres (i.e.; N_2 (g)) to reduce the surface MgO layer for improved sintering [5–7].

Based on this technical background and the large potential application of near shape processed, light weight Mg alloy components, this technical investigation applied the following approach;

- Only commercially available Mg and Mg alloy powders were used,
- The mature near shape fabrication method of press and sinter processing (P + S) was used to consolidate powders, and
- Both solid and liquid phase sintering were used to densify powder compacts.

In this work, commercially available pure Mg and the Mg alloy AZ91D powders both synthesized by inert gas atomization were used. These powders were initially characterized and then consolidated into near shape, cylindrical discs using P + S methods. Sintering was performed in an inert atmosphere at temperatures below and above the melting and solidus temperatures. Green and sintered materials were characterized for density and microstructure. From this investigation, it is intended that the technical understanding of press + sinter processing of Mg and Mg alloy powders will be progressed.

Materials and Experimental Methods

Materials

The Mg materials used in this work are powders of pure Mg and the Mg alloy AZ91D both synthesized by gas atomization and provided by ESM Group, Inc. of Amherst, NY. The Mg powder contains only trace amounts of impurities and has an intended size distribution of −100/+325 mesh but contains approximately 33% −325 mesh powder. AZ91D is a commercially available Mg alloy usually formed by casting processes containing Al, Zn, and Mn as alloying additions and the pre-alloyed powder used in this work has a size distribution of −90/+270 mesh [1]. Compositions of the pure Mg and pre-alloyed AZ91D powders are presented in Table 1. For compaction processing, both Mg and pre-alloyed AZ91D powders were compacted containing 1 wt% ACRAWAX C admixed lubricant and no admixed lubricant with Zn Stearate used as a die wall lubricant.

Experimental Methods

The Mg and pre-alloyed AZ91D powders used in this work were characterized for powder particle morphology, hardness, flow, apparent density, melting temperature, and phase composition. Powder particle morphology determination was performed using a field emission scanning electron microscope (FE-SEM) operated at various accelerating voltages with images acquired using the secondary electron (SE) signal. Hardness of powder particles was determined using Vickers microhardness. Powder samples were mounted at room temperature in 2-part resin plus hardener epoxy and then ground and polished with various grinding and polishing discs and various abrasive polishing suspensions using an automated grinder/polisher. Hardness of the powder particles observed in cross section were then measured using a Wilson VH1202 Vickers/Knoop microhardness tester with a load of 10 gm. and a dwell time of 5 s. Flow and apparent densities of powders were determined generally following MPIF Standards 03 and 04 [8, 9].

Thermal analysis of powders was performed using a TA Instruments SDT Q-600 simultaneous thermal analyzer (STA) collecting both differential scanning calorimetry (DSC) and thermogravimetric analysis (TGA) signals simultaneously. Powder samples for thermal analysis were placed in covered, high purity Al_2O_3 crucibles, heated to and cooled from 700 °C at a rate of 5 °C/min with 100 ml/min. mass flow rate of high purity Ar (99.999%) as a cover gas. Thermal analysis data was analyzed using the TA Universal Analysis software. Phase composition of powders was determined using a Rigaku MiniFlex II x-ray diffractometer. Powder samples were placed in a zero background holder and scanned from approximately 20 to 100° 2Θ at a rate of 1°/min. using Cu K_α radiation with a wavelength of 1.540562 Å. X-ray diffraction data was analyzed using Materials Data Inc. JADE analysis software and the International Centre for Diffraction Data PDF-4 + x-ray diffraction database.

Consolidation of Mg and pre-alloyed AZ91D powders into near shape samples was performed using press and sinter processing. Powder compaction was performed using a 100 ton capacity mechanical compaction press. Generally, a specific mass of powder (ex. 13 gm) was hand fed into a 1.186 in. diameter cylindrical die and compacted to target compaction stress of 300, 400, 500, or 600 MPa. The density of green compacted samples was determined by dividing the green sample mass by the measured dimensions of the green compact. Thermal delubing and sintering of green

Table 1 Compositions of the Mg and pre-alloyed AZ91D powders used in this work

Element	Mg	AZ91D
	Weight %	
Al	–	9.20
Zn	–	0.65
Mn	0.10	0.26
Cu	0.02	–
Sn	0.01	0.01
Other	0.05	0.01
Mg	balance 99.8	balance 89.9

compacts was performed in a 2.75 in. diameter laboratory tube furnace in a high purity Ar (99.999%) atmosphere. All samples here heated at a rate of 10 °C/min., delubed at 400 °C for 20 min., and then sintered at various temperatures for 60 min. Specifically, Mg powder compacts were sintered at 635, 655, and 670 °C for 60 min. and AZ91D powder compacts were sintered at 535, 570, and 585 °C for 60 min.

The density of sintered samples was determined generally following ASTM standards B962-17 or B311-17 depending on the level of porosity in the sintered sample [10, 11]. Hardness of sintered samples was performed using Rockwell hardness testing in the E scale on sections of cylindrical discs. Sintered samples were sectioned for metallographic mounting and grinding/polishing using a slow speed diamond saw in orientations parallel (transverse section) and perpendicular (longitudinal section) to the compaction direction. Sectioned sintered samples were mounted at room temperature in 2-part resin plus hardener epoxy and then ground and polished with various grinding and polishing discs and various abrasive suspensions using an automated grinder/polisher. Imaging and analysis of mounted and polished sintered samples was also performed using a FE-SEM operated at various accelerating voltages and imaged using the backscattered electron (BSE) signal.

Results and Discussion

Powder Characterization Results

Powder particle morphology of pure Mg and pre-alloyed AZ91D powders are presented in Fig. 1. Both Mg and AZ91D powders are spherical in shape as would expected from a gas atomization powder fabrication process.

Powder particle Vickers microhardness, powder flow rate, and powder apparent density of the Mg and AZ91D powders are presented in Table 2. Both powders are relatively soft with the Mg powder slightly softer than the AZ91D. The Mg and AZ91D powders are both free flowing and exhibit similar apparent densities.

Thermal analysis plots of DSC data for pure Mg and pre-alloyed AZ91D powders are presented in Fig. 2 and the

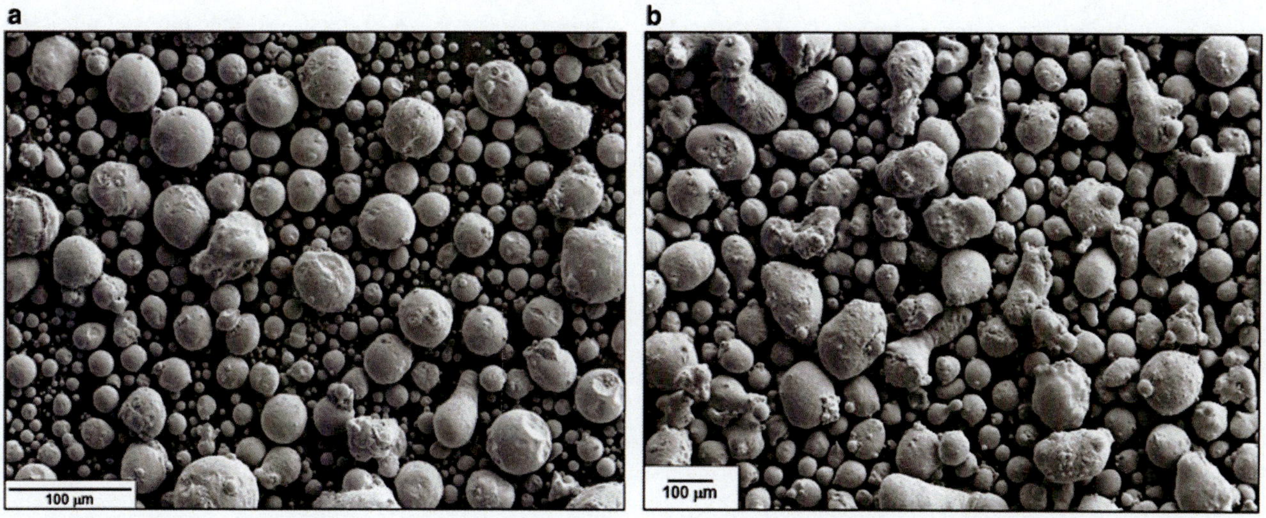

Fig. 1 **a** SE SEM image of pure Mg powder used in this work. **b** SE SEM image of pre-alloyed AZ91D powder used in this work

Table 2 Vickers microhardness, flow rate, and apparent density of the pure Mg and pre-alloyed AZ91D powders used in this work

Powder material	Hardness (HVN)	Flow rate (s/gm)	Apparent density (gm/cm^5)
Mg	39.3 ± 7.1	38.7/50	0.886
AZ91D	58.5 ± 2.9	92/50	0.908

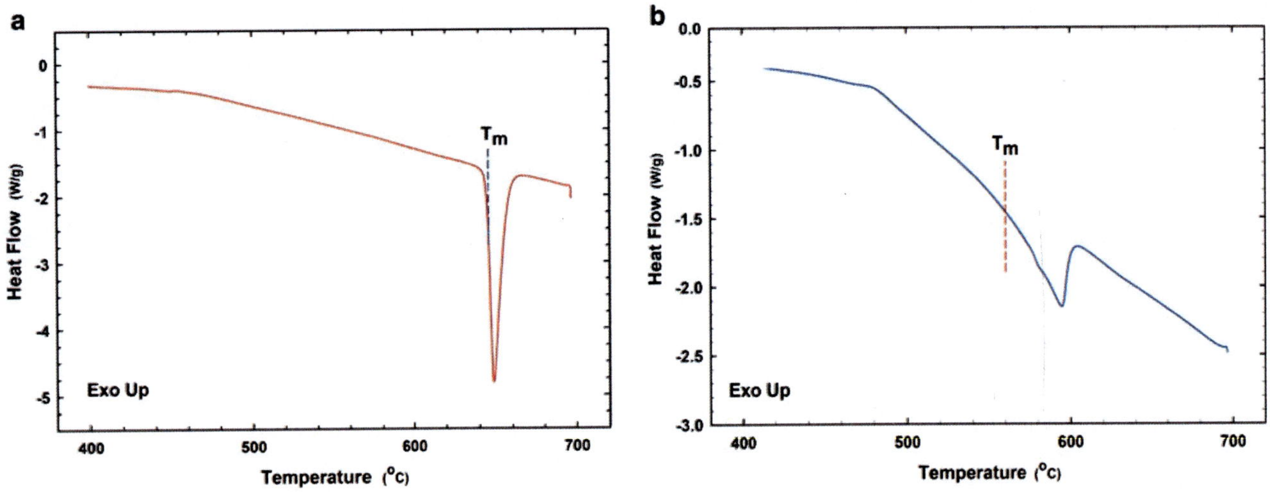

Fig. 2 **a** DSC scan of pure Mg powder during heating to 700 °C. **b** DSC scan of pre-alloyed AZ91D powder. (Color figure online)

Table 3 Experimentally determined onset melting temperature (T_m), peak reaction temperature (peak T), and heat of fusion values for pure Mg and AZ91D powders used in this work

Powder material	T (°C)	Peak T (°C)	Heat of fusion (J/gm)
Mg	644.76 ± 0.37	648.65 ± 0.05	−302.20 ± 40.04
AZ91D	560.50 ± 3.40	594.23 ± 0.22	−189.84 ± 40.64

analytically determined onset melting point (T_m), peak reaction temperature (peak T), and heat of fusion for these powders are presented in Table 3. As shown in Fig. 2a and Table 3, the DSC scan of Mg powder exhibits a single endothermic melting reaction at approximately 645 °C and a heat of fusion of approximately −302 J/gm. The DSC scan and values of T_m, peak T, and heat of fusion for pre-alloyed AZ91D powder are shown in Fig. 2b and Table 3 with this powder also exhibiting a single endothermic melting reaction at approximately 561 °C and a heat of fusion of approximately −190 J/gm.

X-ray diffraction patterns for pure Mg and pre-alloyed AZ91D powders are presented in Fig. 3a and 3b. Both patterns exhibit similar diffraction peaks for the hexagonal close-packed (HCP) α-Mg phase. However, the AZ91D diffraction pattern additionally exhibits peaks associated with the $Mg_{17}Al_{12}$ β-phase as shown in Fig. 3b.

Compaction and Sintering Results

Compaction processing data for pure Mg powder using admixed ACRAWAX C and Zn Stearate die wall lubricants is shown in Fig. 4.a. Similar compaction data for pre-alloyed AZ91D powder also using admixed ACRAWAX C and Zn Stearate die wall lubricants is shown in Fig. 4b. Both figures show an increase in compact green density with increasing compaction stress and generally a nonlinear relationship between green density and compaction stress. The density of fully dense Mg is 1.738 gm/cm^3 and that of wrought AZ91D is 1.82 gm/cm^3 [12].

The density of green and sintered pure Mg and AZ91D powder compacts are presented in Table 4. Generally, sintered density noticeably increased for pure Mg material and little or no density increase occurred for AZ91D material.

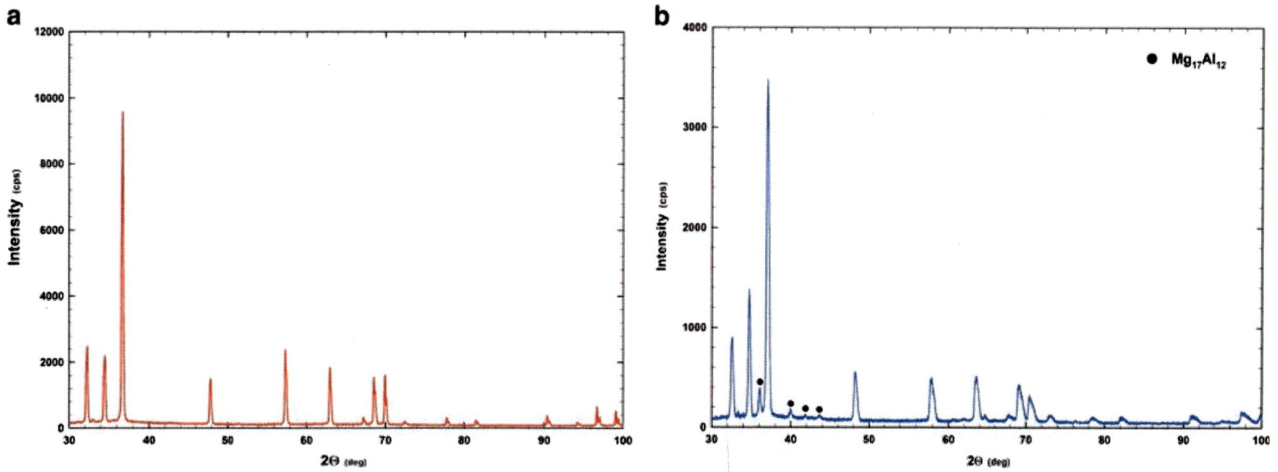

Fig. 3 **a** X-ray diffraction pattern of Mg powder used in this work. **b** X-ray diffraction pattern of AZ91D powder used in this work. (Color figure online)

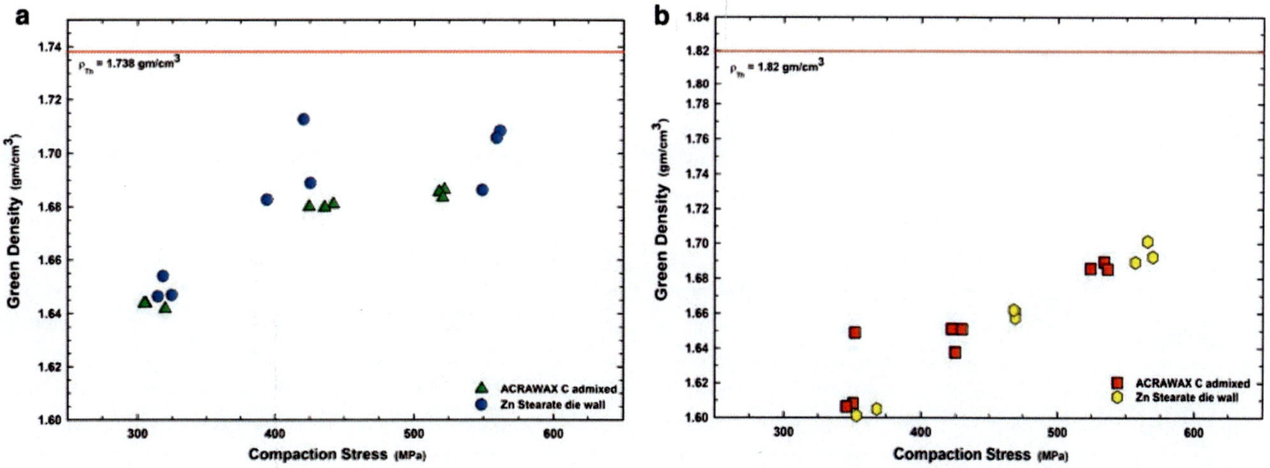

Fig. 4 **a** Compressibility plot of Mg powder with admixed ACRAWAX C and die wall Zn Stearate lubricants compacted at T_R. **b** Compressibility plot of pre-alloyed AZ91D powder with admixed ACRAWAX C and die wall Zn Stearate lubricants compacted at T_R. (Color figure online)

Table 4 Measured values of green and sintered density for pure Mg and AZ91D used in this work

powder material	lubricant	compaction stress (MPa)	green density (% ρ_{Th})	sinter density (% ρ_{Th})
Mg	1 % ACRAWAX C admixed	309.97 ± 8.3	94.54 ± 0.1	95.13 ± 1.3
	Zn Stearate DW	319.13 ± 5.1	94.89 ± 0.3	96.5 ± 1.0
AZ91D	1 % ACRAWAX C admixed	349.50 ± 3.3	89.07 ± 1.3	–
	Zn Stearate DW	352.62 ± 12.3	87.93 ± 0.2	88.10 ± 0.3

Last, representative BSE SEM micrographs of sintered Mg and sintered pre-alloyed AZ91D materials are presented in Fig. 5a and 5b.

Discussion of Results

Compositions of both powders used in this study are presented in Table 1. The Mg powder is shown to have a purity of greater than 99.8% Mg. The pre-alloyed AZ91D powder composition is very close to the accepted nominal composition of Mg-9.5Al-0.6Zn-0.3Mn for the AZ91D casting alloy [1]. Both powders used in this work have particle size distributions that are acceptable for P + S processing with the Mg powder having a smaller size distribution specifically below −325 mesh. As shown in Fig. 1a and 1b, both powders have a spherical morphology as would be expected from rapid solidification of a liquid by gas atomization. Qualitatively the morphology of the AZ91D powder appears slightly less spherical than the pure Mg powder. Both

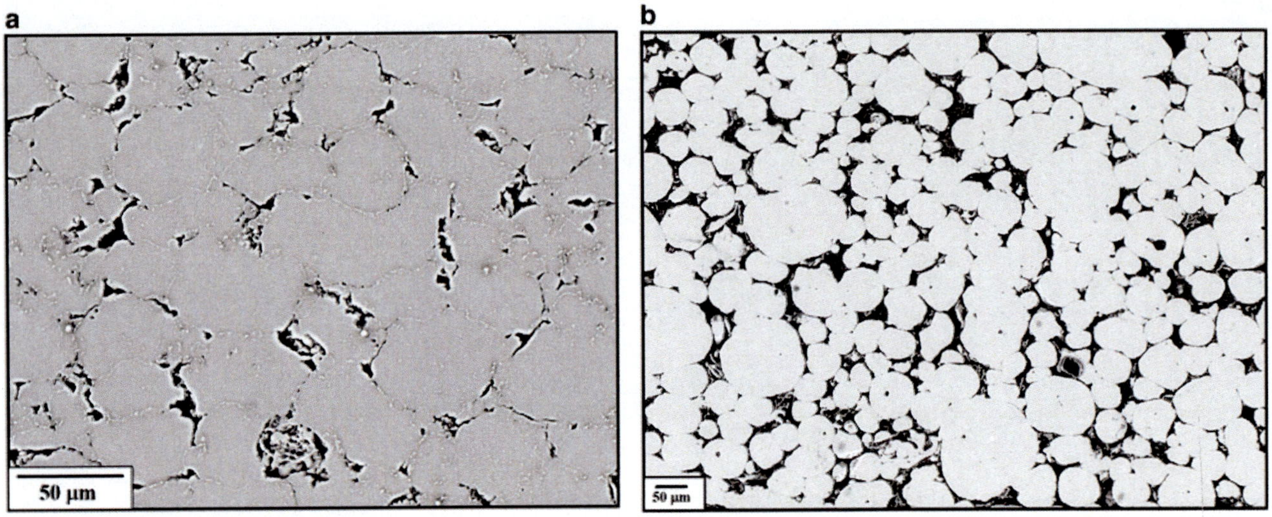

Fig. 5 **a** BSE SEM micrograph of Mg powder compacted with Zn Stearate die wall lubricant sintered at 630 °C. **b** BSE SEM micrograph of sintered AZ91 powder compacted with Zn Stearate die wall lubricant sintered at 545 °C

powders are free flowing and have similar apparent densities as presented in Table 2. Additionally, the pure Mg and pre-alloyed AZ91D powders are both relatively soft with hardness of approximately 39 and 59 HVN respectively as also presented in Table 2. While the AZ91D powder is significantly harder than the Mg powder, hardness values below 100 HVN should be considered very soft. Based on this characterization, these two commercially available Mg/Mg alloy powders appear well suited for consolidation by mechanical compaction and inert atmosphere sintering. Even though these powders have not been engineered specifically for P + S processing, characterization results indicate these powders should have adequate flowability for die filling and exhibit sufficiently compressible for mechanical compaction.

Melting reactions of the pure Mg and pre-alloyed AZ91D powders used in this study are presented as the DSC plots of Fig. 2a and 2b. For Mg powder a single, narrow endothermic melting reaction shown in Fig. 2a results in a T_m of 645 °C with a heat of fusion of −302 J/gm as presented in Table 3. For AZ91D powder, Fig. 2.b presents the single, relatively broad endothermic melting reaction resulting in a T_m of 561 °C with a heat of fusion of −190 J/gm also presented in Table 3. Since the Mg powder of this study is considered a pure metal, the determined value of 645 °C is the single temperature at which melting will occur. However for the pre-alloyed AZ91D powder, the thermal analysis determined value of 561 °C should be taken as the solidus temperature with melting of this alloy occurring over a range of temperature starting at 561 °C. It should be additionally noted that the DSC plots of Fig. 2a and 2b showed no additional reactions during heating to 700 °C other than the melting endotherms. Previous results by Burke and colleagues have shown reactions associated with the decomposition of hydroxide and carbonate compounds during heating of powdered Mg in DSC analyses [13, 14]. For this work, these potential interactions of hydrocarbon compounds with Mg and AZ91D powders is not believed to influence the sintering process. Based on the thermal analysis results, sintering temperatures below and above the Mg and AZ91D melting and solidus temperatures are selected to facilitate solid and liquid phase sintering.

Figure 3a and 3b present the x-ray diffraction patterns of pure Mg and pre-alloyed AZ91D powders respectively. Both diffraction patterns index as Mg as shown in PDF 00-035-0821 [15]. Additionally, in the AZ91D diffraction pattern 4 relatively low-intensity peaks are present around the Mg (101) peak identified with the symbol •. These peaks index as the intermetallic compound $Mg_{17}Al_{12}$ as shown in PDF 04-014-7592 [16]. $Mg_{17}Al_{12}$ has been identified previously in AZ91D and other Mg alloys containing Al [17, 18]. For the work reported here, it will be assumed that the presence of $Mg_{17}Al_{12}$ is in a relatively very small volume fraction and has no influence on the P + S processing of pre-alloyed AZ91D powder.

The room temperature compressibility of pure Mg and pre-alloyed AZ91D powders using 1 wt% admixed ACRAWAX C and Zn Stearate die wall compaction lubricants are presented in Fig. 4a and 4b. Both powders exhibit increased green density with increased compaction stress and a generally nonlinear relationship between green density and compaction stress. For pure Mg, green densities ranging from 94.5 to 98.3% theoretical density (ρ_{Th}) were achieved with circumferential cracking observed in 50% of the samples compacted using ARAWAX C. No circumferential cracking was observed in green compacts fabricated using

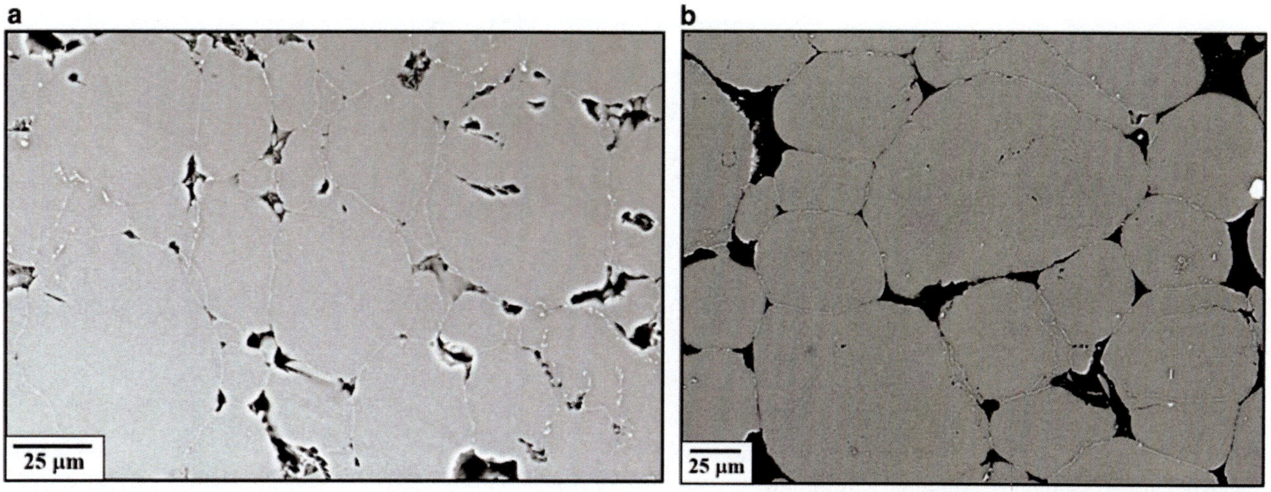

Fig. 6 **a** BSE SEM micrograph of Mg sintered at 670 °C exhibiting liquid phase at particle interfaces. **b** BSE SEM micrograph of AZ91D sintered at 570 °C exhibiting minimal liquid phase at particle interfaces

Zn Stearate die wall lubrication. For pre-alloyed AZ91D, green densities ranging from 87.6 to 93.5% ρ_{Th} were achieved with some degree of circumferential cracking observed in all samples compacted regardless of compaction lubricant used. From this data, it can be concluded that commercially available pure Mg and pre-alloyed AZ91D powders can be consolidated using closed die compaction. The data presented in Fig. 4a and 4b imply that for the same compaction stress, the pure Mg powder is more compressible than the AZ91D powder. However, the presence of circumferential green cracking is undesirable particularly for potential structural application of P + S Mg materials. Future research efforts should be made to eliminate green cracking in Mg and AZ91D compacts.

Based on results of the thermal analysis of pure Mg and pre-alloyed AZ91D powders, compacted samples of both powders were sintered at temperatures below and above the determined T_m values. In this approach, both solid phase and liquid phase sintering were performed as an attempt to overcome the MgO surface layer on powder particles and maximize sintered densities. For the pure Mg powder compacts, sintering was performed at 635, 655, and 670 °C with sintering above T_m termed liquid phase sintering. For pre-alloyed AZ91D powder compacts, sintering was performed at 545, 570, and 585 °C and sintering above T_m is termed supersolidus liquid phase sintering. Sintering results for both pure Mg and pre-alloyed AZ91D green compacts are presented in Table 4 and typical sintered microstructures in Fig. 5a and 5b. These results indicate that some increase in density due to sintering was achieved for pure Mg using both ACRAWAX C and Zn Stearate compaction lubricants with a maximum of 3% density increase for samples compacted using Zn Stearate. However, little or no increase in density due to sintering was achieved for AZ91D regardless of compaction lubricant used. Some insight into the different sintering response can be obtained by analyzing the microstructures of these two materials as presented in the higher magnification BSE SEM images of Fig. 6a and 6b. The Mg material of Fig. 6.a was sintered at 670 °C, 25 °C above the Mg T_m and a temperature where liquid phase sintering should occur. As shown in Fig. 6.a, a light contrast phase decorates the large majority of interparticle boundaries and appears to bond the powder particles in the image. It can be proposed that the interparticle light contrast phase is a transient liquid that has formed, migrated between powder particles, and solidified providing an interparticle bond. The AZ91D material of Fig. 6b was sintered at 570 °C, 10 °C above the AZ91D T_m and a temperature where supersolidus liquid phase sintering should occur. Similarly, a light contrast phase decorates AZ91D interparticle boundaries shown in Fig. 6b, however here the interparticle phase exists at a small fraction of the interparticle boundaries. While liquid phase/supersolidus liquid phase sintering has occurred for both materials, the effect of liquid phase sintering in increasing the materials density appears greater for Mg than for AZ91D. A chemical identification of this interparticle phase has not been performed and is planned for future work.

While experimental evidence shows sintering using a liquid phase has occurred in Mg and AZ91D powder compacts, the efficacy with which liquid phase sintering has resulted in increased density of powder compacts is not clear. For all materials sintered in this work, a delubing treatment of 20 min. at 400 °C prior to sintering was performed to assure removal of all hydrocarbon lubricants and to give all materials the same thermal history prior to actual sintering [19]. Based on the current work, it is possible that modification of the sintering temperature and/or sintering time will be required to greater increase the densities achieved through liquid phase sintering of Mg base materials.

Summary and Proposed Future Work

Commercially available pure Mg and pre-alloyed AZ91D powders were characterized and subsequently processed using press + sinter methods towards fabricating near ρ_{Th} Mg structural materials. The Mg powder used was 99.8% pure and the AZ91D powder had a composition similar to the standard cast form of this alloy. Both powders are relatively soft, free flowing, have similar apparent densities, and are spherical in particle morphology. Additionally, x-ray diffraction showed both powders to be phase pure as α-Mg with the AZ91D containing a small volume fraction of $Mg_{17}Al_{12}$ phase. Thermal analysis was used to determine melting temperatures of both powders as 645 °C for Mg and 561 °C for AZ91D, and these values were used to guide solid and liquid phase sintering. Mechanical compaction in a closed die showed both powders to be reasonably compressible using both admixed and die wall lubricants. Mg powder attained up to 98.6% ρ_{Th} when compacted to 560 MPa and AZ91D up to 93.5% ρ_{Th} when compacted up to 569 MPa, however both powder compacts exhibited some amount of circumferential green cracking. Sintering of these two materials using solid phase and liquid phase mechanisms achieved some increase in density but did not approach ρ_{Th}. Mg powder compacts had up 3% increase in density while AZ91D showed very little to no increase in density both after sintering with formation of some volume of liquid phase. The efficacy of liquid phase sintering of Mg and AZ91D powders is not clear and further investigation is needed to progress towards near ρ_{Th} sintered materials.

Future work on press + sinter processing of Mg and Mg alloy powders for structural applications should focus on attempting to further progress the sintered density towards ρ_{Th}. Specifically work performed determining sintering temperature and time parameters to more efficiently use the liquid phase formed in progressing toward ρ_{Th}. Additionally the interparticle phase observed in BSE SEM images of this study should be compositionally quantified and powder compaction efforts should be made to compact powders without formation of circumferential green compact cracking. Last efforts to reduce or minimize the MgO powder particle surface layer should be made to facilitate greater sinter densification and possible compressibility of Mg base materials.

Acknowledgements This work was partially supported by two CSU/AAUP University Faculty Research Grants Banner Index ARJOHL and ARJOHM, two CCSU Foundation Faculty—Student Research Grants Banner Index AFJONK and AFJOHM, and funding from a CT Next Higher Education Fund grant from the state of Connecticut, Dr. G.J. Gallo, University of Connecticut (UCONN) Principal Investigator.The donation of Mg and AZ91D powders by ESM Group, Inc. Amherst, NY and ACRAWAX C lubricant by Lonza, Inc. Williamsport, PA are gratefully acknowledged. The assistance of National Sintered Alloys, Inc. Clinton, CT in performing compaction processing of powders is greatly appreciated. Access to thermal analysis instrumentation located in the Thermal Analysis Lab of the Institute of Materials Sciences at UCONN, Storrs, CT is acknowledged. Last, electron microscopy was performed using the facilities at the UCONN/Thermo Fisher Scientific Center for Advanced Microscopy and Materials Analysis also in Storrs, CT.

References

1. Polmear, IJ (2006) Light Alloys: From Traditional Alloys to Nanocrystals. 4th edition, Elsevier Butterworth-Heinemann, Massachusetts, p 237–268.
2. Daye, W (2017) Aluminum Powder Metallurgy: A Versatile & Cost-Effective Option for Lightweighting World Jan + Feb 2017:42–52.
3. Liu, R et. al. (2019) Densification of pure magnesium by spark plasma sintering–discussion of sintering mechanism. Advanced Powder Technology 30:2649–2658. https://doi.org/10.1016/j.apt.2019.08.012.
4. Salehi, M *et. al.* (2019) Additive Manufacturing of magnesium-zinc-zirconium (ZK) alloys via capillary-mediated binderless three-dimensional printing. Materials and Design 169:1–15. https://doi.org/10.1016/j.matdes.2019.107683.
5. Burke, P *et. al.* (2009) Sintering Fundamentals of Magnesium Powders. Canadian Metallurgical Quarterly 48(2):123-132.
6. Burke, P et al. (2011) Processing Parameters and post-sintering operations effects in magnesium powder metallurgy. Canadian Metallurgical Quarterly 50(3):240–245. https://doi.org/10.1179/1879139511y.0000000013.
7. Burke, P and Kipouros, GJ (2011) Development of Magnesium Powder Metallurgy AZ31 Alloy Using Commercially Available Powders. High Temp. Mater. Proc. 30:51–61. https://doi.org/10.1515/htmp.2011.007.
8. MPIF standard 03 (2016) Determination of Flow Rate of Free-Flowing Metal Powders Using the Hall Apparatus. Metal Powder Industry Federation, Princeton, NJ.
9. MPIF standard 04 (2016) Determination of Apparent Density of Free-Flowing Metal Powders Using a Hall Apparatus. Metal Powder Industry Federation, Princeton, NJ.
10. ASTM standard 0B962 – 17, "Standard Test Methods for Density of Compacted or Sintered Powder Metallurgy (PM) Products Using Archimedes' Principle", 2017, ASTM International, West Conshohocken, PA.
11. ASTM standard 0B311 – 17, "Standard Test Method for Density of Powder Metallurgy (PM) Materials Containing Less Than Two Percent Porosity", 2017, ASTM International, West Conshohocken, PA.
12. ASM Handbook Online, Volume 2 Properties and Selection: Non-Ferrous Alloys and Special-Purpose Alloys, https://doi.org/10.31399/asm.hb.v02.9781627081627.
13. Burke, P *et. al.* (2011) Thermal Effects of Calcium and Yttrium Additions on the Sintering of Magnesium Powder. Paper presented at TMS 2011 Annual Meeting and Exhibition, San Diego, California, 27 February – 3 March, 2011.
14. Burke, P *et. al.* (2016) DSC and FIB/TEM investigation of calcium and yttrium additions in the sintering of magnesium powder.

Canadian Metallurgical Quarterly 55(1):45–52. https://doi.org/10.1080/00084433.2015.1135535.
15. PDF-4 + 2020; ICDD: Newtown Square, PA, 2020; PDF 00–035-0821 (accessed Sept 7, 2020).
16. PDF-4 + 2020; ICDD: Newtown Square, PA, 2020; PDF 04–014-7592 (accessed Sept 7, 2020).
17. Jabbari Taleghani, MA and Torralba, JM, P (2013) Hot deformation behavior and workability characteristics of AZ91 magnesium alloy powder compacts – A study using processing map. Mater. Sci. Eng. A 580:142–149. https://doi.org/10.1016/j.msea.2013.04.071.
18. Huang, X, *et. al.* (2019) Improvement of mechanical properties of extruded AZX912 magnesium alloy using high-temperature solution treatment. J. Mater. Res. 34(21):3725–3734. https://doi.org/10.1557/jmr.2019.281.
19. Pieczonka, T, et. al. (2018) The effect of nitrogen flow rate on ACRAWAX C decomposition and its removal during sintering of Alumix 431D grade powder. Powder Metallurgy 61(2):149–156. https://doi.org/10.1080/00325899.2017.1410999.

Effect of Sintering Temperature on the Properties of AZ91 Foamed Magnesium Alloy

Hanghang Zhou, Guibao Qiu, Zhenyun Tian, and Qingjuan Li

Abstract

Foamed magnesium alloy has similar porosity and mechanical properties to biological bone, and is often used as a substitute material for biological bone. AZ91 magnesium alloy has high corrosion resistance and is suitable as a substitute material for biological bone. In this paper, AZ91 magnesium alloy powder is used as the raw material and urea is used as the pore-forming agent. The powder is compressed into a cylindrical sample under a pressure of 11 MPa, and sintered at different temperatures. Conduct metallographic observation and mechanical performance test on the sample. The results show that the sintering temperature is between 520 and 550 °C is the most suitable, the compressive strength of the sample is up to 9.14 MPa, the porosity is large and the pore size is uniform, which meets the performance requirements of bone substitute materials.

Keywords

AZ91 magnesium alloy • Sintering temperature • Mechanical properties • Porosity

Introduction

In recent years, magnesium and its alloys have been widely used in aerospace, automotive, biomedicine and other fields due to their strength-to-weight ratio, good mechanical properties and good biocompatibility [1, 2]. Foamed magnesium is an emerging material with excellent performance. It has a special structure and a large number of connected or unconnected pores in the matrix. It not only has good mechanical properties, but also can be used as a functional material [3–5]. As a kind of foam metal, compared with physical metal magnesium, foam magnesium has low density, good energy absorption rate, sound absorption and sound insulation performance [6]. Foamed magnesium has good biomimetic properties and can be used as a biomedical material [7, 8]. Magnesium alloy is a material with very good biodegradability. It has good biocompatibility and bioabsorption. The pore structure of foamed magnesium is similar to that of human bone. The pore structure of foamed magnesium can be controlled to match the human bone tissue, Implanted into the human body instead of human bone [9].

Magnesium is one of the lightest materials in industrial metals, with a density of 1.74 g/cm^3. This value is about 2/3 that of aluminum, 2/5 of titanium, 1/4.5 of steel, and 1/4 of iron. The damping performance is far incomparable to other metals [10]. In addition, the foam material itself has the characteristics of many pores, so that it has a smaller density, greater specific strength, specific stiffness, and good sound absorption performance. In the same pore condition, the foamed magnesium alloy has the same or even better mechanical properties than the widely used aluminum foam by virtue of its lighter weight [11]. At the same time, the foamed magnesium alloy material has very high recyclability and reusability. Because of this, it is also called "twenty-first-century Green Engineering Material."

Although the unique properties of foamed magnesium alloy make it have a very wide development space in many fields, but it has certain defects, such as easy to be oxidized, high activity, poor corrosion resistance, and other defects which severely limit it. The large-scale preparation and commercial use [12, 13]. In addition, people's research on foam metal is relatively late and the research is not sufficient. Therefore, there is still no complete and mature system for preparing foamed magnesium alloy. It is currently almost in the laboratory stage, and magnesium has a porous structure. After that, its easy to be oxidized and easy to corrode defects

H. Zhou · G. Qiu (✉) · Z. Tian · Q. Li
College of Materials Science & Engineering, Chongqing University, Chongqing, 400044, China
e-mail: qiuguibao@cqu.edu.cn

V. M. Miller et al. (eds.), *Magnesium Technology 2021*, The Minerals, Metals & Materials Series,
https://doi.org/10.1007/978-3-030-65528-0_19

will be further expanded, which makes its application greatly restricted. It is because of the emergence of these problems that it is more urgent and important to develop a simple and safe product technology method with good corrosion resistance, lower oxidation rate, high production efficiency, and excellent product performance.

Experimental

In order to obtain a foamed magnesium alloy with better corrosion resistance, AZ91 magnesium alloy powder is selected as the research object in this paper. AZ91 magnesium alloy is a relatively mature magnesium alloy powder with relatively good corrosion resistance. Its main components and properties are as follows shown in Tables 1 and 2.

In this paper, urea is selected as the pore former. Compared with other pore formers, urea has a controllable shape and is more regular. After thermal analysis, it is known that urea starts to melt at 120 °C and can be completely decomposed before 460 °C, so as to avoid the decomposition or reaction of magnesium and its alloys at higher temperatures. The urea used in the experiment was white needle-shaped, with a particle size of 100–500 μm, and a purity of over 99.99%.

The preparation process is as follows: mixing magnesium alloy and urea with a good ratio in a ball mill at a speed of 300 r/min for pretreatment for 1 h to make the raw materials uniformly mixed. Load the mixed powder into a mold, and hold it under a pressure of 11 Mpa for 1 min to obtain a cylindrical sample with a diameter of 16 mm and a height of 10 mm. In order to ensure the repeatability of the experiment, this paper sets 5 sets of variables for the sintering temperature, and each set of experiments is repeated 3 times. The experimental design is shown in Table 3, and the heating curve is shown in Fig. 1. At the first endothermic peak of urea, the heating rate is set to 2 C/min; at the second endothermic peak, the heating rate is set to 4 °C/min; the heating rate set at the third endothermic peak is 6 °C/min. The sintering temperature was set to 490 °C, 520 °C, 550 °C, 580 °C, and 610 °C, respectively. After holding for 1.5 h, it is cooled with the furnace.

Results and Discussion

Figure 1 shows samples at different sintering temperatures. It can be clearly seen that the samples have been burned at 610 °C.

Influence of Sintering Temperature on Porosity

The sintered sample was observed by metallographic microscope and Image J analysis. Three samples in each experiment were observed and calculated, and then the average value was taken to ensure the accuracy of the data. The pore size distribution and porosity of the sample were obtained as shown in the Fig. 3 and Table 4. Figure 2 is the morphology observed after 25 times magnification in the electron microscope As shown in the figure, it cannot be observed due to overburning at 610 °C. It can be found from Fig. 2 that as the sintering temperature increases, the particles inside the green compact adhere to each other. With the movement of atoms and the exchange of substances, the particles change from initial point contact to line contact and surface contact. contact. In this process, the two particles are connected to form a sintered neck, which continuously grows, and finally causes a series of changes in the shape, size and number of pores in the foamed magnesium alloy.

It can be seen from Fig. 3 that as the sintering temperature increases, the proportion of micropores below 0.2 mm in the sample begins to decrease and gradually aggregates into larger pores. When the temperature increases above 550 °C, the pore diameter begins to aggregate to a large pore diameter of 0.6 mm or more. For biological bone materials, it is more desirable for the pore size to be concentrated between 0.2 and 0.6 mm, so the sintering temperature is more appropriate between 520 and 550 °C.

Enter the data in Table 4 into the Origin chart to get the porosity curve of the foamed magnesium alloy with the

Table 1 The main components of AZ91D (%)

Mg	Al	Zn	Mn	Sn	Fe	Ni	Cu
margin	8.5–9.5	0.45–0.90	0.17–0.4	<=0.05	<=0.004	<=0.001	<=0.025

Table 2 The main performance of AZ91D

Melting point (°C)	Density (g/cm^3)	Tensile strength (MPa)	Yield point (MPa)	Specific strength	Elongation (%)	Young's modulus (GPa)
596	1.82	250	160	7	154	44.8

Table 3 Pressing pressure, sintering parameters and green components

Sample number	Pressure (MPa)	Sintering temperature (°C)	Sintering time (h)	Urea content (vol.%)
1#	11	490	1.5	60
2#		520		
3#		550		
4#		580		
5#		610		

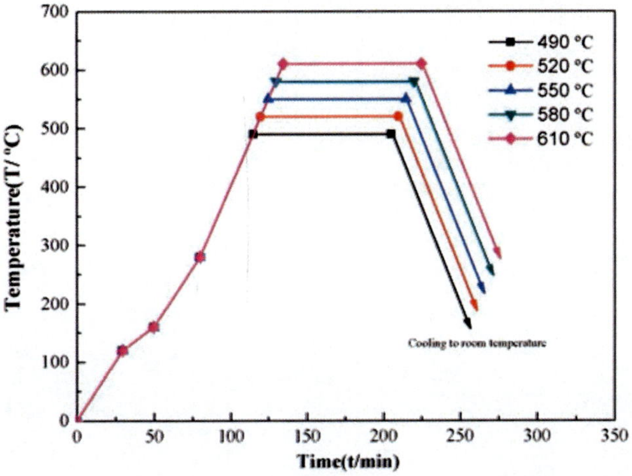

Fig. 1 The heat treatment for samples. (Color figure online)

Table 4 The porosity of materials at different sintering temperatures

Sintering temperature (°C)	490	520	550	580	610
Porosity(%)	49.98	49.21	48.36	48.02	–

sintering temperature at different sintering temperatures, as shown in Fig. 4. After data simulation, it is found that there is roughly a linear function relationship of $y = 60.894–0.022 \times$ between sintering temperature and porosity. We can see from the figure that as the sintering temperature increases, the porosity shows a downward trend. The analysis in this paper believes that it is caused by the closing of microscopic pores and the further shrinkage of macroscopic pores during the sintering process.

The Influence of Sintering Temperature on Mechanical Properties

Perform compression analysis on 3 samples of each group of experiments and take the average value to reduce the error of the experiment and ensure the repeatability of the experiment. The stress-strain curve of the sample is shown in Fig. 5. After calculation, the mechanical properties of the sample are shown in Table 5. It can be seen that the compressive strength, yield strength, and elastic modulus are

Fig. 2 Samples at different sintering temperatures (1#–5# from left to right). (Color figure online)

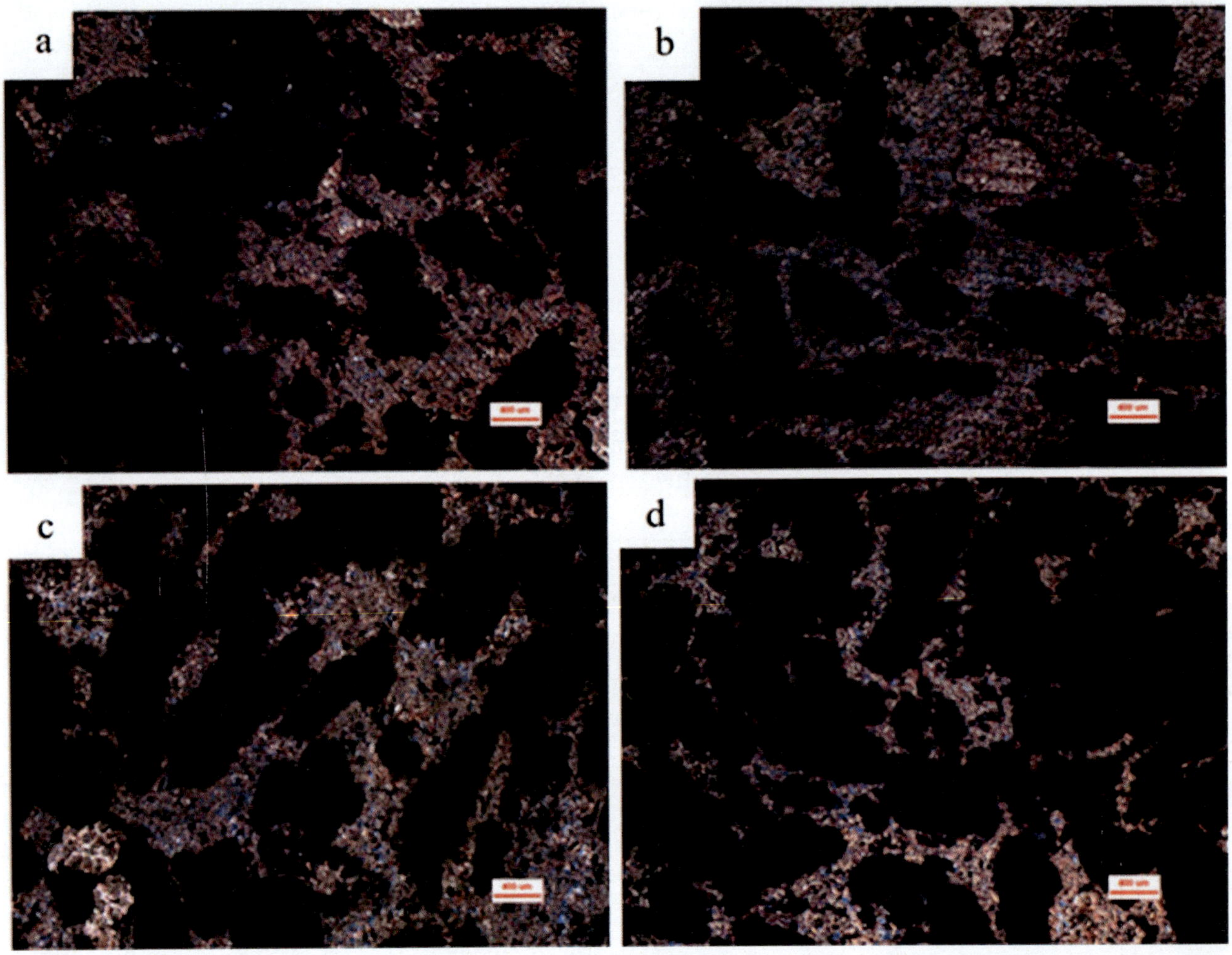

Fig. 3 The metallographic diagram of foamed magnesium alloy at different sintering temperatures. (Color figure online)

between 0.74 MPa and 9.14. MPa, 0.59 MPa–7.31 MPa, 0.07 MPa–0.21 MPa. Both the compressive strength and the yield strength showed a trend of increasing first and then decreasing. At 520 °C, the maximum values are 9.14 MPa and 7.31 MPa, respectively.

The main reason may be that as the sintering temperature increases, the diffusion capacity of atoms becomes stronger, and at the same time, the rate of formation and growth of the sintering neck increases. There will be more metallurgical joints between particles. At the same time, the shape of the hole is also changing, gradually becoming spherical. The mechanical properties of the sintered body are guaranteed by the bonding surface between the particles. The increase in temperature can increase the alloying strength of the sintered body. In a suitable temperature range, increasing the sintering temperature as much as possible can improve the mechanical properties of the material. However, if the temperature is too high, overburning will occur, so it is necessary to comprehensively consider the appropriate sintering temperature (Fig. 6).

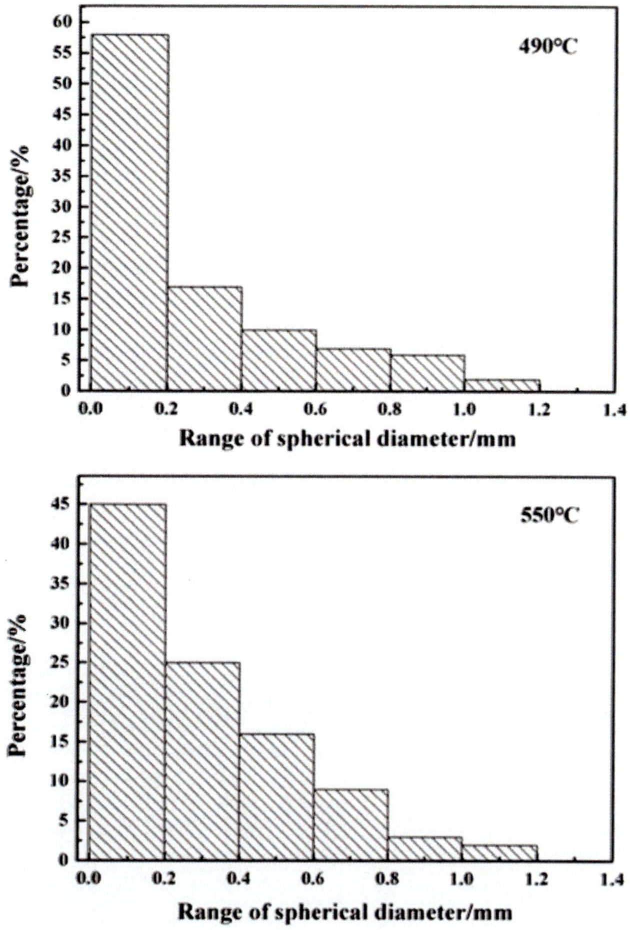

Fig. 4 The pore size distribution

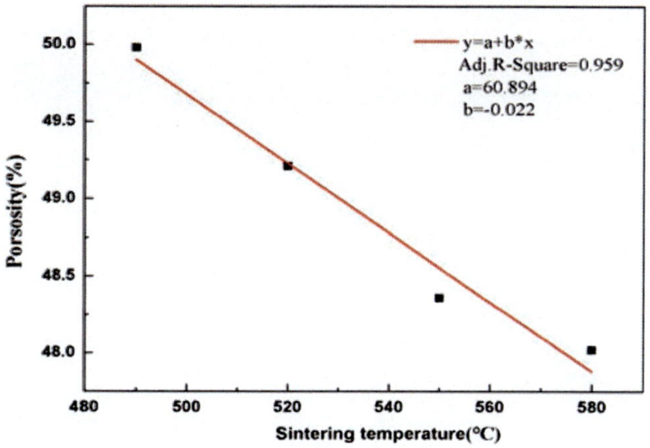

Fig. 5 The porosity of foam magnesium alloy under different sintering temperature. (Color figure online)

Conclusions

(1) With the increase of the sintering temperature, the contact between the particles and the material exchange become stronger, resulting in a decrease in the porosity of the material and a larger pore diameter.

(2) As the sintering temperature increases, the compressive strength, and yield strength both show a trend of increasing first and then decreasing. At 520 °C, the maximum values are 9.14 MPa and 7.31 MPa, respectively.

(3) Considering porosity and pore size, the sintering temperature in this experiment is the most suitable between 520 and 550 °C, which can ensure larger porosity and suitable pore size, and at the same time optimize the mechanical properties of the material.

Table 5 The statistics of porosity and mechanical properties

Sintering temperature (°C)	σ_{bc}/MPa	σ_s/MPa	E/GPa
490	7.90	6.32	0.15
520	9.14	7.31	0.16
550	8.08	6.46	0.21
580	7.58	6.06	0.09
610	0.74	0.59	0.07

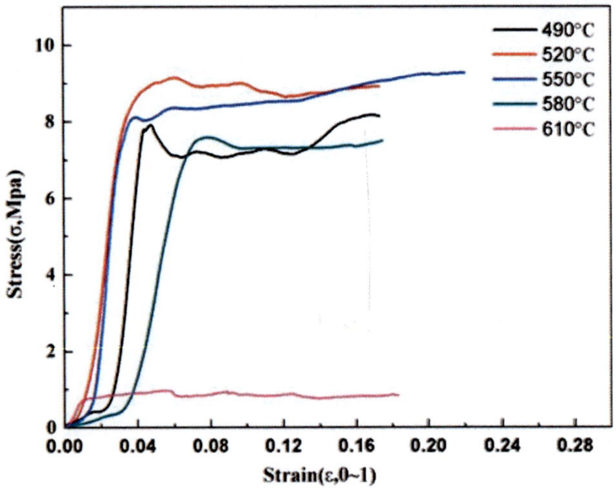

Fig. 6 The stress-strain curves under different sintering temperatures. (Color figure online)

References

1. Wang Yuecun, Zhou Fan, Ge Yanfeng, Zhou Genshu, Jiang Bailing, Shan Zhiwei. Research progress on surface modification and protection of magnesium alloys[J]. Progress in Chinese Materials, 2020, 39(02): 100–112.
2. Liu Lanyun. Research progress of biomedical metal materials[J]. Journal of Zhejiang Vocational College of Industry and Trade, 2008(01):47–50.
3. Yuan Yuanping, Chen Ge, Chen Leping, Zhou Quan. Preparation and application prospects of magnesium foam[J]. Shanghai Aerospace, 2019, 36(02): 78–83.
4. Huang Anbin, Hu Zhiliu, Wen Shikun. Research on the structure and properties of foam aluminum materials and their applications [J]. Metallic Functional Materials, 2010, 17(04): 62–65.
5. Cai Delong, Chen Fei, He Fengmei, Jia Dechang, Kuang Ning, Miao Lei, Qiu Haipeng, Wang Hongsheng, Xu Nianxi, Yang Zhihua, Yu Changqing, Zhang Junwu, Zhang Weiru, Zhou Yanchun. Research progress of high-temperature wave-transparent ceramic materials[J]. Modern Technical Ceramics, 2019, 40(Z1): 4–120.
6. Metal and powder metallurgy[J]. New Material Industry, 2019 (11):93–95.
7. Yang Xuejuan, Liu Ying, Li Meng, Tu Mingjing. Preparation and application of porous metal materials[J]. Materials Guide, 2007 (S1): 380–383.
8. Li Xinfang, Jiang Wufeng, Hao Suju, Zhang Yuzhu. Research on preparation methods and applications of porous metal materials[J]. Southern Metal, 2016(02): 11–15 + 27.
9. Huang Haoyan, Liu Shenghui, Lang Meidong. Research progress of biomaterials for tissue engineering[J]. Journal of Tissue Engineering and Reconstructive Surgery, 2005(03):175–178.
10. Xu Hongxia, Zhang Xiuli. Green environmental protection material in the 21st century-magnesium alloy [J]. Journal of Shanghai University of Engineering Science, 2007(04): 322–325.
11. Zhang Zan, Chen Xiaowei, Xia Xingchuan, etc. Research status of porous magnesium alloys[J]. Special Casting and Nonferrous Alloys, 2015, 35(06):580–585.
12. Guo Quanfen. Preparation and properties of foamed magnesium alloy [D]. Xi'an University of Technology, 2018.
13. Tan Weimin, Shen Xiaodong, Zhang Zhenzhong, Lu Chunhua, Xu Zhongzi. Research progress and prospects of damping properties of magnesium-based materials[J]. Materials Review, 2006(08): 127–130 + 139.

Effects of Hot Isostatic Pressing on the Microstructure and Properties of Mg-Gd-Y-Zn Alloys

Janet M. Meier, Josh Caris, and Alan A. Luo

Abstract

Hot isostatic pressing (HIP) treatment after solution treatment has been investigated on its effects on the mechanical and corrosion properties of Mg-Gd-Y-Zn alloys with high long-period stacking order (LPSO) 14H phase fractions. Plate and cylinder samples of three different compositions were cut from permanent mold castings. The alloy samples were solution treated for 25 h at 500 °C and processed with the following HIP conditions: 485–490 °C, 100–200 MPa, for 2 h. The mechanical properties generally increased with HIP. Under accelerated corrosion conditions (submersion in 3% KCl at a temperature of 90 °C), the corrosion rate was observed to increase after HIP. HIP was found not to change the density of the samples, indicating micro-porosity was not an issue. The change in corrosion rate and mechanical properties were a result of microstructural changes due to the HIP thermal cycle. CALPHAD (CALculation of PHAse Diagrams) modeling of the three compositions as well as scanning electron microscopy (SEM) microstructural observations of the alloys are provided.

Keywords

Magnesium alloys • CALPHAD modeling • Alloy development • Hot isostatic pressing (HIP) • Long period stacking order (LPSO)

This material is based upon work supported by the Army Contracting Command—Adelphi, MD under Contract No W911QX-18-P-0038. Any opinions, findings, and conclusions or recommendations expressed in this material are those of the author(s) and do not necessarily reflect the views of the Army Research Laboratory (ARL).

J. M. Meier (✉) · J. Caris
The Ohio State University, Columbus, OH, USA
e-mail: meier.129@osu.edu

A. A. Luo
Terves LLC, Euclid, OH, USA

Introduction

The effects of rare earth (RE) elements on the properties of Mg alloys have been an exciting source of advancement in the study of magnesium alloys in recent years [1]. Among the unique phases formed in these RE-containing systems are long period stacking order (LPSO) phases in Mg-RE-transition metal (TM) alloys [2, 3]. The stacking of several hexagonal Mg layers between face centered cubic RE and TM enriched layers produces this LPSO structure [4–6]. The resulting stacking faults play an important role in kink banding and the suppression of twinning [5–9].

In our previous work, we explored the ability of two existing CALPHAD software packages to accurately predict the phases present in Mg-Gd-Y-Zn alloys [10]. It was found that the major phases were predicted, but the accuracy of the volume fraction predictions of the LPSO and other secondary phases varied, primarily in the Scheil condition (assumption that diffusion is infinite in the liquid and there is no diffusion in the solid). Overall, the equilibrium predictions could be used to maximize the LPSO phase fraction in the alloys. This work explores the effect of high LPSO phase fraction on mechanical and corrosion properties. A hot isostatic pressing (HIP) treatment was also evaluated for improved mechanical properties, since HIP has been reported to simultaneously reduce porosity and increase mechanical properties in many alloy castings [11].

Experimental Methods

Based on the previous CALPHAD analysis, three alloy compositions were down-selected and cast by Terves, LLC (Euclid, OH). Commercially pure Mg, Zn, Gd, and Y were melted in a 74 lb gas-fired furnace, with a pneumatic shear mixer, under a CO_2-SF_6 cover gas. The cover gas was also applied to the mold riser of a 7" × 7" × 4" steel permanent mold into which the samples were cast. Inductively coupled

V. M. Miller et al. (eds.), *Magnesium Technology 2021*, The Minerals, Metals & Materials Series,
https://doi.org/10.1007/978-3-030-65528-0_20

plasma mass spectroscopy (ICP) and spark optical emission spectroscopy (OES) were used to determine alloys chemistry (Table 1). Tensile bars were machined from the castings and were solutionized for 25 h at 500 °C and air cooled. Tensile tests all followed the ASTM E8 standard.

A plate (0.5" × 4" × 7") and three cylinders (16 mm diameter × 25 mm length) were cut from each of the three castings. The plates and cylinders were solutionized for 25 h at 500 °C and the water weight density was determined for the cylinders. A HIP treatment was performed on the cylinders and plates at 485–490 °C with a pressure of 100–200 MPa for 2 h by American Isostatic Presses, Inc. (Columbus, OH). HIP parameters were selected based on the calculated solidus temperature determined by the CALPHAD method. Thus, the HIP temperature is approximately 50 °C below the solidus of the alloys. Water-weight density measurements were taken on the HIP samples to determine if a density change had occurred. Tensile bars were machined from the plates and tested by the ASTM E8 standard. Accelerated corrosion tests were performed on the as cast, solutionized, and HIP samples. Samples were submerged in 3wt% KCl brine solution at 90 °C for 24 h.

Scanning electron microscopy (SEM) was done on the as cast, solutionized, HIP samples using a FEI Apreo FEG microscope with EDAX Octane Elect energy dispersive spectroscopy (EDS) capabilities. Samples were ground and polished up to 0.05 μm colloidal silica polish. Phase fraction measurement was carried out using the thresholding method in ImageJ and were averaged over 10 images from each sample.

Results and Discussion

Microstructure

Macroscopic porosity was observed in the casting, primarily along the edges of the casting, but was not obvious in the microscopy samples. Microstructural examination revealed that the as cast alloys consisted of the α-Mg matrix and the LPSO 14H phase with a blocky morphology (Fig. 1a, d, and g). As the Y and Zn content increase, as in MC21, the W-phase ($Mg_3Y_2Zn_3$) with eutectic-like morphology becomes more prominent. In the MC14 sample (Fig. 1d) the W-phase has a more block-like morphology while in the MC21 samples (Fig. 1g) it has a eutectic-like morphology due to the larger volume fraction and higher Y content. In both cases, the W-phase is in contact with the blocky LPSO 14H phase and the α-Mg matrix. Phase fraction analysis indicated that the LPSO 14H phase fraction was about the same in the MC07 and MC14 samples. The phase fraction increases significantly in the MC21 sample.

After solution treatment, more of the lamellar LPSO 14H phase morphology is observed in all the alloys (Fig. 1b, e, and h). The W-phase observed in solution treated MC14 (Fig. 1e) samples is mostly surrounded by LPSO 14H. In the MC21 sample (Fig. 1h), the W-phase is disconnected and is mainly either surrounded by blocky LPSO 14H or has the lamellar LPSO 14H extending from it. This change in morphology indicates that the W-phase is dissolving into the matrix at the solution temperature. The increase in lamellar LPSO 14H and the decrease in W-phase fraction is supported by the previous CALPHAD results. The lamellar LPSO phase forms due to the diffusion of the RE and Zn into the matrix [6], [9], [12]–[15]. The W-phase is predicted to be less stable at 500 °C and is currently predicted to dissolve completely. As addressed in our previous work, there are some issues with the current prediction of the W-phase [10]. This is supported by DICTRA simulations, which indicate that equilibrium is reached within 5 to 15 h depending on composition (Fig. 2).

After HIP, the lamellar LPSO 14H is thicker and more prevalent in the interior of the grains (Fig. 1c, f, and i). The W-phase in the MC14 (Fig. 1f) and MC21 (Fig. 1i) samples is smaller than in the solution treated samples and is primarily surrounded by LPSO 14H. This suggests there is a continued dissolution or transformation of the W-phase. Of the three compositions, the MC14 sample showed the least change between the three conditions. In all three conditions, there is a small quantity of a bright, cuboidal phase that is stable at high temperatures and has been identified as yttria.

Density

The average density was taken for each alloy composition in the as cast and HIP conditions. Due to the presence of macro-porosity observed near the edges of the castings, there was some concern that there was a large amount of

Table 1 Composition of the three samples used in this study (determined using ICP and OES)

Sample	Mg (wt%)	Gd (wt%)	Y (wt%)	Zn (wt%)
MC07	82.98	10.4	3.62	3.00
MC14	81.73	10.57	5.25	2.45
MC21	79.95	10.2	4.74	5.1

Table 2 Tensile properties collected using the ASTM 8E standard

	As Cast			Solutionized 500°C 25h			HIP		
Sample	UTS (MPa)	YS (MPa)	e_f (%)	UTS (MPa)	YS (MPa)	e_f (%)	UTS (MPa)	YS (MPa)	e_f (%)
MC07	160	141	1.5	150	130	1.0	173	135	5.0
MC14	168	157	1.5	169	156	1.7	168	158	2.3
MC21	151	107	2.5	150	113	2.0	183	116	2.4

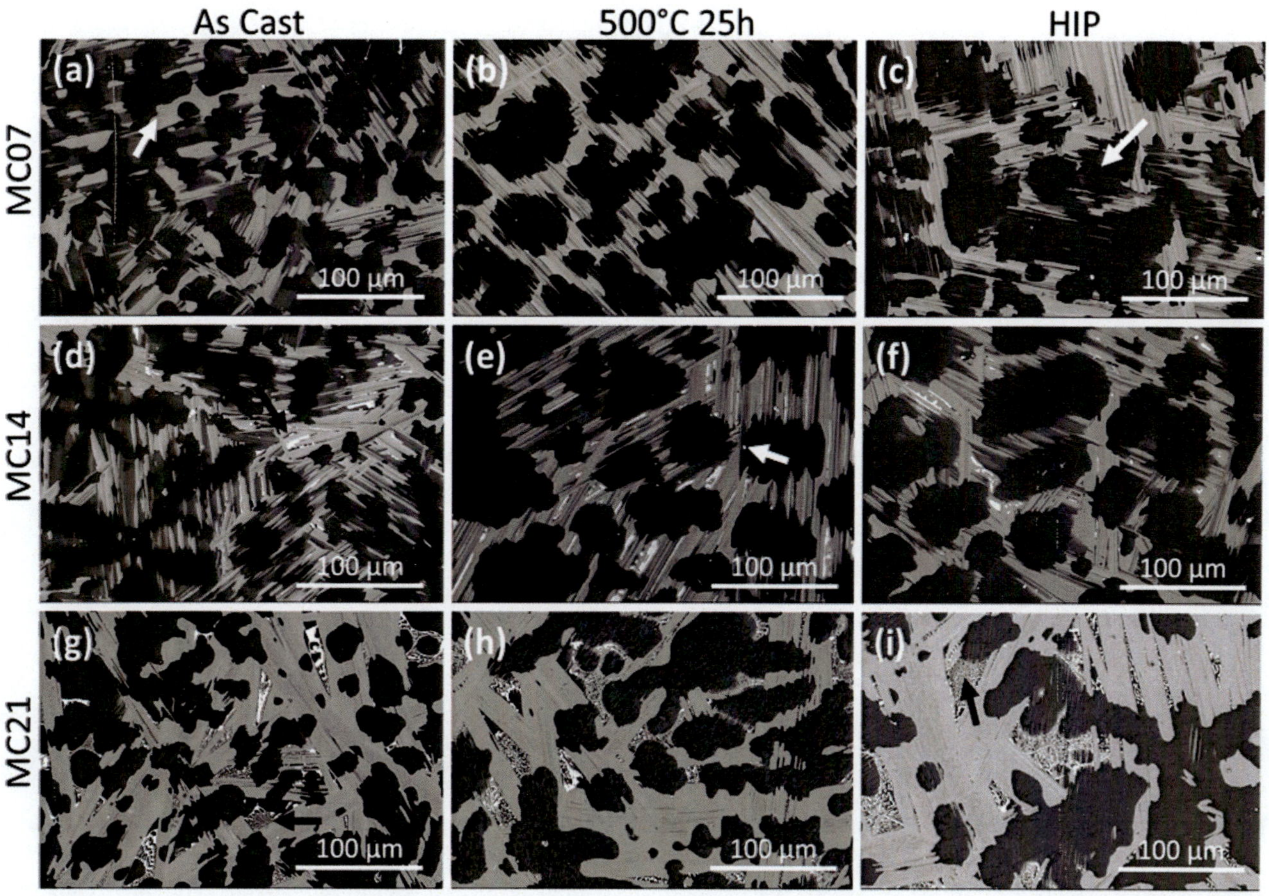

Fig. 1 SEM images of the MC07 (a, b, c), MC14 (d, e, f), and MC21 (g, h, i) compositions in the as cast (a, d, g), solution treated at 500 °C (b, e, h), and HIP conditions (c, f, i). Examples of the following phases are indicated by an arrow: **a** white arrow indicating bulking LPSO 14H, **c** white arrow indicating lamellar LPSO 14H, **d** black arrow indicating bulky W-phase, **e** white arrow indicating cuboidal yttria, **g** black arrow indicating eutectic like W-phase, and **i** black arrow indicating partially dissolved eutectic-like W-phase

micro-porosity in the samples (Fig. 3). Figure 4 shows the theoretical (circle), solution treated at 500 °C for 25 h (triangle, before HIP), and HIP (square, after HIP) sample densities. After HIP the density was shown to change within one standard deviation of the solution treated sample value. This indicates that the difference in density observed between the two conditions may be a result of measurement error. The decrease in density for the MC07 sample after HIP is likely the result of measurement error. As a result, it was concluded that the HIP process did not have much effect on the sample density and that micro-porosity was not an issue in the bulk of the material. It should be noted that the theoretical densities of the MC14 and MC21 samples, calculated based on elemental compositions, are lower than the measured densities. This difference in density may be attributed to the differences in the microstructure. Further work is needed to determine a more accurate theoretical density based on alloy phase constituents.

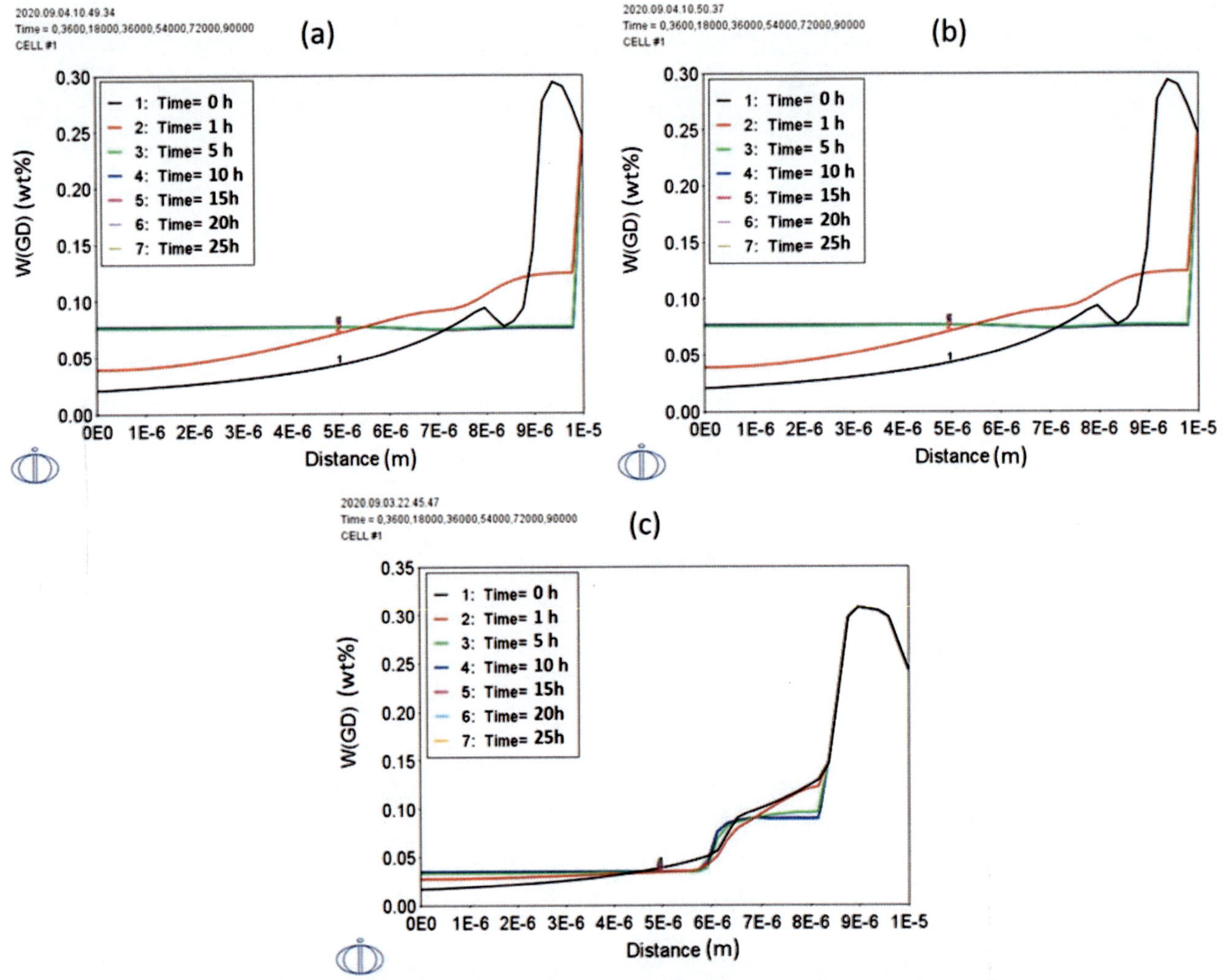

Fig. 2 DICTRA simulation results for the diffusion of Gd in **a** MC07, **b** MC14, and **c** MC21 compositions from the Scheil model segregation to 25 h. Equilibrium is reached when there is no longer a difference in the curve. Each plot represents half of a 20 μm wide dendrite, with the centerline of the dendrite at 0 m and the center of the interdendritic region at 1E-5 m. (Color figure online)

Tensile Properties

Tensile results (Table 2) were averaged over two samples for the as cast condition and three samples for the solutionized and HIP conditions. For the MC07 sample, we see the expected drop in yield and tensile strength commonly associated with solution treatment. After HIP the yield strength remains lower, but there is a 13 MPa increase over the as cast properties. Most notable however is the improvement in ductility from 1.5% in the as cast to 5.0% after HIP. Based on the microstructural analysis, these improvements in properties are attributed to the increase and thickening of the lamellar LPSO phase.

This trend is not observed in the MC14 samples. Both yield stress and ultimate tensile stress were consistent across the three conditions. Elongation improved slightly, but not to the extent seen in the MC07 composition. One possible explanation for the lack of improvement is the presence of the W-phase. The W-phase is a brittle phase that can concentrate stress and lead to failure [16]. Although there was some change in the morphology of the W-phase, it was not significant and the stress concentration could not be compensated for by the increase in the lamellar LPSO 14H phase fraction.

In the MC21 sample there is little change in mechanical properties between the as cast and solution treated conditions. In the HIP conditions, there is improvement in the strength but the elongation remains similar to the other conditions. Based on the microstructural changes, these improvements are attributed to the dissolution of the W-phase and the increase and thickening of the lamellar LPSO 14H. The lack of improvement in elongation in the MC21 sample when compared to the MC07 is likely due to the incomplete dissolution of the W-phase.

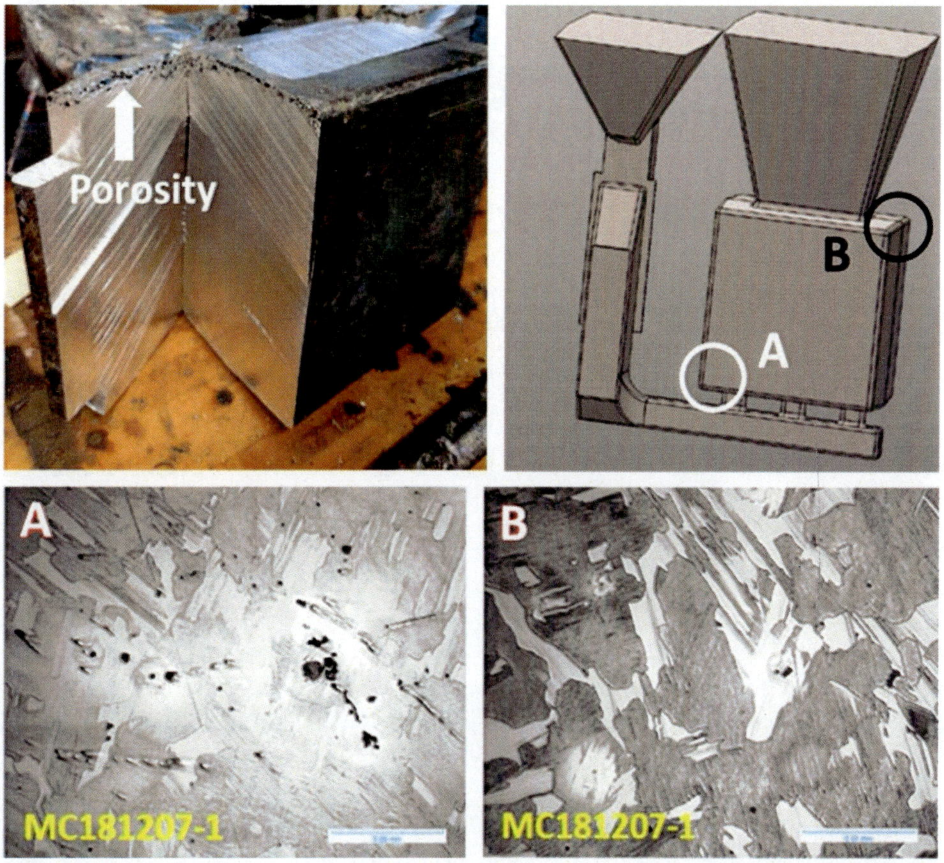

Fig. 3 Examples of porosity in the as cast MC07 sample. Optical micrographs show the porosity form regions A (white) and B (black) of the casting. Macro-porosity is indicated by the white arrow in the picture of the casting. (Color figure online)

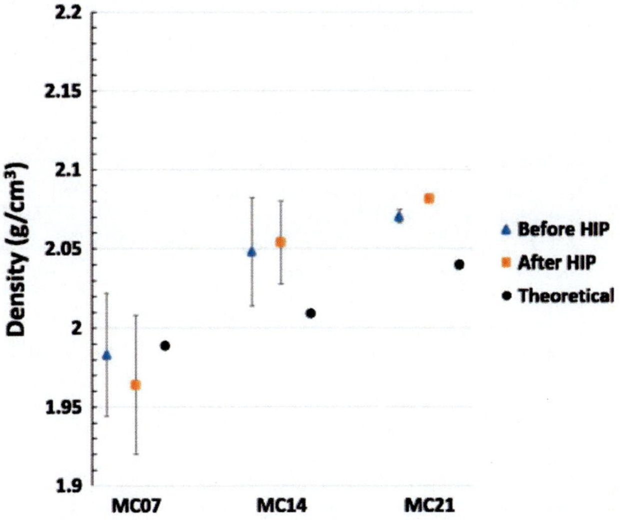

Fig. 4 Water-weight density measurements before HIP (solution treated 500 °C 25 h) and after HIP compared to the calculated theoretical density. Error bars correspond to one standard deviation. (Color figure online)

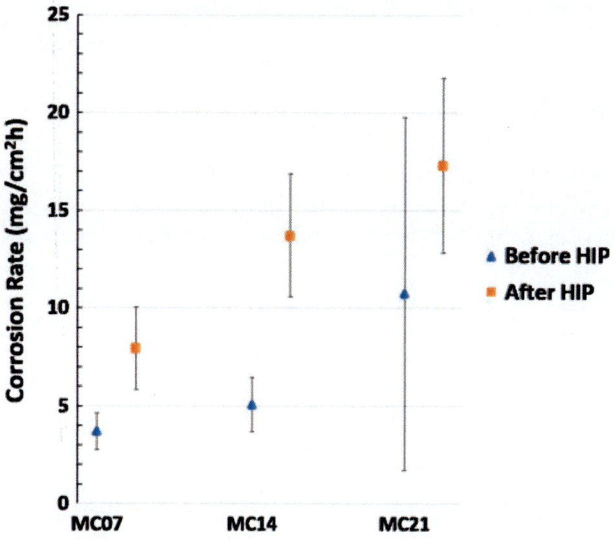

Fig. 5 Accelerated corrosion (3wt% KCl, 90 °C, 24 h) results for each alloy composition before and after HIP. Error bars correspond to one standard deviation. (Color figure online)

Corrosion Properties

Accelerated corrosion tests on these alloys showed that corrosion increased with increasing Zn content in the alloys and with HIP (Fig. 5). Previous work has indicated that the LPSO phases can act as a cathodic phase in the microstructure and thus increased corrosion if it forms a connected network that breaks up the Mg matrix [17]. Solution treatment and HIP resulted in an increase in the lamellar LPSO 14H, which may explain the increase corrosion rate as it continued to break up the Mg matrix and any protective films. Interestingly, the MC14 sample showed the largest increase in corrosion rate despite having the least obvious microstructural change. This indicates there are other factors that are increasing the corrosion rate other than the increase in cathodic reaction sites.

Conclusions

HIP processing results in an increase of lamellar LPSO 14H phase fraction and the dissolution of the W-phase in the as cast and heat treated Mg-Gd-Y-Zn alloys. Although porosity was a concern in these casting due macro-porosity along the casting edges, density measurements indicated that the HIP did not result in a density change since micro-porosity was not an issue in these castings. The microstructural change due to HIP generally lead to the improvement of both strength and elongation in the studied alloys. The exception to this was the MC14 sample, which showed little change in mechanical properties with solution treatment or HIP. It is speculated that this was due to the lack of change in the morphology and phase fraction of the W-phase in this composition. Accelerated corrosion testing showed that HIP increases the corrosion rate of these alloys. This is likely due to the increase in the lamellar LPSO phase which serves as a cathodic reaction site. HIP does offer some mechanical property improvement, but this is at the expense of increased corrosion.

Acknowledgements This work was funded by the Army Research Laboratory (ARL) and Terves LLC. The authors would like to acknowledge Dr. Vincent Hammond with ARL, Dr. William Meier of Oak Ridge National Laboratory, and the Light Metals and Manufacturing Laboratory members at The Ohio State University for their insightful discussions. This material is based upon work supported by the Army Contracting Command—Adelphi, MD under Contract No W911QX-18-P-0038. Any opinions, findings and conclusions or recommendations expressed in this material are those of the author(s) and do not necessarily reflect the views of ARL.

References

1. J. F. Nie, "Precipitation and hardening in magnesium alloys," *Metall. Mater. Trans. A Phys. Metall. Mater. Sci.*, vol. 43, no. 11, pp. 3891–3939, 2012, https://doi.org/10.1007/s11661-012-1217-2.
2. E. Abe, Y. Kawamura, K. Hayashi, and A. Inoue, "Long-period ordered structure in a high-strength nanocrystalline Mg-1 at% Zn-2 at% Y alloy studied by atomic-resolution Z-contrast STEM," *Acta Mater.*, vol. 50, no. 15, pp. 3845–3857, 2002, https://doi.org/10.1016/s1359-6454(02)00191-x.
3. K. Hagihara, N. Yokotani, and Y. Umakoshi, "Plastic deformation behavior of Mg12YZn with 18R long-period stacking ordered structure," *Intermetallics*, vol. 18, no. 2, pp. 267–276, 2010, https://doi.org/10.1016/j.intermet.2009.07.014.
4. Y. Kawamura and M. Yamasaki, "Formation and mechanical properties of Mg 97 Zn 1 RE 2 alloys with long-period stacking ordered structure," *Mater. Trans.*, vol. 48, no. 11, pp. 2986–2992, 2007, https://doi.org/10.2320/matertrans.mer2007142.
5. F. Lu, A. Ma, J. Jiang, D. Yang, and Q. Zhou, "Review on long-period stacking-ordered structures in Mg-Zn-RE alloys," *Rare Met.*, vol. 31, no. 3, pp. 303–310, 2012, https://doi.org/10.1007/s12598-012-0510-y.
6. J. K. Kim, L. Jin, S. Sandlöbes, and D. Raabe, "Diffusional-displacive transformation enables formation of long-period stacking order in magnesium," *Sci. Rep.*, vol. 7, no. 1, pp. 1–8, 2017, https://doi.org/10.1038/s41598-017-04343-y.
7. D. Xu, E. H. Han, and Y. Xu, "Effect of long-period stacking ordered phase on microstructure, mechanical property and corrosion resistance of Mg alloys: A review," *Prog. Nat. Sci. Mater. Int.*, vol. 26, no. 2, pp. 117–128, 2016, https://doi.org/10.1016/j.pnsc.2016.03.006.
8. T. Honma, T. Ohkubo, S. Kamado, and K. Hono, "Effect of Zn additions on the age-hardening of Mg-2.0Gd-1.2Y-0.2Zr alloys," *Acta Mater.*, vol. 55, no. 12, pp. 4137–4150, 2007, https://doi.org/10.1016/j.actamat.2007.02.036.
9. L. Luo, Y. Liu, and M. Duan, "Phase Formation of Mg-Zn-Gd Alloys on the Mg-rich Corner," *Materials (Basel).*, vol. 11, no. 8, p. 1351, 2018, https://doi.org/10.3390/ma11081351.
10. J. M. Meier, J. Caris, and A. A. Luo, "CALPHAD modeling and microstructure investigation of Mg–Gd–Y–Zn alloys," in *Magnesium Technology*, 2020, pp. 61–69, https://doi.org/10.1007/978-3-030-36647-6.
11. A. G. Padalko and V. A. Baklan, "Transformation in metal materials during hot isostatic pressing," *Inorg. Mater.*, vol. 48, no. 13, pp. 1226–1242, 2012, https://doi.org/10.1134/s0020168512130031.
12. S. Zhang, G. Y. Yuan, C. Lu, and W. J. Ding, "The relationship between (Mg,Zn)3RE phase and 14H-LPSO phase in Mg-Gd-Y-Zn-Zr alloys solidified at different cooling rates," *J. Alloys Compd.*, vol. 509, no. 8, pp. 3515–3521, 2011, https://doi.org/10.1016/j.jallcom.2010.12.136.
13. Y. Hu, C. Zhang, T. Zheng, F. Pan, and A. Tang, "Strengthening Effects of Zn Addition on an Ultrahigh Ductility Mg-Gd-Zr Magnesium Alloy," *Materials (Basel).*, vol. 11, no. 10, p. 1942, 2018, https://doi.org/10.3390/ma11101942.
14. J. Wu, Y. L. Chiu, and I. P. Jones, "Microstructure of as-cast Mg-4.2Zn-0.8Y (at.%) alloys containing Gd," *J. Phys. Conf. Ser.*, vol. 522, no. 1, pp. 8–12, 2014, https://doi.org/10.1088/1742-6596/522/1/012033.

15. K. Wang *et al.*, "Enhanced mechanical properties of Mg-Gd-Y-Zn-Mn alloy by tailoring the morphology of long period stacking ordered phase," *Mater. Sci. Eng. A*, vol. 733, no. July, pp. 267–275, 2018, https://doi.org/10.1016/j.msea.2018.07.050.
16. H. S. Jiang, X. G. Qiao, C. Xu, M. Y. Zheng, K. Wu, and S. Kamado, "Ultrahigh strength as-extruded Mg-10.3Zn-6.4Y-0.4Zr-0.5Ca alloy containing W phase," *Mater. Des.*, vol. 108, pp. 391–399, 2016, https://doi.org/10.1016/j.matdes.2016.06.116.
17. J. Zhang, J. Xu, W. Cheng, C. Chen, and J. Kang, "Corrosion behavior of Mg-Zn-Y alloy with long-period stacking ordered structures," J. Mater. Sci. Technol., vol. 28, no. 12, pp. 1157–1162, 2012, doi: 10.1016/S1005-0302(12)60186-8.

Low-Cost Magnesium Primary Production Using Gravity-Driven Multiple Effect Thermal System (G-METS) Distillation

Madison Rutherford, Armaghan Ehsani Telgerafchi, Gabriel Espinosa, Adam Powell, and David Dussault

Abstract

Vapor compression distillation (VCD) in a gravity-driven multiple-effect thermal system (G-METS) distiller can reduce the energy use and cost of magnesium distillation refining by up to 90% versus today's batch distillation processes. This could potentially provide a key unit operation for efficient primary production of magnesium from MgO, by molten salt electrolysis using a reactive cathode, e.g., liquid tin followed by VCD separation. This work presents a techno-economic model of cost, energy consumption, and emissions associated with magnesium primary production by this reactive cathode molten salt electrolysis process with a G-METS distiller. The model includes a mass balance with 17 elements, electrolysis process energy balance with carbon or solid oxide membrane anodes, and detailed operating and capital cost estimates. Based on the properties of magnesium and expected operating conditions, the cost of magnesium production using this process could be comparable to or lower than that of aluminum production.

Keywords

Magnesium • Electrometallurgy • Extraction and processing • Technical cost modeling

Introduction

The Hall–Héroult cell turned aluminum from a precious metal into a low-cost commodity. Since its introduction in the late nineteenth century, its principles have improved the cost and quality of other metals, such as rare earths. An equivalent electrolysis cell using magnesium oxide instead of anhydrous chloride would eliminate chloride drying under HCl or carbochlorination, and chlorine gas handling at high temperature, though it would increase carbon anode consumption and energy use. Unfortunately, magnesium metal density is below that of all molten salts such that it floats and shorts the cell, and it dissolves into some salts leading to electronic current and lower efficiency.

Yerkes proposed an MgO electrolysis cell with a reactive cathode in the 1940s to solve both of these problems [1] and recently Kang et al. developed the concept further [2]. The cell uses a dense liquid metal such as lead, tin, or antimony as the cathode, which absorbs reduced magnesium metal and creates an alloy with high density and low magnesium metal activity. But the process requires distillation to separate the magnesium product from the cathode metal, and distillation has been a slow batch process with high energy consumption of 5–7 kWh/kg Mg product [3].

A new magnesium distillation technology called Gravity-Assisted Multiple Effect Thermal System (G-METS) could dramatically reduce the cost and energy use of magnesium distillation [4], as another paper in this Proceedings describes in detail. This work presents a techno-economic analysis of that new distillation technology in the context of a primary production plant, using capital and operating cost estimate methodologies similar to those of Moudgal et al. [5].

M. Rutherford (✉)
Environmental Engineering Program, Worcester Polytechnic Institute, Worcester, MA, USA
e-mail: marutherford@wpi.edu

A. E. Telgerafchi · G. Espinosa
Department of Mechanical Engineering, Worcester Polytechnic Institute, Worcester, MA, USA

A. Powell
Materials Science and Engineering Program, Worcester Polytechnic Institute, Worcester, MA, USA

D. Dussault
Elemental Beverage Company, Watertown, MA, USA

V. M. Miller et al. (eds.), *Magnesium Technology 2021*, The Minerals, Metals & Materials Series,
https://doi.org/10.1007/978-3-030-65528-0_21

Techno-Economic Analysis Methods

The operational expenditure and capital investment models were developed using a combination of the U.S. Department of Energy ARPA-E METALS Tool techno-economic analysis spreadsheet v1.0 [6] and a novel capital cost model for electrowinning processes developed by Stinn and Allanore [7], respectively. The capital investment cost model correlates the direct capital cost required for the electrolysis plant with production capacity, process temperature, electrode surface area, and power requirements, in the following equations:

$$C = F + E + R \tag{1}$$

$$F = \frac{51010}{1 + e^{-(3.823*10^{-3}*(T-631))}} P^{0.8} \tag{2}$$

$$E = \frac{5634000}{1 + e^{(-7.813*10^{-3}*(T-349))}} * \left(\frac{pzF}{jA\varepsilon M}\right)^{0.9} \tag{3}$$

$$R = 750,000QV^{0.5}N^{0.5} \tag{4}$$

where C is the total capital investment for the process, F is the front-end processing cost, E is the electrolytic and metal recovery process costs, and R is the cost of the rectifiers. The front-end processing cost is broken down in Eq. 2, where P is the production capacity in metric tons per year and T is the process temperature in Celsius. E is broken down in Eq. 3, where T is the temperature again in Celsius, p is the production rate in kilograms per second, z is the moles of electrons reacting per mole of product, F is Faraday's constant, j is the current density in amps/m^2, A is the cathode surface area in m^2, ε is the current efficiency, and M is the product molar mass. The rectifier cost is broken down in Eq. 4, where Q is the total installed power capacity in megawatts, V is the voltage required, and N is the number of rectifier lines.

Within the ARPA-E METALS tool, variable costs included raw materials, utilities, and miscellaneous materials, determined by information pulled from a full material and energy balance developed and added onto the ARPA-E METALS Tool.

The mass balance includes 16 elements total, including those found in common impurities in caustic calcined magnesium oxide, various fluoride salts for the molten salt bath, and tin as the reactive cathode. All streams, including recycle streams, are calculated based on their relationships to each other with the major overall inputs and outputs being determined by a linear equation system. Other parameters, such as product recovered per distillation pass, concentrations of impurities in the magnesium oxide feed, target electrolyte bath composition, and impurities recovered from recycled tin, are adjustable parameters. A switch facilitates quick comparison between carbon and solid oxide membrane (SOM) anodes. This material balance feeds into the main ARPA-E METALS Tool stream composition balances, which verifies perfect closure (within the spreadsheet accuracy) of the mass balance for all 16 elements, and calculates the value flow of each input or output stream based on its compounds. Values for each compound were found from various online vendors and used as possible best and worst-case scenarios.

Voltage and heat requirements were calculated using the added energy balance. The total voltage required was calculated based on the free energy from the dissociation of MgO, the free energy from the formation of CO_2 if a carbon anode is being used, the ohmic resistance of the molten salt bath, the ohmic resistance of a zirconia membrane if a carbon anode is not being used, and electrical energy loss in the electrodes, as follows:

$$V = \sum \frac{\Delta G}{nF} + \sum \frac{J_i L_i}{\sigma_i} + 2\sqrt{L_{el} T \Delta T} \tag{5}$$

where ΔG is free energy of given reaction i (MgO dissociation and CO_2 formation), n the number of electrons transferred in reaction i, F Faraday's constant, J_j, L_j, and σ_j are current density, resistivity, and thickness of component j (YSZ SOM and molten salt), L_{el} the Wiedmann–Franz constant, T absolute temperature, and ΔT temperature difference. Molten salt conductivity is approximate; values are used from a similar salt bath consisting of MgF_2, CaF_2, YF_3, CaO, SiO_2 [8].

Major heat uses for the electrolysis process were calculated as kilowatt hours per kilogram and included the dissociation and formation enthalpy for MgO, and CO_2, if there is a carbon anode, the electrode heat losses, and the energy required to heat MgO. This is summarized in the following equation:

$$Energy = \sum -H_f + 4nF\sqrt{L_{el} T \Delta T} + \int c_p dT \tag{6}$$

where ΔH_f is the enthalpy of formation for MgO and optionally CO_2, and c_p is the specific heat of MgO. Heat losses through the walls and to process gases are unknown but should be similar to those of comparable Hall–Héroult cells.

The energy required for the VCD process was estimated at 1.0 kWh/kg based on the thermal model described in the G-METS distillation paper in this Proceedings and rounded up as a contingency.

The ARPA-E METALS Tool calculated utility costs based on the energy requirement. Data for average electricity costs and emissions per kWh for each state from the Energy Information Administration were added as well, and used to inform assumptions for best and worst case scenario electricity costs.

Other operating cost parameters, such as plant overheads, laboratory costs, and maintenance were calculated through the ARPA-E METALS Tool spreadsheet. Estimates for operating labor were based on an interview with century Aluminum Company about their aluminum plant in Mt. Holly, South Carolina [9]. The production capacity of the aluminum plant was compared with the capacity of the theoretical magnesium plant, and possible operating labor values were calculated proportionally from this to create the reference case.

The list of assumptions for the best, worst, and reference cases are summarized in Table 1.

Cost calculations for raw material feeds other than MgO depend on the MgO feed composition. Assumptions for feed compositions by mass are summarized in Table 2.

Model Results

The proportions of each material flow compared to the product flow are shown in Fig. 1. Thus, for a magnesium plant producing 20,000 metric tons of product per year, about 33,400 metric tons of 97% magnesium oxide feed would be required. The bath feed is completely dependent on the amount of calcium oxide in the magnesium oxide feed and the amount of yttrium oxide in the bath; with the feed consisting of 0.8% calcium oxide, around 680 tons of magnesium fluoride bath would be required with a carbon anode. With a SOM anode, yttrium oxide is added to the bath to stabilize the solid oxide membrane.

Given these flow rates, the raw material cost could be calculated based on each stream's value. Table 3 gives all raw material cost streams for 20,000 t/a magnesium production in the reference case assuming a carbon anode.

The energy requirement for the process is shown in Fig. 2. In order for the electrolysis process to be self-heating at this current density and anode-cathode distance, the cells will need to operate at temperatures under 1050 °C with carbon anodes, or under 1150 °C with a SOM anodes. Heat requirements increase while voltage requirements decrease with increasing temperature. The ohmic resistance for the flux causes the largest voltage drop and changes the most with temperature, while the dissociation enthalpy of magnesium oxide requires the most heat.

Assuming the best case energy cost is $0.04 USD/kWh and the worst case is $0.09 USD/kWh, the annual electricity costs amount from $6.8 million to $17.5 million, respectively, amounting to roughly 14–25% of total annual operating costs. The total raw materials cost ranges from $12.6 million to $25.4 million, or 31–36% of the total annual operating costs (Fig. 3).

	Total annual cost	Cost per kg Mg
Best case	$41,310,629.9	$2.04
Reference case	$48,684,454.3	$2.43
Worst case	$72,249,727.6	$3.69

The total capital cost for electrolysis, including the cost of a facility housing only the electrolysis process and its associated equipment, was calculated to be almost $250,000,000 with 180 kA cells.

Front-end processing	$136,044,186
Electrolytic capital costs	$75,682,532
Rectifier costs	$11,398,818
Total capital cost for electrolysis	**$223,125,536**

Discussion

Electrolysis paired with a VCD system has the potential to cost as little as $2.04USD per kilogram magnesium, making it a viable option for sustainable magnesium production. For comparison, aluminum is currently selling for $1.72USD per kilogram. Taking into account the amount of magnesium needed to have a similar rigidity compared to aluminum, the price ratio for magnesium to aluminum should be no more than 1.8 for magnesium to compete in automotive markets [11], i.e., $3.00/kg. This process meets that criterion in both the best and reference cases.

Table 1 Assumptions informing cost calculations for the best, worst, and reference cases. FTE is full time equivalent, OL operating later, FCI fixed capital investment, VC variable costs, and FC fixed costs

	Best case	Reference case	Worst case
MgO feed cost	$250 per ton	$250 per ton	$500 per ton
Electricity	$0.035 per kWh	$0.035 per kWh	$0.09 per kWh
Operating labor	30 FTEs	30 FTEs	45 FTEs
Other labor	91.5% OL	91.5% OL	91.5% OL
Maintenance	3.75% FCI	7.5% FCI	7.5% FCI
Indirect costs	10% of VC and FC	12.8% of VC and FC	12.8% of VC and FC
Insurance	1.0% FCI	1.0% FCI	1.0% FCI
Royalties	1.0% FCI	1.0% FCI	1.0% FCI

Table 2 Magnesium Oxide Feed compositions for the best, reference, and worst case cost scenarios

	Best case	Reference case	Worst case
MgO	0.98	0.97	0.9544
CaO	0.015	0.008	0.008
Fe_2O_3	0.001	0.0044	0.0068
SiO_2	0.003	0.0132	0.024
Al_2O_3	0.001	0.0044	0.0068

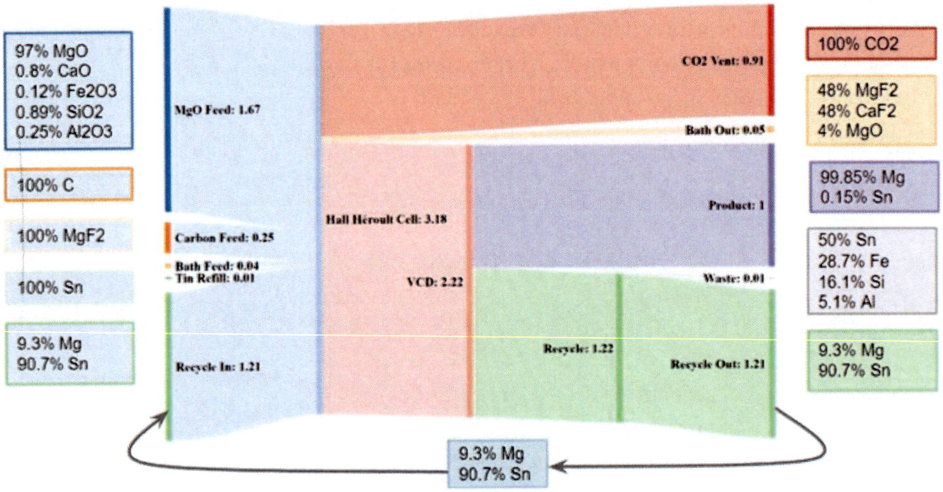

Fig. 1 Material balance flow ratios for the process with a carbon anode. Using SOM anodes would eliminate the carbon anode input, but require a small amount of Y_2O_3 to balance Y activity in YSZ and prevent leaching. (Color figure online)

Table 3 Process input streams, flow rates, and values assuming a 20,000 t/a magnesium plant with prices for magnesium fluoride [10] and tin [LME June 2020]

Stream	Flow rate (kg/hr)	Cost (USD/kg)	Annual cost (USD)
MgO Feed	3,888	$0.25–0.50	$8,514,700–$17,029,400
Carbon anode feed	581	$0.58	$2,949,000
MgF_2 bath feed	78	$1.88	$1,290,600
Tin refill	53	$16.50	$7,728,000
Total	4,600		$20,482,300–$28,997,000

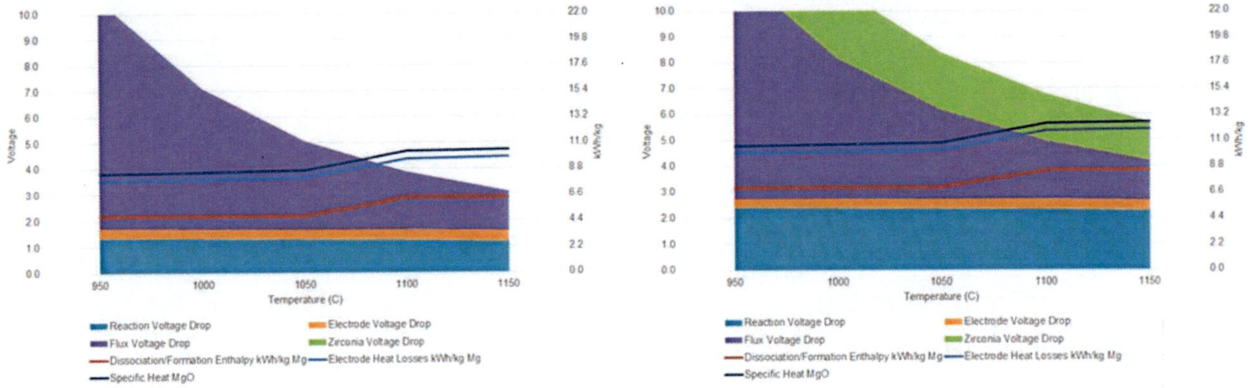

Fig. 2 Energy requirements for the process with a carbon anode (left) and solid oxide membrane anode (right). The stacked areas correspond with voltage requirements while stacked lines correspond with the heat requirements in kWh/kg. The process is self-heating where the top line is below the top area. The sudden increase in heat requirements at 1090 °C shows the effect of the magnesium boiling point. (Color figure online)

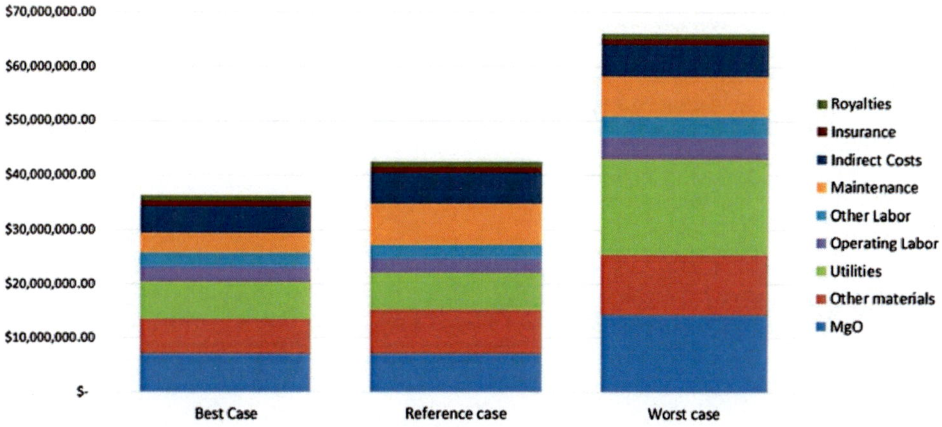

Fig. 3 Three annual operating cost stack scenarios for 20,000 t/a magnesium production. (Color figure online)

This cost can be minimized by reducing the amount of tin lost throughout the recycle process; tin is the most expensive raw material used, and while relatively small amounts are needed, it constitutes 27–39% of the total raw material cost. Overall operating costs can be reduced in the long term with a higher investment in technology to recover as much tin as possible in the recycle stream. Lead could also serve as a cheaper alternative for a liquid cathode in place of tin, but is undesirable due to its related negative health effects. Further material costs could be recovered by separating and recycling the magnesium fluoride from the salt bath.

Replacing the typical carbon anode with a SOM anode would save close to 20 ¢/kg Mg spent on electrodes, but cost an additional 2.2–2.5 kWh/kg plus the anode cost. It would also eliminate direct CO_2 emissions from the process, though life cycle emissions effects will not be clear until there is a good estimate for SOM anode lifetime.

The Yerkes patent on this flow sheet [1] suggested lead rather than tin as the cathode. This would have higher density for better pad stability, and lead costs about 90% less than tin. Health and safety issues would need to be addressed, particularly for very hot metal exiting the distiller.

Comparing with aluminum, the 33% higher productivity by mass of a magnesium cell at the same current (assuming similar current efficiency) should lead to about 25% lower processing cost per kg for electrolysis, including electricity, anodes, and operating labor. In this flow sheet, if the cost of distillation and extra raw materials (tin, MgF_2) can then be kept below that margin, in the range of 25–40¢/kg, then magnesium production cost is lower than that of tin. The low energy use, simplicity, and potential robustness of G-METS distillation could achieve this.

Furthermore, to lower the capital cost, it may be possible to retrofit an existing aluminum production plant to make magnesium. This could reuse the current bus, anode production and loading, raw material distribution, off-gas scrubbing, analysis, and under some circumstances casting systems already in the plant.

Conclusions

Techno-economic analysis of magnesium production using an updated Hall–Héroult process with reactive liquid metal cathode, with a new distillation operation, shows potential for production cost and emissions at or below those of aluminum. Using an updated capital cost model, this indicates potential profitability at a scale as small as 20,000 t/a. The model is sufficiently detailed to include effects of bath chemistry, raw material impurities, anode type, and system energy balance on material consumption and overall cost. The biggest technical risks are likely in G-METS distiller feasibility and scale-up. If this scale-up is successful, then the process could remove a major barrier to more widespread magnesium use in transportation, and potentially new applications such as energy storage.

References

1. L. A. Yerkes, "Electrolytic Method for Producing Magnesium," US 2,342,723, Dec. 02, 1947.
2. J. Kang, J. Lee, and T. Lee, "Metal refining method by using liquid metal cathode," Korean Patent 10-2004902, Jul. 29, 2019.
3. A. Ditze and C. Scharf, *Recycling of Magnesium*. 2008.
4. A. C. Powell, D. M. Dussault, M. R. Earlam, A. Tajima, and C. Raymes, "Method and Apparatus for Efficient Metal Distillation and Related Primary Production Process," U.S. Patent Application 16/944,147, Jul. 31, 2020.
5. A. Moudgal *et al.*, "Finite Element Analysis and Techno-Economic Modeling of Solar Silicon Molten Salt Electrolysis," *JOM*, vol. 73, no. 1, Jan. 2021.
6. D. Matuszak, *ARPA-E METALS Tool*. U.S. Department of Energy, 2013.

7. C. Stinn and A. Allanore, "Estimating the Capital Costs of Electrowinning Processes," *Electrochem. Soc. Interface*, vol. 29, no. 2, pp. 44–49, Jun. 2020, https://doi.org/10.1149/2.f06202if.
8. T. Villalón Jr, "Zero-Direct Emission Silicon Production via Solid Oxide Membrane Electrolysis," Ph.D., Boston University, Boston, MA, USA, 2018.
9. C. Ritter, "Interview regarding operation and economics of Century Aluminum's Goose Creek smelter," Mar. 09, 2020.
10. T. Handy and T. Tack, "LEEDS Business Unit, Magnesium Fluoride Procurement Cost," Sep. 23, 2020.
11. S. Das, "Primary magnesium production costs for automotive applications," *JOM*, vol. 60, no. 11, pp. 63–69, Nov. 2008, https://doi.org/10.1007/s11837-008-0151-7.

Efficient Low-Cost Gravity-Driven Multiple Effect Thermal System (G-METS) Distillation of Magnesium

Armaghan Ehsani Telgerafchi, Gabriel Espinosa, Madison Rutherford, Adam Powell, and David Dussault

Abstract

Vapor compression distillation (VCD) can reduce the energy use and cost of magnesium distillation refining by up to 90% versus today's batch distillation processes. This work describes a new continuous gravity-driven multiple effect thermal system (G-METS) process for magnesium VCD with just one internal moving part. The distiller will likely use less than 1 kWh/kg magnesium product, and high throughput of the continuous process can lower capital cost considerably. A detailed thermal model of the system presented here describes multicomponent evaporation, and batch distiller experiments validate key components of the model. There are multiple alloy distillation challenges with potential solutions described here, including liquid diffusion resistance, aerosol carry-over, and removal of volatile elements such as zinc. This efficient low-cost process could play a key role in multiple new flow sheets, from magnesium alloy recycling to rare earth magnet recycling to primary magnesium production.

Keywords

Magnesium • Recycling and secondary recovery • Modeling and simulation • Distillation

A. E. Telgerafchi · G. Espinosa
Department of Mechanical Engineering, Worcester Polytechnic Institute, Worcester, MA, USA

M. Rutherford
Environmental Engineering Program, Worcester Polytechnic Institute, Worcester, MA, USA

A. Powell (✉)
Materials Science and Engineering Program, Worcester Polytechnic Institute, Worcester, MA, USA
e-mail: acpowell@wpi.edu

D. Dussault
Elemental Beverage Company, Watertown, MA, USA

Introduction

Magnesium (Mg) is a low-density metal whose alloys have outstanding stiffness/weight and excellent strength/weight. It is proposed for use in very lightweight high-efficiency motor vehicles. For example, in 2015 Ford and Magna demonstrated an aluminum-intensive vehicle, which they called "Mach 1", with 25% weight reduction vs. a steel baseline. Their magnesium-intensive "Mach 2" vehicle design achieved 50% lower weight, though in practice would probably be 40–45% lighter than the steel baseline [1]. Two of the barriers to more widespread magnesium use are: low-cost low-emissions primary production method suitable for OECD or similar labor and environmental regulatory framework for supply and cost stability; and a recycling method capable of producing low-iron low-copper low-nickel magnesium from post-consumer scrap.

Magnesium metal distillation can address both of these barriers, enabled by its relatively low boiling point of 1091 °C—indeed, about 15% of the more than 2000 patents in US classification C22B26/22 "Obtaining magnesium" involve distillation. For recycling, it can use post-consumer scrap and produce pure magnesium with copper and nickel concentrations below even the demanding specifications of low-corrosion alloys such as AZ91E [2]. For primary production, distillation enables direct reduction of magnesium oxide (MgO) in a Hall–Héroult type cell with a dense reactive liquid metal cathode such as tin to make an alloy, then separation of pure magnesium from the alloy and returning the host metal to the electrolysis cell [3, 4], as discussed further in a separate paper in these Proceedings. However, the high vapor pressure of magnesium at its melting point makes it challenging to condense as a liquid, particularly when partially diluted by an inert gas [5–7]. In addition, high temperature generally leads to low energy efficiency, with distillers using 5–7 kWh/kg Mg product, cf. 1.5 kWh/kg enthalpy of vaporization.

V. M. Miller et al. (eds.), *Magnesium Technology 2021*, The Minerals, Metals & Materials Series,
https://doi.org/10.1007/978-3-030-65528-0_22

Vapor compression distillation uses the heat released in condensation to provide much of the heat required for evaporation, and can dramatically reduce the energy use and cost of distillation. Pumping magnesium vapor at high mass flow rate is very challenging, but multiple effect thermal system (METS) distillers overcome this by pumping liquid rather than vapor [8–12], and there are entire patent classifications based on this principle (in USA: B01D1/26 "Multiple effect evaporating", B01D1/0088 "Cascade evaporators"). Either way, each condenser operates at higher pressure than an adjacent evaporator, such that the boiling point is higher in the condenser, and heat flows from the hotter condenser to the cooler evaporator.

Here we present a new METS magnesium distiller concept which uses gravity to create this pressure difference with no moving parts, called G-METS. The weight of a column of liquid metal raises the pressure in the evaporating chamber below it, and thus raises the pressure in the condenser connected with the lower evaporating chamber, as shown in Fig. 1. Heat flows upward from the main heater, and each condenser feeds the heat needed for the evaporator above it. Standpipes control liquid level in each evaporator. This is similar to the water distiller of Petrek [10], which uses vertical drain pipes between effects to achieve the desired pressure difference. But in this magnesium distiller, a barrier and tube separate each evaporator from the condenser in the same effect, leading to better thermal efficiency and segregation of volatiles such as zinc. Other features can potentially further enhance zinc separation.

This work first describes the G-METS distiller, and features which overcome specific challenges. We then present:

- Results of steel corrosion tests and one- and two-effect batch distiller experiments;
- A reduced-order thermal model which estimates temperatures, pressures, compositions, and flow rates;
- Solid modeling of an industrial distiller design which illustrates the distiller scale, and structural Finite Element Analysis (FEA) using the solid model estimates maximum stress;
- FEA with fluid flow and heat transfer estimates vapor velocity distribution in the evaporator and condenser.

Distiller Description

Figure 1 shows a schematic of a four-effect magnesium G-METS distiller. For recycling, solid alloy is introduced into the top melter. Melted Mg alloy flows at a controlled rate downward into the top evaporator, where a portion of the Mg alloy evaporates. That portion which evaporates rises to the top condenser, where it condenses. The remaining magnesium which does not evaporate here flows down through the liquid standpipe into the next evaporator, where a portion of it evaporates, and so on. Each evaporator-condenser pair is called an "effect". Heat flows upward from effect to effect, from lower condensers to the evaporators above them. Liquid Mg flows downward between evaporators from effect to effect. Mg vapor flows upward within each effect from its evaporator into its condenser above it. Standpipes control the height of liquid Mg alloy in each evaporator.

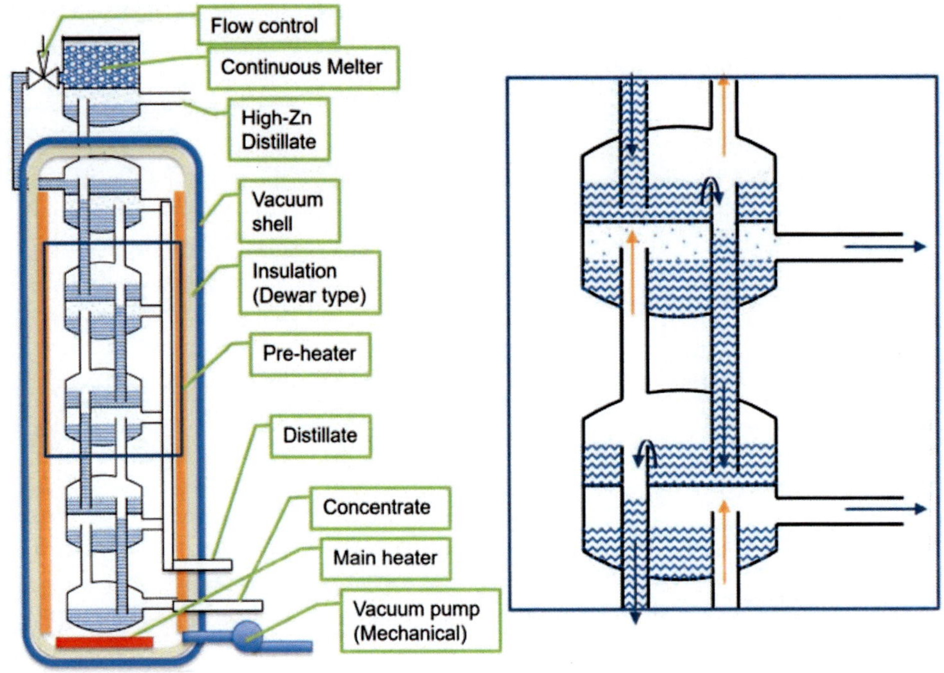

Fig. 1 Schematic diagram of a four-effect magnesium G-METS distiller, with detailed view of a single effect. (Color figure online)

Several issues complicate the operation of a G-METS distiller. First, if the main heater is run at constant power, the distiller is unstable, in that more liquid metal flow downward from evaporator to evaporator leads to lower temperatures, reducing evaporation rate, and increasing liquid metal flow rate further. Using a proportional integral derivative (PID) or similar algorithm to control temperature in the bottom evaporator avoids this instability: faster flow leads to higher heat input and evaporation rate, slowing down concentrate flow. One can further enhance stability by regulating flow of influent metal to control the temperature at the top of the distiller.

Second, in recycling it is necessary to heat and melt the incoming magnesium alloy metal, which uses energy. This distiller can in some cases use the top effect condensation enthalpy as part or all of the energy required to heat and melt incoming magnesium, reducing the overall energy required. Third, incoming magnesium scrap can have zinc or other volatile metals with higher vapor pressures than magnesium, which concentrate in the distillate. In addition to the barrier between evaporator and condenser described above, a counter-flow evaporator further segregates zinc into the top condenser; and a vented counter-flow condenser can separate most zinc out of a condenser [13].

Fourth, magnesium alloy evaporation sometimes leads to aerosol creation and carry-over, bringing low-volatility impurities to the condenser. Flow devices which create high velocity and streamline curvature, such as cyclone separators and even angled vapor flow channels, can separate such aerosols and reduce distillate contamination.

Steel Corrosion and Batch Distiller Experiments

Production of magnesium distillate requires the use of nickel-free steels which neither corrode nor contaminate liquid magnesium. For this reason, a 200-hour corrosion test compared five stainless alloys in an inert environment with liquid magnesium: 310, 316L, 347, 410, and 440C. Figure 2 shows the apparatus consisting of a graphite separator in a mild steel container, a welded steel lid and tube for evacuation are not shown. A mechanical vacuum pump reduced air pressure below 10 Pa, then the tube was crimped to seal it. A kiln heated the apparatus to 1100 °C for 200 h. After cooling, cutting open the apparatus revealed the steel coupons embedded in magnesium.

The 3xx series alloys showed considerable corrosion and porosity, with SEM EDS indicating at least 1 wt% Mg in the centers of each of those coupons. Solid Mg discs surrounding the 3xx alloy coupons showed 11–13 wt% Ni. In contrast, 410 and 440C steel samples showed negligible corrosion with no nickel in the magnesium around them. Based on this, the distiller structure should use nickel-free stainless alloys. Strength of these alloys is low, but operating in an inert gas-filled vessel with equal pressure to that of the bottom effect (10–15 kPa) can minimize hoop stress in the hottest part of the distiller, and supports in the lowest effect can be made of ceramic materials.

23 one- and two-effect batch distillation experiments examined behavior of magnesium and a synthetic scrap alloy. A summary of five two-effect experiments follows. Synthetic

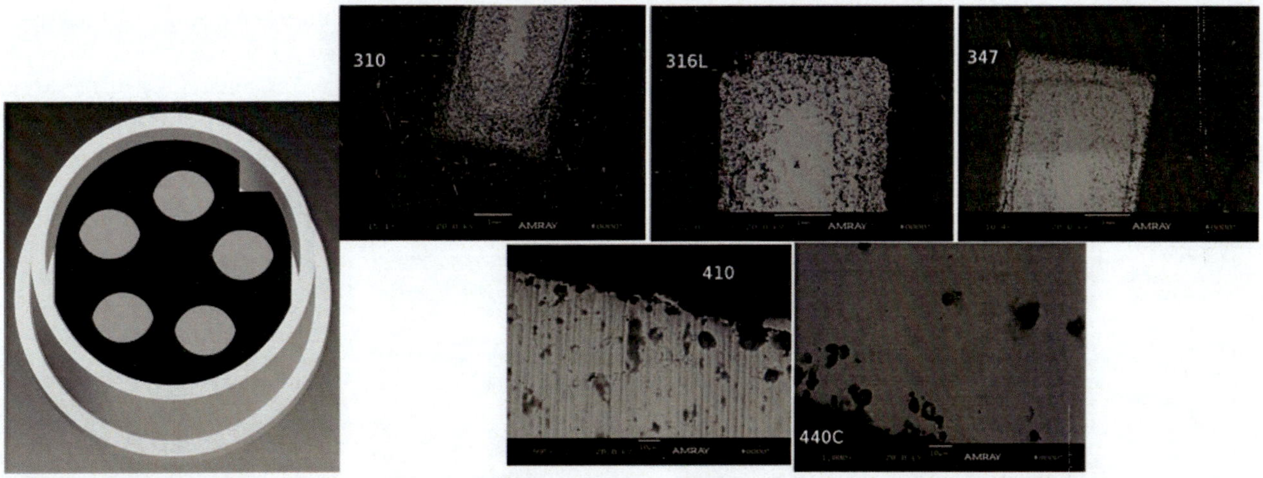

Fig. 2 (Left) mild steel vessel with graphite separator used in corrosion experiments (welded lid not shown); (right) SEM backscatter micrographs of stainless coupon cross section after 200-hr Mg exposure showing alloys 310, 316L, 347, 410 and 440C. Note 1 mm scale bars in 3xx steel samples, and 10 μm scale bars in 4xx samples

scrap consisted of AZ31 with addition of ~4 wt% Fe, 1 wt% Cu, and 1 wt% Ni. Side heaters raised the temperature in the whole apparatus to start, these then switched off and only a bottom heater ran during distillation. Most experiments used welded 6-inch schedule 80 pipe and plate with 316 stainless in the bottom evaporator and 1018 steel in the remainder of the apparatus. Figure 3 shows all five setups cut open after the experiments. Table 1 summarizes the parameters and composition data of the following five experiments.

Experiment I used pure magnesium in an apparatus made of welded 4-inch schedule 80 stainless pipe cap sections. The top and bottom effects each began with 500 g pure magnesium. The side heater remained on throughout the experiment. All of the magnesium in both effects transferred to the condensers, likely because of the side heater.

Experiment II began with 4.0 kg synthetic scrap in the bottom effect and 2.6 kg in the top. The side heater extended from the bottom of the bottom evaporator to about halfway up the top condenser, leaving the top of the top condenser open. As a result, the top of the top condenser was too cold, so it clogged. Bottom heater power was 2.0 kW, and significant non-volatile impurities entered the bottom condenser.

Experiment III also began with 4.0 and 2.6 kg synthetic scrap in the two effects. The side heater heated the entire apparatus to 650 °C, then shut off. Bottom heater power was only 1.0 kW, and low-volatility impurity concentrations were half or less of those in Experiment II. But in 6.6 h only 2 kg Mg distilled in the bottom effect, and the low resulting heat flux did not distill any of the top effect. Our hypothesis was that at this low power and distillation rate, an Mg-depleted layer formed in the evaporator liquid surface, reducing the evaporation rate.

Experiment IV also began with 4.0 and 2.6 kg of synthetic scrap in the two effects. The side heater heated the apparatus to 650 °C, then shut off, after which the bottom ran at 2.0 kW. Our hypothesis was that as with Experiment II, high heat flux led to bubbling in the bottom evaporator, increasing distillation rate while also creating aerosol carry-over into the condenser. Condensation rate in the bottom effect created enough heat to distill 1.3 kg Mg in the top effect. But with the side heater off during distillation, the tall and narrow apparatus lost a considerable amount of heat to radiation from the sides; this agreed well with reduced-order thermal model predictions described next.

Experiment V began with 3.5 and 2.2 kg of synthetic scrap in the two effects. A piece of angle iron was welded over the bottom effect conduit entrance, requiring vapor to turn 90° twice with just ~1 cm maximum turn radius. The side heater heated the apparatus to 750 °C then shut off, after which the bottom heater ran at 1.5 kW. Distillation was slower than in experiments II and IV, but faster than Experiment III, yet Al, Cu, and Ni in distillate were very low. This supports the aerosol formation hypothesis, with centrifugal separation preventing condenser contamination.

Reduced-Order Thermal Model

A reduced-order continuous flow thermal model estimates the temperatures, vapor, and total pressures, heat and mass flow rates, and distillate and concentrate compositions in a multi-effect distiller. It uses the General Solution Model [14] with parameters from Zivkovic for Al–Mg–Zn ternary liquid [15] to estimate vapor pressures. It solves a set of equations simultaneously using a matrix-free Newton–Krylov nonlinear solver.

Fundamentally, distillation rate is limited by heat transfer up through the distiller. This in turn is mostly limited by heat conduction through the steel evaporator support and liquid magnesium in the evaporator. Magnesium vapor flow requires only a very small pressure difference, so within an effect magnesium vapor pressure at the evaporator surface is very close to vapor pressure at the condenser surface—though evaporator temperature is higher due to the lower magnesium activity in liquid solution. This model assumes equal evaporator and condenser vapor pressures.

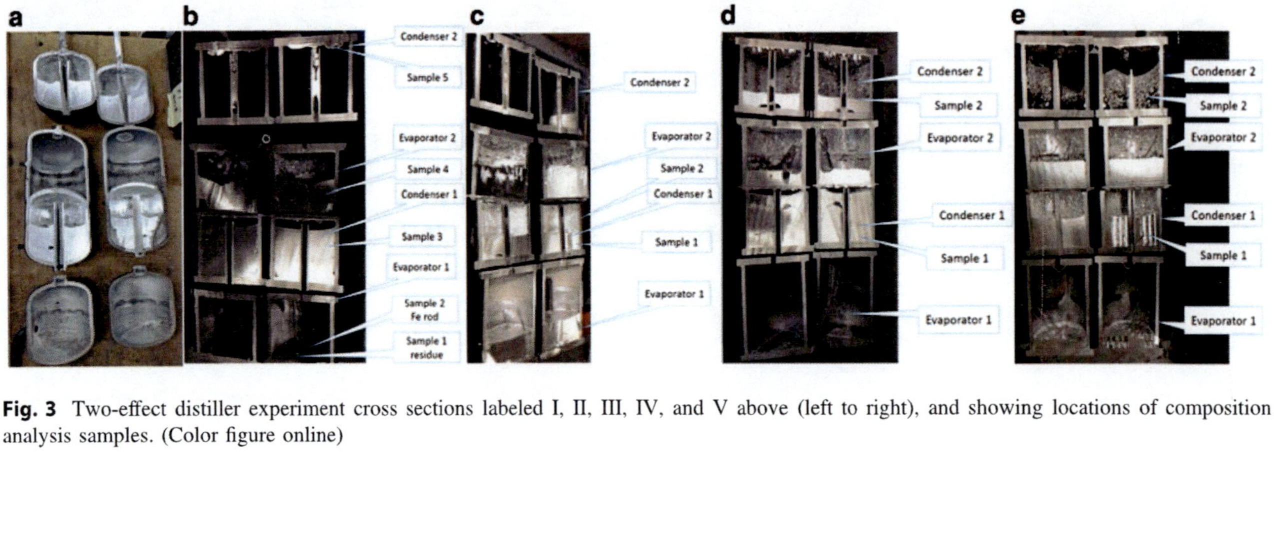

Fig. 3 Two-effect distiller experiment cross sections labeled I, II, III, IV, and V above (left to right), and showing locations of composition analysis samples. (Color figure online)

Table 1 Distillation experiment durations, distillation rates, heater power levels, and compositions

Experiment	I	II*	III	IV	V
Duration at temperature, h	1.5	5.0	6.6	5.0	7.0
Bottom effect rate, kg/h	3	0.8	0.3	0.8	0.5
Top effect rate, kg/h	unk.	0	0	0.25	0
Sample 1 Al wt%		1.3	0.32	0.86	–
Sample 1 Zn wt%		0.8	1.76	1.07	0.012
Sample 1 Cu wt%		0.4	0.13	0.49	–
Sample 1 Ni wt%		0.2	0.12	0.36	–
Sample 2 Al wt%		99	0.01	0.02	
Sample 2 Zn wt%		1	1.33	1.11	
Sample 2 Cu, wt%		–	–	0.05	
Sample 2 Ni, wt%		–	–	0.01	

*Sample 3 and 5 compositions, not 1 and 2. Concentrations below detection limits indicated by "–"

Figure 4 shows a schematic and a result of this thermal model applied to a continuous large laboratory distiller with five effects and total influent flow rate of 2.7 kg/h. Influent composition is Mg-5 wt% Al-1 wt% Zn. Temperatures at the bottom evaporator and top melter are set to 950 °C and 800 °C, respectively. Final concentrate flow rate is set to 10% of influent rate; distillate flow rates in each effect are determined by heat flow balances. Maximum pressure difference between effects is 4 kPa, corresponding to 28 cm of liquid Mg height in a standpipe. Evaporation rate in the higher effects is lower than the lower ones because radiation loss out the sides reduces heat flux up the distiller.

Figure 5 shows results of the same model for 2–7 effect pilot-scale distillers with 1.14 m inner diameter and a target total Mg distillate production rate of 100 kg/hour (500–700 t/a). Bottom evaporator temperature is again 950 °C, and melter temperature is 750 °C. Evaporators are 10 cm high with 2.5 cm liquid Mg depth, and condensers are 17.5 cm high with 0–15 mg liquid Mg depth. Each condenser holds about 230 kg Mg, and must be tapped every 2.5–11 h depending on the number of effects (more effects have lower production rate per effect and need less frequent tapping). When using counter-flow evaporators [13], only the top effect distillate has higher Zn than the influent, production rate of this high-Zn distillate is shown in orange on the stacked graph on the left and the remainder with low Zn is shown in blue. The graph on the right shows that required bottom heater power to achieve these flow rates (shown in

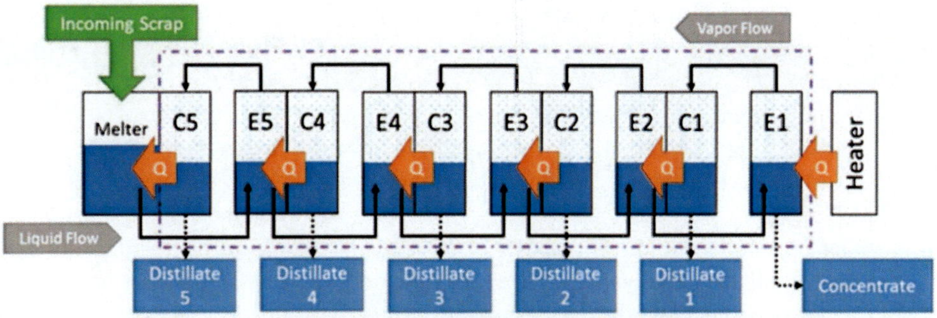

	Effect 5		Effect 4		Effect 3		Effect 2		Effect 1	
	C5	E5	C4	E4	C3	E3	C2	E2	C1	E1
Temperature (°C)	810	814	825	831	843	851	864	877	892	950
Pressure (psia)	0.68		0.83		1.00		1.30		1.90	

	Distillate 5	Distillate 4	Distillate 3	Distillate 2	Distillate 1	Concentrate
Total (kg/hr)	0.40	0.44	0.49	0.54	0.59	0.27
Magnesium (%)	97.8%	98.3%	98.9%	99.3%	99.7%	50.0%
Zinc (%)	2.2%	1.7%	1.1%	0.7%	0.3%	0.0%
Aluminum (%)	0%	0%	0%	0%	0%	50%

Fig. 4 Five-effect METS schematic, and output of reduced-order thermal model of a 2.4 kg/h large laboratory G-METS magnesium distiller showing distillate flow rates and compositions starting from Mg-5%Al-1%Zn. (Color figure online)

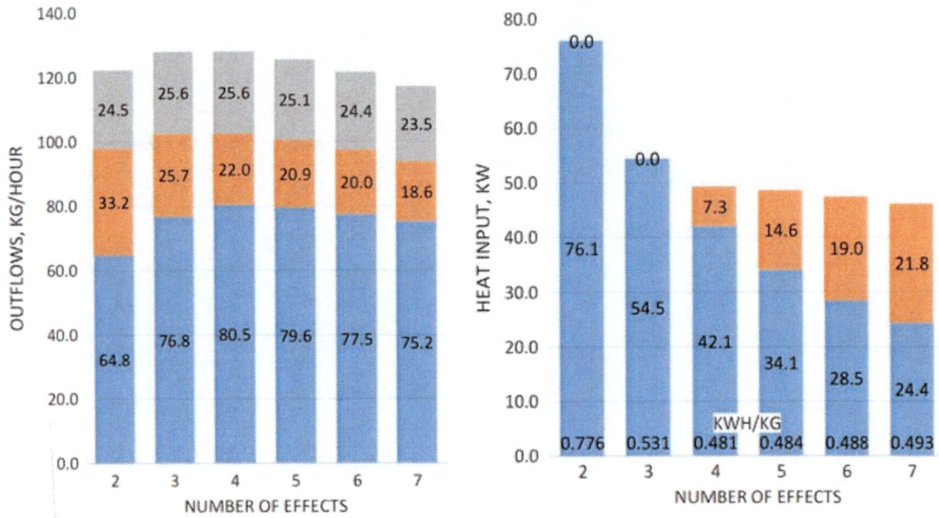

Fig. 5 Reduced-order model estimates of ~100 kg/h pilot distiller outflows (left) consisting of, from the top, concentrate, high-Zn distillate, and <1% Zn distillate; and heat input (right) showing, from the top, melter heater power, bottom heater power, and total specific energy use in kWh per kg of magnesium distillate product. (Color figure online)

blue) falls with the number of effects, as is expected, but for solid Mg alloy scrap input and more than three effects, heating and melting Mg requires additional heat input to the melter (shown in orange).

Based on the model output, the three-effect distiller was selected for scale-up design and further study, as it is simpler than the larger distillers but with little additional energy input and high-Zn distillate.

First-Cut Industrial Distiller Solid Model and Structural Finite Element Analysis

Figure 6 shows a first-cut scaled G-METS magnesium distiller design with three effects and ~1000 kg/h (5000–7000 t/a) pure Mg production rate. Internal diameter is 3.6 m, and the cross-section area is 10 m^2; this is ten times the area of the 100 kg distiller described above. In the cross section, liquid Mg is shown in dark gray, with a thin 2.5 cm layer in each evaporator and up to 15 cm of distillate storage (~2300 kg) in each condenser. There is a vessel at the bottom where concentrate accumulates. Tapping can be done by melting frozen plugs in the drain pipes.

The outer shell maintains inert gas pressure roughly equal to Mg vapor pressure in the lowest evaporator in order to minimize hoop stresses in that hottest part of the distiller. Both 410 and 440 nickel-free stainless alloys were found to exhibit negligible corrosion in experiments which exposed them to Mg liquid and vapor for 500 h, but both are relatively weak at high temperature. The structural Finite Element Analysis shown in Fig. 7 with gravitational and pressure forcing shows stress in the bottom effect supports around 5–8 MPa, with maximum stress in a few spots up over 27 MPa. It may be necessary to make some of these out of a non-reactive ceramic material with high compressive strength at 950 °C such as AlN, MgO, or CaO. Higher effects will have additional inward hoop stresses due to lower internal pressure, but those regions will be at a lower temperature such that steel will be stronger.

Flow and Heat Transfer Finite Element Analysis

In this study of a G-METS magnesium distiller, FEA of flow and heat transfer has two major goals: model heat conduction in steel and magnesium which limits the performance of the distiller, and model coupled evaporation and flow in gas phases for later use to design counter-flow evaporator and condensers.

Fig. 6 1000 kg/hour G-METS magnesium distiller design cross section (left) and approximate scale (right). (Color figure online)

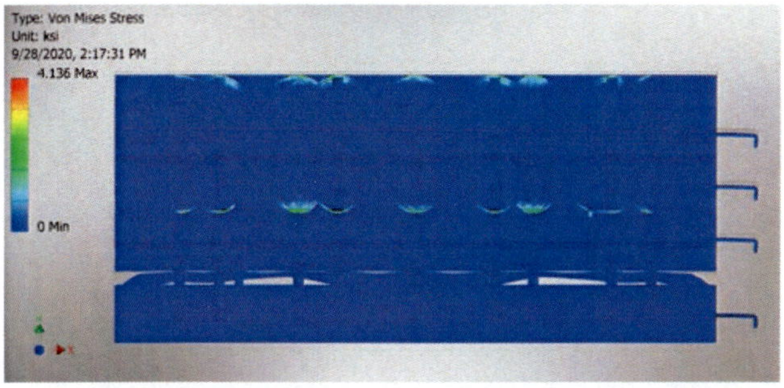

Fig. 7 Structure FEA results showing stresses ~2 ksi = 13.8 MPa in lower supports, and 28 MPa max stress. (Color figure online)

This model uses Elmer Version 8.4 with the Heat Model and Navier-Stokes Model for gas phase fluid flow. The geometry is based on the 1.14 m inner diameter ~100 kg/hour 3-effect distiller and meshes were made with Salome Program version 7.7.1 and the NETGEN 1D-2D-3D algorithm. The mesh has 92.480 nodes with element size from 0.001 to 0.05 m. In this geometry, each effect includes steel as a body of the bottom effect, liquid magnesium in the evaporator, steel as body between evaporator and condenser, and a condenser boundary condition on top of the top of effect. Vapor pressure is calculated from temperature using the Clausius–Clapeyron equation with liquid Mg constants [16] and the heat flux boundary condition at the liquid-gas interface of liquid and gas magnesium includes evaporation enthalpy. The model uses temperature-dependent thermal conductivities [17], viscosities [18, 19] and densities of magnesium in the liquid and vapor phases.

Figure 8 shows the distiller model geometry, velocity magnitude throughout the distiller, and velocity and relative pressure along the vertical axis of the vapor conduits. Maximum velocity is about 8 m/s, leading to a Reynolds number in the conduit around 250, so these simulations ran without a turbulence model.

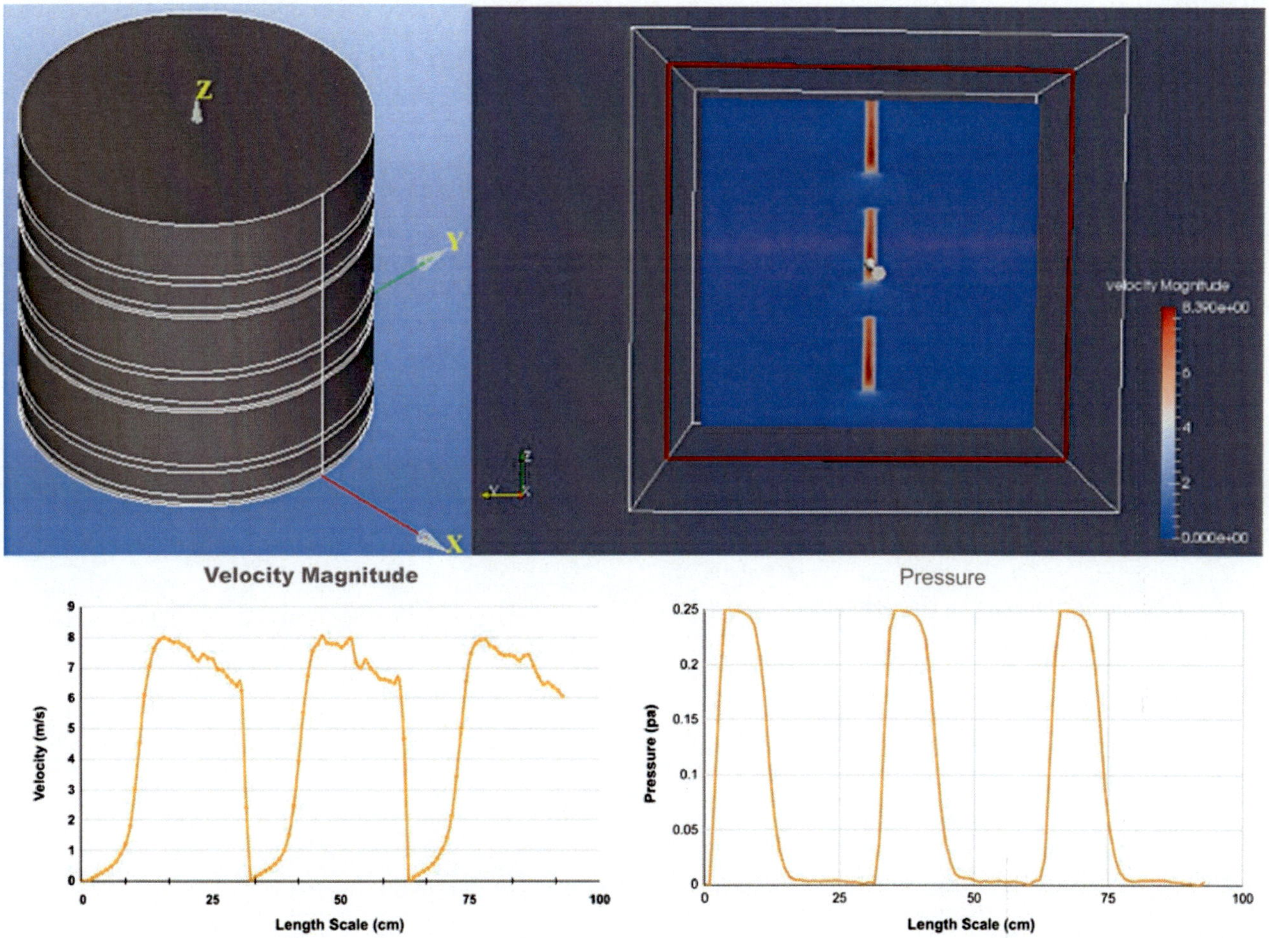

Fig. 8 FEA model geometry, cross-section velocity magnitude, and center axis velocity and pressure profiles. (Color figure online)

Discussion

The biggest potential issue with the G-METS distiller is the kinetics of alloy evaporation. Experimental work showed that although pure Mg distillation rate very closely followed the reduced-order thermal model predictions, when an alloy was used, the rate was either much slower than the pure Mg model predicted with low heat flux, or with sufficient heat flux, matched the pure Mg distiller—but with product contamination by less volatile impurities such as Al. The current hypothesis for this behavior is formation of a surface layer depleted of magnesium which impedes evaporation at low heat flux, and disruption of that layer by bubbling with aerosol creation at higher heat flux. Though early experiments have supported this hypothesis as described above, characterization of this phenomenon requires further study, as does its remedy by cyclone or other flow curvature separation.

Conclusions

This work presented results from batch magnesium distillation experiments, a validated reduced-order thermal model applied to a pilot-scale 100 kg/h distiller, a 1000 kg/h distiller design study with an estimation of stress due to weight of the distiller and pressure difference, a thermal-fluid FEA model of vapor flow, and techno-economic analysis. These show that a G-METS magnesium distiller can potentially reduce the energy use from 5 to 7 kWh/kg today to as little as 0.5–1 kWh/kg, a nearly 90% reduction. The distiller is continuous and compact, and can potentially enable new recycling processes for low-grade post-consumer magnesium alloy scrap. Further study is needed to clarify the kinetics of alloy evaporation, to validate the reduced-order and FEA thermal-fluid models, to identify potential limiting mechanisms in scale-up, and to engineer key subsystems in the distiller.

If it succeeds, this distiller could enable electrolytic production of magnesium using a Hall–Héroult type cell with reactive dense liquid metal cathode e.g., tin, as first envisioned in the 1940s. The lower valence of Mg than Al leads to higher productivity in a plant with the same potline current, potentially producing Mg at or below the cost of Al.

References

1. D. Wagner, J. L. Conklin, T. Skszek, and M. Zaluzec, *The Multi-Material Lightweight Vehicle (MMLV) Project*. Warrendale, Pennsylvania, USA: SAE International, 2015.
2. A. Ditze and C. Scharf, *Recycling of Magnesium*. 2008.
3. L. A. Yerkes, "Electrolytic Method for Producing Magnesium," US 2,342,723, Dec. 02, 1947.
4. J. Kang, J. Lee, and T. Lee, "Metal refining method by using liquid metal cathode," KR 10-2004902, Jul. 29, 2019.
5. W. Schmidt, "Condensation of Magnesium Vapors," US 3,505,063, Apr. 07, 1970.
6. A. F. S. Schoukens, T. R. Curr, M. Abdellatif, and R. M. Whyte, "Metal Vapour Condensation and Liquid Metal Withdrawal," US 7,641,711, Jan. 05, 2010.
7. A. C. Powell *et al.*, "Apparatus and Method for Condensing Metal Vapor," US 8,617,457, Dec. 31, 2013.
8. B. Thoens and S. S. Shears, "Multiple Still," US 1,069,829, Aug. 12, 1913.
9. C. W. Lotz, "Multiple Effect Distillation Apparatus," US 3,536,591, Oct. 27, 1970.
10. J. P. Petrek, "Multiple-Effect Evaporation System and Process," US 5,094,721, Mar. 10, 1992.
11. H. H. Sephton, "Desalination of Seawater by Evaporation in a Multi-Stack Array of Vertical Tube Bundles, with Waste Heat," US 5,853,549, Dec. 29, 1998.
12. M. Schrems, "Method and Device for Treating Liquids by Partial Evaporation," US 6,080,273, Jun. 27, 2000.
13. A. C. Powell, D. M. Dussault, M. R. Earlam, A. Tajima, and C. Raymes, "Method and Apparatus for Efficient Metal Distillation and Related Primary Production Process," U.S. Patent Application 16/944,147, Jul. 31, 2020.
14. K.-C. Chou, "A general solution model for predicting ternary thermodynamic properties," *Calphad*, vol. 19, no. 3, pp. 315–325, Sep. 1995, https://doi.org/10.1016/0364-5916(95)00029-e.
15. D. Zivkovic, Y. Du, L. Balanovic, D. Manasijevic, D. Minic, and N. Talijan, "Prediction of the Thermodynamic Properties for Liquid Al-Mg-Zn Alloys," *Materials and Technology*, vol. 46, no. 5, pp. 477–482, 2012.
16. W. F. Gale and T. C. Totemeier, *Smithells Metals Reference Book*, Eight Edition. Elsevier, ASM International, 2004.
17. C. Y. Ho, R. W. Powell, and P. E. Liley, *Thermal conductivity of selected materials (Part 2)*. National Bureau of Standards, 1968.
18. S. Chapman and T. G. Cowling, *The Mathematical Theory of Non-Uniform Gases*, 2nd Edition. 1953.
19. M. Guanbao, A. V. Okhapkin, N. Yu. Konstantinova, A. A. Sabirzyanov, P. S. Popel', and L. Pytsze, "Kinematic viscosity of Mg-Al melts in the magnesium-rich portion," *Russ. Metall.*, vol. 2013, no. 2, pp. 90–95, Feb. 2013, https://doi.org/10.1134/s0036029513020079.

Industrial Practice of Extracting Magnesium from Serpentine

Huimin Lu and Neale R. Neelameggham

Abstract

There are 140 million tons of nickel-containing serpentine mines in Tuquan County, Inner Mongolia, China. There is no industrial practice in the world to extract magnesium metal from magnesium silicate minerals. Laboratory research shows that it is difficult to extract magnesium metal from serpentine. Based on our laboratory research, an industrial demonstration production line with an annual processing capacity of 21,000 tons of ore has been constructed. Especially the continuous vacuum metal magnesium reduction furnace developed by ourselves is stable and reliable, which proves its feasibility. From ore crushing, grinding, ball pressing, reduction, to metal magnesium refining and casting ingots, the entire production line is fully continuous and automated. The production process has no carbon dioxide emission, clean production, high production efficiency, and low cost, which lays the foundation for large-scale industrialization of extracting magnesium metal from serpentine.

Keywords

Industrial demonstration production line • Aluminum-silicon thermal reduction • Magnesium • Nickel-containing serpentine • Nickel iron alloy

H. Lu (✉)
Beijing Oufei Jintai Technology Co. Ltd, Miyun Economic Development Zone, No. 12 Xingyuan Street, Beijing, 101500, China
e-mail: lhm0862002@aliyun.com

N. R. Neelameggham
IND LLC, 9859 Dream Circle, South Jordan, UT 84095, USA
e-mail: neelameggham@gmail.com

Introduction

Magnesium is the lightest metal of the commonly used metals and can be applied widely. For example, there are metallurgical applications, chemical applications, structural applications, and automobile parts production [1, 2]. In 2016 the global production of magnesium reached 900 thousand tons. The production of magnesium is based on the Pidgeon process and electrolytic process, and more than 90% production of magnesium was produced by the Pidgeon process from China, and all magnesium producers use dolomite as magnesium reduction feedstock.

Magnesium production in the world is currently dominated by the Pidgeon process which uses silicon, in the form of ferrosilicon, to reduce magnesia from calcined dolomite under vacuum. The overall reaction of the process can be written as follows [3]:

$$2MgO(s) + 2CaO(s) + (xFe)Si(s) = 2Mg(g) + Ca_2SiO_4(s) + xFe(s) \quad (1)$$

The reaction is performed in batch mode within steel retorts that operate around 1200 °C and a vacuum of 10–20 Pa to produce approximately 20 kg of Mg over an eight to ten hour period. The process suffers from high energy usage and low productivity.

In Tuquan County, Inner Mongolia, China, magnesium resources are not the traditional dolomite but serpentine, huge reserves (about 1.4 billion tons), extraction of magnesium with the type of serpentine ore, there is no precedent all over the world, it is a great challenge. In 2017, we have completed the research of extracting magnesium metal from serpentine in a laboratory vertical continuous vacuum induction furnace.

In this paper, based on our laboratory research, an industrial demonstration production line with an annual processing capacity of 21,000 tons of ore has been constructed. Especially the continuous vacuum metal magnesium reduction furnace developed by ourselves is stable and reliable,

V. M. Miller et al. (eds.), *Magnesium Technology 2021*, The Minerals, Metals & Materials Series,
https://doi.org/10.1007/978-3-030-65528-0_23

which proves its feasibility. From ore crushing, grinding, ball pressing, reduction, to metal magnesium refining and casting ingots, the entire production line is fully continuous and automated. The production process has no carbon dioxide emission, clean production, high production efficiency, and low cost, which lays the foundation for large-scale industrialization of extracting magnesium metal from serpentine.

Serpentine Reduction Principle and Experimental Procedure

Reduction Principle

Magnesium production in the laboratory is carried on which uses aluminum-silicon alloy as a reduction agent, to reduce magnesium silicate from serpentine under vacuum. In the Al–Si alloy reduction process of magnesium, due to the presence of calcium oxide, resulting in a more stable 12 $CaO{\cdot}7\,Al_2O_3$, making the theoretical minimum temperature of the reaction reduced by 249.2 °C, showing the presence of calcium oxide makes the reaction easier. The overall reaction of the process can be written as follows:

$$\begin{aligned} &54CaO(s) + 14[2MgO \cdot SiO_2(s)] + 7[MgO \cdot SiO_2(s)] + 14Al(s) + 7Si \\ &\quad = 12CaO \cdot 7Al_2O_3(s) + 14[2CaO \cdot SiO_2(s)] \\ &\quad + 35[Mg](g) + 14[CaO(s) \cdot SiO_2(s)] \end{aligned} \quad (2)$$

Industrial Test Equipment

Industrial experiments for producing magnesium were conducted in a 2400 kW serpentine continuous reduction of magnesium test furnace, as pictured in Fig. 1. Process conditions were as follows: the reaction is performed in continuous feeding charge 3000 kg each within magnesium test furnace that operate around 1200 °C and a vacuum of 10–20 Pa over a four-hour period.

1-feeding device, 2-material isolation valve, 3-first vacuum valve, 4-temperature control meter, 5-control system, 6-heating power supply, 7-fifth vacuum valve, 8-sixth vacuum valve, 9-the second vacuum unit, 10-slag discharge

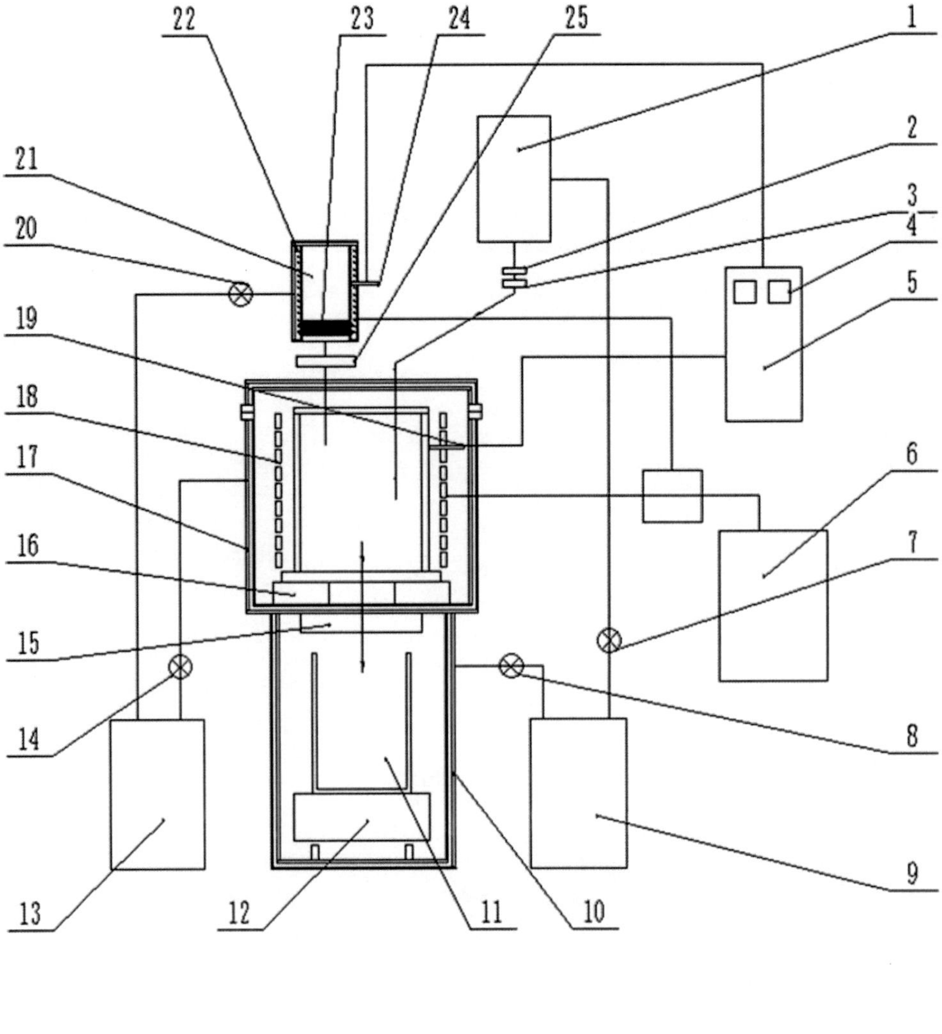

Fig. 1 2400 kW industrial continuous reduction of magnesium test furnace

chamber, 11-slag collecting tank, 12-mobile truck, 13-first vacuum unit, 14-fourth vacuum valve, 15-second flap isolation valve, 16-first turn plate isolation valve, 17-reduction chamber, 18-electromagnetic induction heating device, 19-reduction temperature measuring device, 20-third vacuum valve, 21-mold, 22-copper tube, 23-graphite filter, 24-crystal test temperature device, 25-second vacuum valve.

Raw Materials

Serpentine. Serpentine ore powder was taken from Inner Mongolia Xintai Construction and Installation Group Co., Ltd, the size was smaller than 0.5 mm. The main minerals in the serpentine ore are 96 mass% serpentine, and with a small amount of olivine, chrome spinel, and metal minerals including magnetite, hematite, chromite, nickel pyrite, and chalcopyrite. The assay of the serpentine ore is shown in Table 1 (Fig. 2).

The main mineralogical characteristics of the serpentine ore by mineralization, SEM/EDS, and XRD analysis of serpentine ore are as follows:

(1) Serpentine is the main ore minerals in the serpentine ore, with a small amount of olivine and chrome spinel, metal minerals, including magnetite, hematite, chromite, nickel pyrite, and chalcopyrite. Serpentine ore is fiber crystal, granular crystal, fiber interwoven structure, substituted structure, and Substituted residual structure, block structure.
(2) Serpentine fiber length is generally in the tens—hundreds of microns, fiber length changes in a large number of individual mm.

High-aluminum bearing coal fly ash. High-aluminum bearing coal fly ash (HACFA) was taken from Hequ Second Power Co. Ltd.; its sizes were smaller than 120 μm. The main composition in HACFA is 88.16 mass% ash content, 2.66 mass% volatile content, 1.54 mass% moisture content, 5.91 mass% solid carbon content. The chemical compositions of the ash content are as follows: 38.05 mass% Al_2O_3, 41.12 mass% SiO_2, 4.66 mass% Fe_2O_3, 4.07 mass% CaO, 0.34 mass% MgO, $Na_2O{\cdot}K_2O$ 0.32 mass%, and 11.44 mass % balance.

Bitumenite. Bitumenite (BT) was from Shanxi Datong Washing Coal Plant. Prior to use it was ground to <0.5 mm. Its composition was 12.64 mass% ash content, 28.92 mass% volatile content, 3.70 mass% moisture content, and 54.74 mass% solid carbon content. The chemical compositions of the ash content are as follows: 19.52 mass% Al_2O_3, 49.84 mass% SiO_2, 6.52 mass% Fe_2O_3, and 11.72 mass% CaO.

Table 1 Analysis of serpentine ore used in this work/mass%

MgO	Al_2O_3	CaO	Fe_2O_3	FeO	SiO_2	Na_2O
K_2O	Mn	Ni	Cr	H_2O	P	S
35.54	0.64	3.36	5.23	1.47	37.17	0.0097
0.055	0.048	0.22	0.12	0.48	0.0022	0.0009

Fig. 2 Photo of serpentine used for industrial extraction of magnesium metal. (Color figure online)

Petroleum coke. Petroleum coke (PC) was from Tianjin Dagang Petroleum Chemical Industrial Company. Before experiments it was ground to <0.5 mm. Its composition was 0.28 mass% ash content, 8.04 mass% volatile content, 0.39 mass% moisture content, and 91.29 mass% solid carbon content.

Charcoal. Charcoal (CL) used in experiments was purchased from market, its sizes were smaller than 0.5 mm. Its composition was 1.03 mass% ash content, 2.99 mass% volatile content, 3.94 mass% moisture content, and 92.04 mass% solid carbon content.

Adhesive paper industry wastewater. Adhesive paper industry wastewater was from Jizhou Paper Mill. Its composition was 27.04 mass% ash content, 46.74 mass% volatile content, 5.26 mass% moisture content, and 20.96 mass% solid carbon content.

Iron oxide powder. Iron oxide powder was from Tangshan Iron and Steel Plant. Its composition was 54.12 mass% Fe_2O_3, 12.74 mass% CaO, 14.02 mass% MgO, 6.11 mass% Al_2O_3, and 9.01 mass% SiO_2.

Aluminum-silicon alloys. The content of the furnace charge is 65 mass% HACFA in which the impurities are quartz, calcium oxide, and magnesia, etc. and the total impurities amount is not in excess of 20 mass%; 22 mass% reducing agent (bitumenite and petroleum coke, mixing ratio of bitumenite and petroleum coke 70:30) 7 mass% iron oxide powder and 6 mass% adhesive paper industry wastewater. First, all the raw materials are mixed uniformly, briquetted and dried, then the carbothermal reduction experiments are conducted in the 1800 kVA DC submerged arc furnace with reducing temperature 2200 °C and reducing time 4 h, the aluminum-silicon alloys containing 35.00 mass% aluminum, 48.06 mass% silicon, and 17% mass% iron are obtained with aluminum recovery rate 80% and silicon 70%. Table 2 lists the results of chemical analysis for the Al–Si alloys.

Test Methods

The raw materials for producing magnesium were mixed uniformly in the given proportions in a stirring machine, dry pressure briquetted into ovals with the major axis 40 mm and the minor axis 20 mm. The average density of briquettes was 1.63–1.65 g/cm^3, the porosity was 45–48%. The briquettes were charged into the 2400 kW industrial continuous reduction of magnesium test furnace with temperature 1150–1250 °C, vacuum of 10–20 Pa and time 4 h. The Mg was discharged. The samples were analysed by XRD, SEM analysis and chemical analysis.

Industrial Experiment for Producing Magnesium

Description of Industrial Continuous Reduction Equipment for Magnesium Metal

Figure 1 of the industrial continuous production equipment for metallic magnesium provided in this article includes a feeding device 1, a reduction chamber 17, a crystallizer 21 and a slag discharge chamber 10. The feeding device 1, a reduction chamber 17, a crystallizer 21, and a slag discharge chamber 10 are all equipped with the vacuum detector also connected with a vacuum device. The feeding device 1 is located above the reduction chamber 17 and communicates with the reduction chamber 17, and a material isolation valve 2 is provided between the feeding device 1 and the reduction chamber 17. In order to isolate the reduction chamber 17 and the feeding device 1, it is convenient to maintain the vacuum state of the reduction chamber 17. The slag discharge chamber 10 is set directly below the reduction chamber 17, and is fixedly connected by bolts to form a vertical upper and lower double chamber structure. The outer walls of the reduction chamber 17 and the slag discharge chamber 10 are both double-layer structures used to pass in cooling water. The reduction chamber 17 and the slag discharge chamber 10 are connected with a slag discharge isolation valve. The slag discharge isolation valve is closed during the reaction. After the reaction in the reduction chamber 17 is completed, the slag discharge isolation valve 25 needs to be opened to release the reduced slag. A crystallizer 21 is connected above the reduction chamber 17, so that magnesium vapor can enter the crystallizer 21 upward for crystallization. The inner wall of the reduction chamber 17 is a graphite crucible, and the surface of the graphite crucible is plasma sprayed with titanium boride, and the titanium boride does not react and bond with the reduction slag, thereby avoiding the occurrence of slag sticking. The reduction chamber 17 is provided with an electromagnetic induction heating device 18. The electromagnetic induction heating device 18

Table 2 The results of chemical analysis for the Al–Si alloys/mass%

Al	Si	Fe	Cu	Ti	Mg
Zn	Mn	Ni	Ca		
35.00	48.06	16.31	0.022	0.011	0.021
0.13	0.13	0.023	0.35		

includes a spirally wound electromagnetic coil. The electromagnetic coil is arranged in the reduction chamber 17, and the electromagnetic coil is electrically connected to the heating power source 6. The electromagnetic induction heating device 18 enables the reduction chamber 17 heated uniformly, and the graphite crucible acts as a heating element to conduct heat, so that the reduction chamber 17 is heated rapidly, thereby realizing the rapid and uniform heating function in the reduction chamber 17, which can significantly save the reaction time.

Specifically, the feeding device 1 is connected to the reduction chamber 17 through a feeding pipe, and the material isolation valve 2 is set on the 5 feeding pipe, a first vacuum valve 3 is also provided on the feeding pipe.

The reduction chamber 17 communicates with the crystallizer 21 through a crystallizer tube. A graphite filter 23 is provided at the front end of the crystallizer 21 where the crystallizer tube is connected to filter dust and prevent temperature diffusion. The magnesium vapor in the reduction chamber 17 can enter the crystallizer 21 through the crystallizing tube and the graphite filter 23 in sequence. The crystallizing tube is provided with a second vacuum valve 25, and the lower end of the crystallizer 21 is connected with a collecting tank. The crystallizer 21 has a double-layer outer wall structure, and a spirally wound electromagnetic heating wire is arranged between the double-layer outer walls of the crystallizer 21. The electromagnetic heating coil is the copper tube 22, which is electrically connected to the heating power source 6, and the copper tube 22 is used to pass cooling water. And a crystallization temperature measuring device 24 is provided on the inner and outer wall of the crystallizer 21 to ensure that the crystallization temperature is controlled at 650 °C under the control of heating or water cooling. Magnesium vapor will crystallize when cold in the crystallizer 21 to form a liquid into the collector. When the receiving tank is full, the second vacuum valve 25 can be temporarily closed to replace the empty receiving tank. The temperature on the graphite filter 23 is 1000 °C, so that metallic magnesium can crystallize here. A certain amount of metallic titanium and zirconium is placed in the receiving tank. At this time, the iron in the liquid magnesium reacts with the titanium and/or zirconium and deposits on the bottom to reach the high purity level of metallic magnesium, which is as high as 99.9%.

There is a slag collecting tank 11 in the slag discharging chamber 10, and a mobile truck 12 is arranged under the slag collecting tank 11, and a mobile truck 12 can move in the slag discharging chamber 10. The slag isolation valve includes a first flap type isolation valve 16 and a second flap plate type isolation valve 15. The first flap type isolation valve 16 is set at the bottom of the reduction chamber 17, and the second flap type isolation valve 15 is set on the top of the slag discharging chamber 10. The first flap type isolation valve 16 and the second flap type isolation valve 15 are positioned to match. With a flap structure, the bottom of the entire reduction chamber 17 is basically fully opened, and the reduction slag in the reduction chamber 17 is discharged instantaneously, avoiding the phenomenon that the slag is slow, easy to block and the reduction slag at the bottom of the reduction chamber 17 is difficult to discharge. The first flap type isolation valve 16 and the second flap type isolation valve 15 are both connected to a hydraulic system, and the hydraulic system can control the opening and closing of the first flap type isolation valve 16 and the second flap type isolation valve 15. The electromagnetic induction heating device 18 is arranged on the first flap type isolation valve 16.

The vacuum device includes the first vacuum unit 13 and the second vacuum unit 9, the first vacuum unit 13 through the third vacuum valve 20 communicates with the crystallizer 21, the fourth vacuum valve 14 communicates with the reduction chamber 17, the second vacuum unit 9 communicates with the feeding device 1 through the fifth vacuum valve 7, and the sixth vacuum valve 8 communicates with the slag discharging chamber 10. The vacuum detector is connected to the control system 5, the reduction temperature measuring device 19 is provided on the outer wall of the inner layer of the reduction chamber 17, the crystal temperature measuring device 24 is provided on the outer wall of the inner layer of the crystallizer 21, the crystal temperature measuring device 24, the reduction temperature measuring device 19 is connected to the control system 5, and the measured temperature is displayed by the temperature control meter 4 on the control system. The control system 5 can send control signals to the vacuum device and the heating power source 6, respectively, to realize the vacuum state and the reduction temperature, maintenance of crystallization temperature. At the same time, the hydraulic system and all the vacuum valves in this embodiment are connected to the control system 5 and the control system 5 controls their opening and closing states.

Description of Industrial Test Process

According to the condition test parameters, the industrial experiment was carried out on a 2400 kW continuous magnesium reduction furnace. Each filling 3000 kg serpentine small briquetts into the furnace, keeping vacuum 10–20 Pa, temperature 1200 °C, to maintain the reduction time 4 h.

The industrial test process is as follows:

(1) The dried serpentine ore (grain size 1 mm), additive lime, catalyst fluorite and reducing agent silico-aluminum alloy (120 mesh) are mixed and

Table 3 The chemical analysis results of the crude magnesium/mass%

Si	Cr	Ni	Cu	Zn	Ca	Fe	Al	Mg
0.001	0.001	0.001	0.002	0.002	0.003	0.005	0.002	99.9

ground in a ball mill according to the proportions, and then dry pressed into serpentine ellipses with the major axis 40 mm and the minor axis 20 mm;

(2) Add the serpentine oval ball to the feeding device 1, open the first vacuum valve 3 and the material isolation valve 2, the serpentine oval ball enters the reduction chamber 17, when the serpentine oval ball enters the reduction chamber 17, close the first vacuum valve 3 and the material isolation valve 2 isolate the reduction chamber 17 and the feeding device 1;
(3) Start the first vacuum unit 13 and the second vacuum unit 9 to vacuum the reduction chamber 17, the slag chamber 10 and the crystallizer 21 to the working vacuum;
(4) Start the heating power supply 6 to supply power to the electromagnetic induction heating device 18 to heat the reduction chamber 17. The control system 5 controls the reduction chamber 17 to heat up to the process temperature, and the magnesium in the serpentine ellipsoidal ball rises to the crystallizer in the form of vapor in 21, the crystallizer 21 is controlled by the control system 5 to maintain the crystallization temperature (600–650 °C), the magnesium vapor in the crystallizer 21 is crystallized when it is cold, and solid crude magnesium is formed in the crystallizer 21;
(5) After the serpentine ellipsoid reaction is completed, close the second vacuum valve 25, take out the magnesium crystallizer, replace with an empty crystallizer, evacuate the crystallizer 21 to the working vacuum degree, and open the second vacuum valve 25;
(6) Through the control of the hydraulic system, the second flap isolation valve 15 and the first flap isolation valve 16 are sequentially opened, the reduced slag falls into the slag collecting tank 11, and the second flap isolation valve 15 and the first flap isolation valve 16 are sequentially closed. Open the slag discharging chamber 10, take out the slag collecting tank 11, replace it with an empty slag collecting tank 11, and evacuate the slag discharging chamber 10 to the working vacuum through the second vacuum unit 9;
(7) Add the serpentine ellipsoid to the feeding device 1, vacuumize the feeding device 1 to the working vacuum through the second vacuum unit 9, and then cycle step 2 to step 6 in sequence.

In the end, the current efficiency of the industrial test process reached 90%, the recovery rate of magnesium was 85%. The chemical analysis results of the crude magnesium obtained in the test is shown in Table 3. Visible, as crude magnesium, its quality is very good. Further refined, high quality Mg products can be obtained. The reduction slag can be used as a cement raw material due to the transformation of calcium dichromate crystal in form of expansion and self-crushing loose white powder, the size smaller than 50 μm, the chemical analysis results were as follows: 61.09 mass% CaO, 14.45 mass% Fe_2O_3, 15.44 mass% SiO_2, 3.39 mass% Al_2O_3, and 4.03 mass% MgO. The obtained residue was subjected to magnetic separation to obtain a nickel iron block. The chemical analysis results were as follows: 2.42 mass% Ni, 97.58 mass% Fe. Nickel iron recovery rate was about 85%.

Conclusions

(1) The industrial process of extracting magnesium by the aluminum-silicon alloy heat-reducing from serpentine is reasonable. The Al–Si alloy can extract the metallic magnesium from the serpentine at 10–20 Pa of vacuum and from 1150 to 1250 °C. In the process of magnesium extraction, at the same time these nickel iron alloy and dicalcium silicate and aluminum silicate slag used as raw materials for the preparation of cement can be obtained.
(2) Industrial scale of experimental studies has shown that serpentine extraction of magnesium metal is feasible, and magnesium recovery rate reaches 85%. This is a cleaning process with no carbon dioxide emissions.
(3) Industrial continuous reduction furnace improves magnesium production efficiency and reduces costs, it is worth further study, and it has a bright future. Raw materials for production of Al–Si alloy are obtained easily, such as coal gangue, fly ash, and so on.
(4) From ore crushing, grinding, ball pressing, reduction, to metal magnesium refining and casting ingots, the entire production line is fully continuous and automated. The production process has no carbon dioxide emission, clean production, high production efficiency, and low cost, which lays the foundation for large-scale industrialization of extracting magnesium metal from serpentine.

References

1. D. Eliezer, et al., "Magnesium Science, Technology and Applications," Advanced Performance Materials, Vol. 5(1998), 201–212.
2. M. Halmann, et al., "Magnesium Production by Pidgeon Process Involving Dolomite Calcinations and MgO Silicothermic Reduction: Thermodynamic and Environmental Analyses," Industrial & Engineering Chemistry Research, Vol. 47(2008), 2146–2154.
3. G. Hanko, H. Antrekowitsch, P. Ebner, "Recycling Automotive Magnesium Scrap", JOM, Vol. 54 (2002), 51–54.

Research on Extracting Magnesium from Carbide Slag and Magnesite in Flowing Argon Atmosphere

Junhua Guo, Daxue Fu, Jibiao Han, Zonghui Ji, Zhi'he Dou, and Ting'an Zhang

Abstract

Carbide slag is a kind of industrial waste produced from the calcium carbide industry, which is difficult to treat and cannot be used reasonably. In this paper, carbide slag as a calcium source for extracting magnesium was proposed, i.e., carbide slag was directly mixed with low-grade magnesite instead of dolomite as raw material for extracting magnesium without pretreatment. The process is simple to operate and can obtain metal magnesium while reducing nearly 50% of CO_2 emissions in the process of extracting magnesium. In the work, the strength of pellets, the effects of Ca/Mg ratio, and calcination temperature on the recovery rate of Mg were investigated. The results showed that low temperature calcination was beneficial to reduction and the recovery rate of Mg increased with the increase of Ca/Mg ratio, and the reduction rate was 87.95% when the Ca/Mg ratio was 1.2.

Keywords

Carbide slag • Silicothermic process • Calcination temperature • Reduction rate

Introduction

Magnesium has excellent properties and wide application, and has become the third largest metal engineering material after steel and aluminum [1]. The Pidgeon process is the main method of extracting magnesium at present [2]. The raw material is dolomite, which is calcined to obtain dolime. Then it is used to produce magnesium by vacuum thermal reduction with fluorite and ferrosilicon. At present, researchers have three opinions on the reaction mechanism of this reaction process: (1) During the reaction process, volatile gas SiO is generated, which participates in the reaction as an intermediate compound [3]; (2) Intermediate compounds CaSi and $CaSi_2$ are generated, and then participate in the reduction reaction [4, 5]; (3) Intermediate compound $CaSi_2$ is generated. When the temperature exceeds 890 $\sim$ 1020, $CaSi_2$ decomposes into silicon, and then Ca vapor participates in the reaction [6, 7]. However, no matter which reaction mechanism, CaO is involved in the reaction process. Therefore, CaO plays an important role in the silicothermic reaction.

According to Eq. (1), 10 tons of dolomite (about 3.04 tons of CaO) is consumed to produce 1 ton of magnesium. Although there are abundant dolomite reserves in China, dolomite used in silicothermic process should not only meet the process requirements in terms of chemical composition, but also meet the requirements of the process in terms of mineral structure, because the chemical composition and mineral structure of dolomite have a great influence on each smelting process of the process [4]. If the MgO content in dolomite is too high (more than 22% MgO), the CaO content is lower (less than 30% CaO), and $2CaO{\cdot}SiO_2$ slag cannot be formed in the reduction process, so CaO should be added. On the contrary, if the content of CaO in dolomite is too high (more than 33% CaO), the content of MgO is lower (less than 19% MgO), and the output rate of magnesium in the reduction process is low, so MgO should be added to the raw material. Therefore, dolomite for the extraction of magnesium by silicothermic process has high requirements on composition and limited resources to meet the conditions. Considering the advantages of abundant magnesite resources in China and the large storage of calcium carbide slag, this paper proposed to use carbide slag as calcium source, and mix it with magnesite in proportion to replace dolomite for magnesium smelting. Then, magnesium is produced by pre-prepared pellets technology, that is, firstly, it is mixed with ferrosilicon and fluorite for fine grinding, and then the magnesium is directly extracted in flowing argon, this

J. Guo · D. Fu · J. Han · Z. Ji · Z. Dou · T. Zhang (✉)
Northeastern University Key Laboratory of Ecological Metallurgy of Multi-Metal Intergrown Ores of Ministry of Education, Special Metallurgy and Process Engineering Institute, Shenyang, Liaoning 110819, China
e-mail: zta2000@163.net

V. M. Miller et al. (eds.), *Magnesium Technology 2021*, The Minerals, Metals & Materials Series,
https://doi.org/10.1007/978-3-030-65528-0_24

process can not only make comprehensive use of magnesite in industrial production, but also solve the problem of carbide slag storage.

$$2(CaO \cdot MgO) + Si = 2Mg + 2CaO \cdot SiO_2 \quad (1)$$

Carbide slag is a kind of industrial waste residue produced by the acetylene process in the chemical industry. Its main component is $Ca(OH)_2$ [8, 9]. However, calcium hydroxide in carbide slag reacts with carbon dioxide to form calcium carbonate due to long-term stacking of carbide slag in the air. Therefore, the main characteristics of the deposited carbide slag are: (1) high calcium content; (2) high alkaline slurry; (3) large amount of storage; and (4) difficult purification, which may cause groundwater pollution, soil alkalization, and air dust pollution, and have an adverse impact on the ecological environment [10–12]. How to turn carbide slag into treasure, the most reported is that carbide slag is used to prepare light or nano calcium carbonate, capture CO_2, and fix sulfur, and has achieved remarkable results [13–18]. However, the above-mentioned treatment method needs to pretreat the carbide slag to remove impurities to increase its effect. Carbide slag is used as calcium source to produce magnesium without pretreatment, which has large capacity and simple operation. According to statistics, there are millions of tons of calcium carbide slag piled up every year in China [8]. Therefore, using carbide slag to produce metal magnesium has the advantages of low cost and wide source. It can not only reduce environmental pollution, but also turn waste into treasure and reduce the production cost of magnesium smelting. In this paper, the strength of pellets, the effects of Ca/Mg ratio, and calcination temperature on magnesium recovery were studied, which provided a theoretical basis for carbide slag used in the extraction of magnesium by silicothermic process.

Experiments

Raw Materials

The experimental raw materials are low-grade magnesite, carbide slag, and ferrosilicon as well as analytical-grade calcium fluoride. Among them, ferrosilicon comes from Henan, China, with 74.13% silicon content. Magnesite comes from Liaoning, China. The chemical compositions are shown in Table 1. Carbide slag is the waste of PVC production by acetylene method.

The XRD pattern of magnesite and carbide slag are shown in Figs. 1 and 2, respectively. The main phase is $MgCO_3$ with a small amount of $CaCO_3$ in magnesite. Table 1 shows a small amount of SiO_2, Fe_2O_3, and Al_2O_3.

Table 1 Chemical composition of magnesite

Ore	Ignition loss	SiO_2	Al_2O_3	Fe_2O_3	CaO	MgO
Magnesite	51.92	0.75	0.19	0.74	0.90	45.50

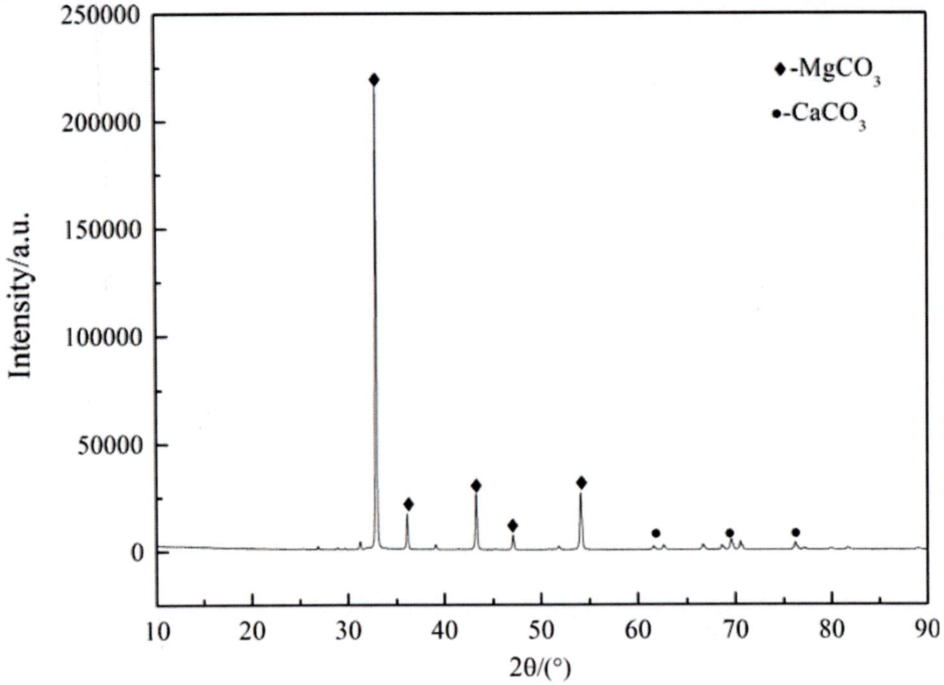

Fig. 1 XRD pattern of magnesite

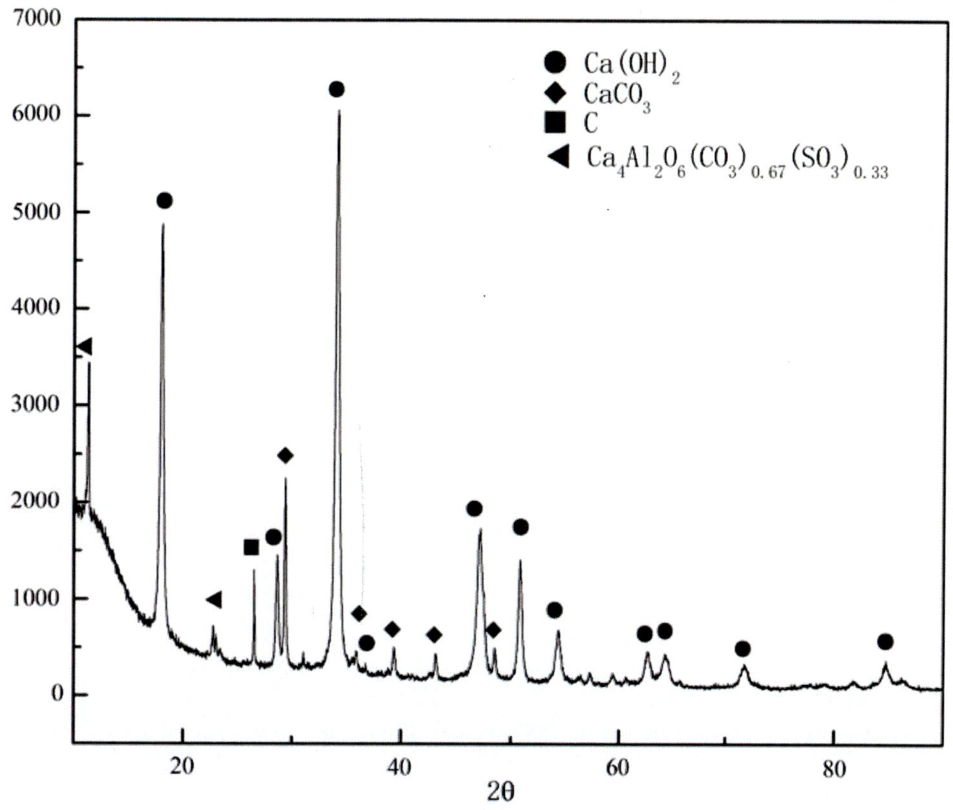

Fig. 2 XRD pattern of carbide slag

The main phase is $Ca(OH)_2$ with a small amount of $CaCO_3$ in carbide slag.

Experimental Equipment

The calcination and reduction process was carried out in a vertical tube furnace with the silicon molybdenum rod as the heating body. The equipment schematic diagram is shown in Fig. 3.

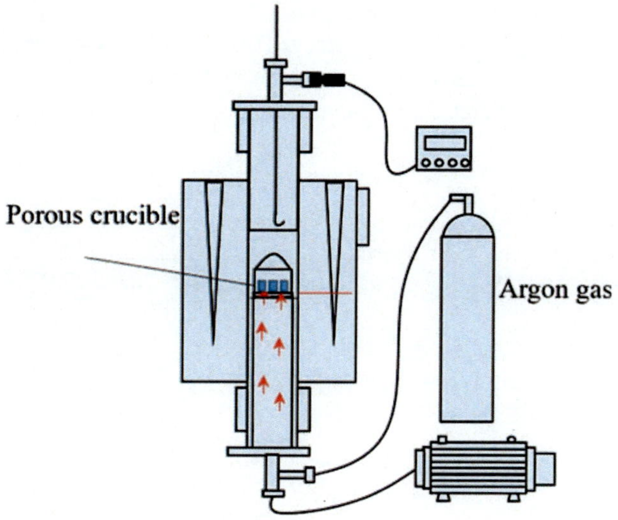

Fig. 3 Schematic diagram of vertical tube furnace. (Color figure online)

Experimental Procedures

It can be seen from Fig. 2 that due to long-term storage and absorption of carbon dioxide, a small amount of $Ca(OH)_2$ will be converted into $CaCO_3$, and a small amount of water may remain. Therefore, it was necessary to determine its component content first. Through XRF analysis, the impurities such as SiO_2, Al_2O_3, MgO, and Fe_2O_3 in calcium carbide slag were less and not considered. Since the decomposition temperature of $Ca(OH)_2$ and $CaCO_3$ is different, $Ca(OH)_2$ could decompose at about 650 °C, while $CaCO_3$ could decompose at 850 ~ 900 °C. Therefore, the constant weight method was used to test the composition. The carbide slag was calcined at 150, 650 °C, and 900 °C until the weight loss was stable in each stage. The experiment was repeated three times. The average value was taken as the content of $Ca(OH)_2$ and $CaCO_3$ in carbide slag. The measurement results were shown in Table 2.

After the content of the carbide slag was determined, it was mixed with magnesite, ferrosilicon, and fluorite in proportion to finely grind and mix evenly before briquetting. Then the pellets were put into corundum crucible with holes and put into a vertical tubular furnace with flowing argon for

Table 2 Chemical composition of carbide slag

Number	CaO (%)	$Ca(OH)_2$ (%)	$CaCO_3$ (%)	H_2O (%)
1	3.37	45.35	49.55	1.73
2	2.88	45.36	49.92	1.84
3	2.85	45.21	50.13	1.81
Average value	3.03	45.31	49.87	1.79

calcination and reduction. After the experiment, weighing and chemical analysis were carried out to calculate the weight loss rate of calcination and the recovery rate of magnesium.

Results and Discussion

Effect of Briquetting Pressure on Pellet Properties After Calcination

In order to ensure the quality of pellets after loading into the reduction tank, the effect of briquetting pressure on the properties of calcined pellets was investigated. Calcium hydroxide, the main component of calcium carbide slag, was used as the calcium source. As calcium hydroxide easily deteriorates, raw materials were used and prepared on site to ensure the accuracy of experimental data as far as possible. The pressed pellets were packaged in a vacuum for standby. Because the diameters of briquetting equipment and pellet were not different, the equipment display pressure and actual pellet pressure were also different. When the equipment pressure indication is 5 MPa, 10 MPa, 15 MPa, 20 MPa, and 25 MPa, the corresponding actual pellet pressure was 69.69 MPa, 139.38 MPa, 209.07 MPa, 278.76 MPa, and 348.45 MPa, respectively. After the furnace temperature reached the set temperature, the pellet was put into the furnace for calcination experiment. The experimental temperature was 1000 °C, calcined for 1 h, and the ratio of Ca to Mg was 1. The weight loss rate and hydration activity of pellets were measured after calcination, and the compressive strength of pellets was measured by an electronic universal testing machine. The experimental results were shown in Figs. 4 and 5.

It could be seen from Fig. 4 that the briquetting pressure had little effect on the weight loss rate of pellets. The weight loss rate fluctuated around 33.33%, and the hydration activity fluctuated around 17%. When the briquetting pressure was 20 MPa, the hydration activity was the highest. Compared with magnesite–calcium carbonate pellets, the hydration activity was lower. Through theoretical calculation, the theoretical weight loss rate of magnesite–calcium

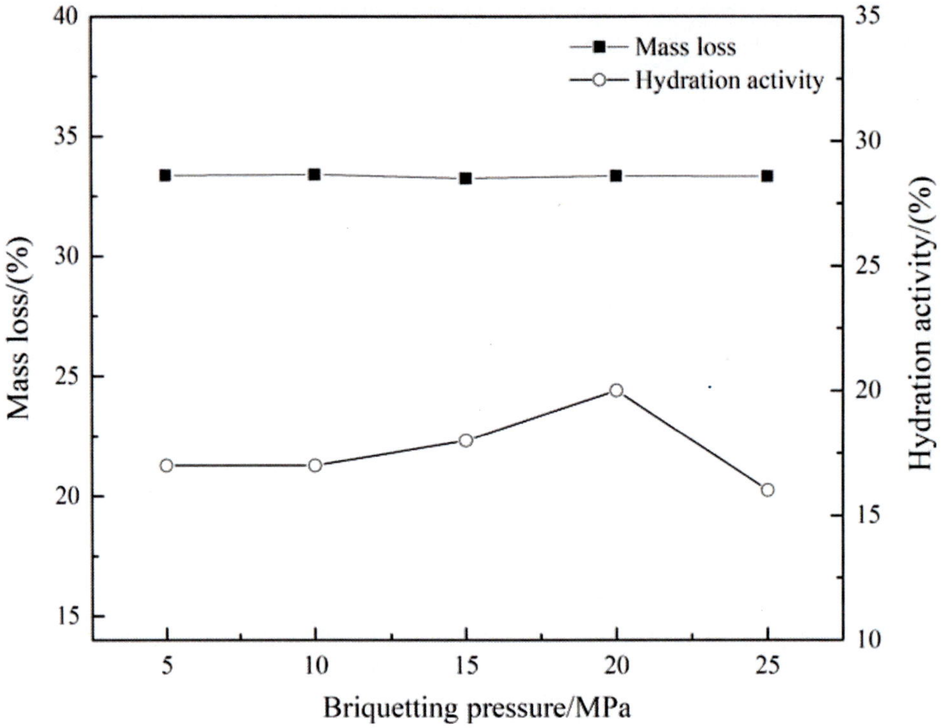

Fig. 4 Mass loss and hydration activity after calcination under different briquetting pressures

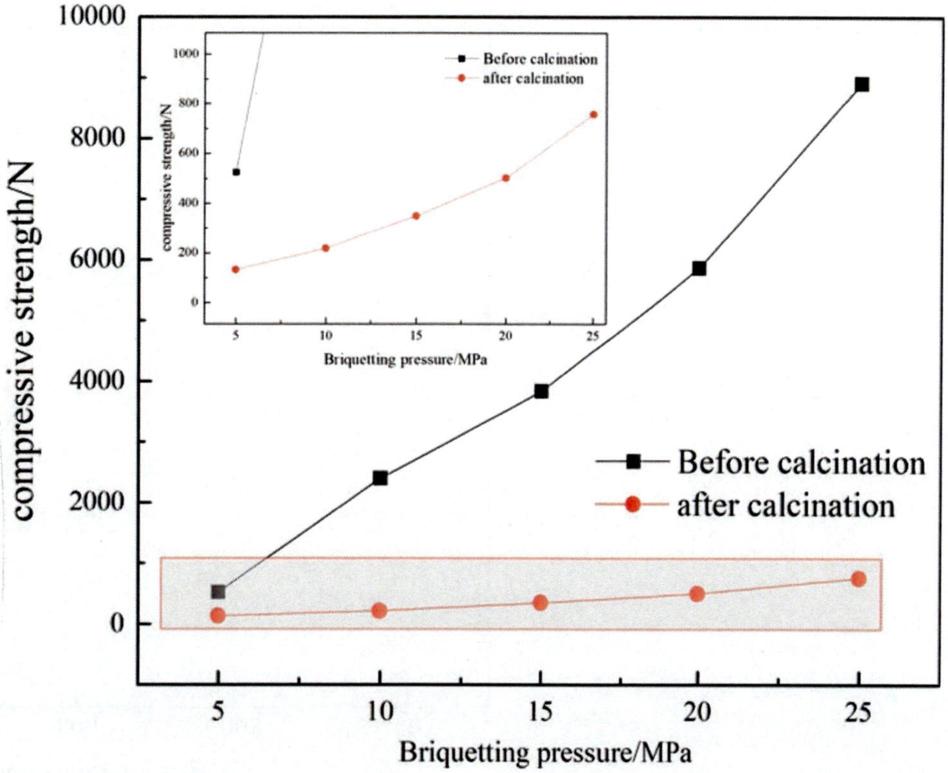

Fig. 5 Compressive strength before and after calcination under different briquetting pressures. (Color figure online)

hydroxide pellets was 33.32%, and the actual weight loss rate of pellets was higher than the theoretical weight loss rate. It could be inferred that calcium hydroxide absorbed CO_2 and deteriorated in the process of pellet preparation, which made the weight loss rate higher and the measured hydration activity lower. It could be seen from Fig. 5 that the compressive strength of pellets before calcination and after calcination increased with the increase of briquetting pressure. The compressive strength of pellets before calcination was 8913 N and that of pellets after calcination was 1000 N at 25 MPa. Since the ballability of calcium hydroxide was greater than that of calcium carbonate, the pressed pellets were more firmly bonded. Therefore, compared with magnesite–calcium carbonate pellets, the load-bearing capacity of magnesite–calcium hydroxide pellets increases nearly twice [19].

Effect of Calcination Temperature on Magnesium Reduction Rate

In order to only consider the influence of calcination temperature on reduction rate and exclude the influence of impurities, $Ca(OH)_2$ was used as calcium source, mixed with magnesite in proportion of Ca/Mg = 1 to replace dolomite. The pellets were prepared by pre-prepared pellets technology, calcined at different temperatures (700 °C ~ 1100 °C) for 1 h, and then reduced at 1300 °C for 2 h. The argon flow rate was 0.2 m^3/h. The experimental results were shown in Fig. 6.

The results showed that the reduction ratio of magnesium decreased with the increase of calcination temperature. The reduction ratio was 78.53% for material calcined at 700 °C and 74.67% at 1100 °C. According to the report [20, 21], at low temperature, the gas escaped and the material formed a porous structure. At this time, CaO and MgO were just formed, with small grain size, many defects, and large specific surface area. Therefore, the activity was high and the silicothermic reaction speed was fast, and the recovery rate of metal magnesium was high in the same time. With the increase of temperature, CaO and MgO grains grew rapidly. The higher the temperature, the faster the grain growth, and even sintering occurred. The activity decreased, resulting in the reduction of magnesium recovery. In addition, the reduction slag was analyzed by XRD, as shown in Fig. 7. The results showed that in addition to Ca_2SiO_4 and MgO, there were also CaMg $(SiO_3)_2$ in the slag. With the increase of calcination temperature, the peak strength of the phase $CaMg(SiO_3)_2$ increased. This may be due to the sintering of CaO and MgO at high temperature, which made MgO enter into the slag, resulting in the reduction rate decreasing, finally existed in the form of $CaMg(SiO_3)_2$ in the slag.

Fig. 6 Effect of calcination temperature on reduction ratio

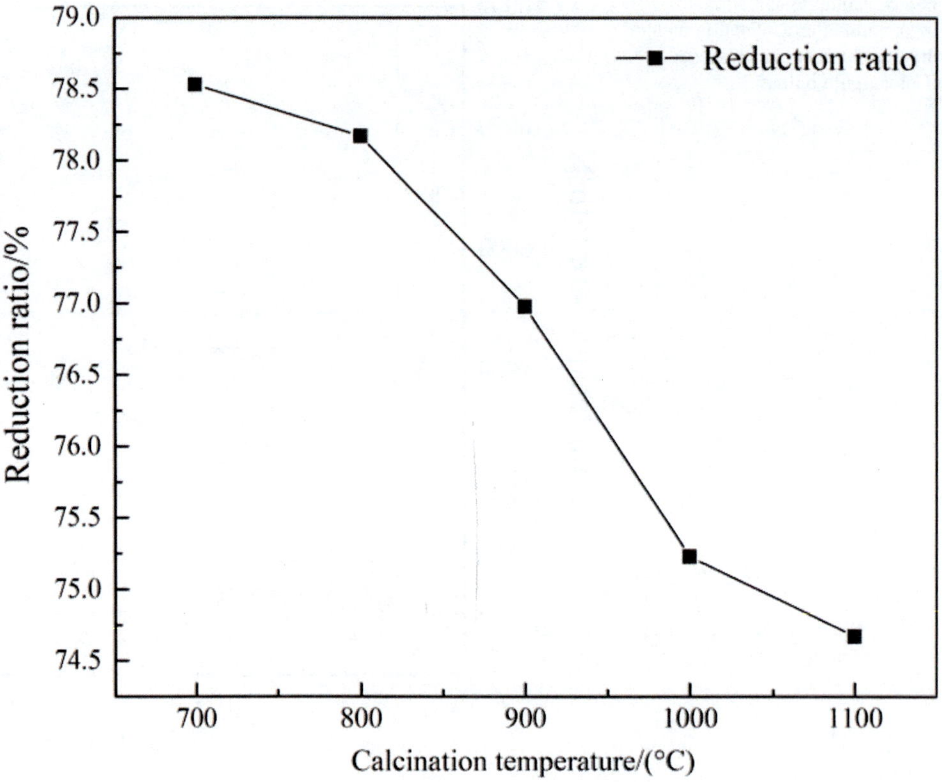

Effect of Ca/Mg Ratio on Magnesium Reduction Rate

The content of calcium and magnesium in dolomite had a great influence on the magnesium smelting process. The ratio of carbide slag to magnesite should be explored when carbide slag and magnesite were used as raw materials to extract magnesium. Therefore, the effect of Ca/Mg ratio on reduction rate was investigated. Firstly, the pellets with the molar ratio of calcium to magnesium of 0.8, 0.9, 1.0, 1.10, and 1.20 were prepared. Then the pellets were calcined at 1000 °C for 1 h, and reduced at 1300 °C for 2 h. The whole experimental process was carried out in flowing argon atmosphere with the argon flow rate of 0.2 m^2/h. The surface morphology of the pellets before and after calcination was observed by metallographic microscope, as shown in Fig. 8. In the metallographic pictures, it could be seen that the surface of the original pellet was flat and dense, while the surface of the pellet after calcination was rough and loose, which was similar to that of the preformed pellet with dolomite as raw material after calcination. It was due to the decomposition of CO_2 from carbonate and the formation of pores on the surface of pellets.

The mass loss rate of the pellet after calcination was calculated by weighing, as shown in Fig. 9. The mass loss rate of the pellets with calcium magnesium molar ratio of 0.8, 0.9, 1.0, 1.10 and 1.20 was 36.8%, 36.5%, 36.5%, 36.2%, and 36.0%, respectively, while the theoretical mass loss was 36.8%, 36.7%, 36.6%, 36.6%, and 36.5%, respectively. It was found that the actual mass loss rate was basically consistent with the theoretical mass loss rate. The results of reduction rate were shown in Fig. 9. The experimental results showed that the reduction rate of magnesium increased with the increase of the molar ratio of calcium to magnesium in the range of 0.8 $\sim$ 1.2, and the reduction rate was 87.95% when the molar ratio of calcium to magnesium was 1.2. However, with the increase of Ca/Mg molar ratio, the increase range decreased. When the molar ratio of Ca to Mg was 1.0, the reduction rate of Mg was over 80%. It showed that the content of CaO was slightly higher than that of MgO, which was beneficial to the improvement of reduction rate.

Conclusions

(1) The effect of briquetting pressure on the mass loss rate of pellets was not obvious, and the compressive strength of pellets before and after calcination increased with the increase of briquetting pressure. The compressive strength of pellets before calcination was

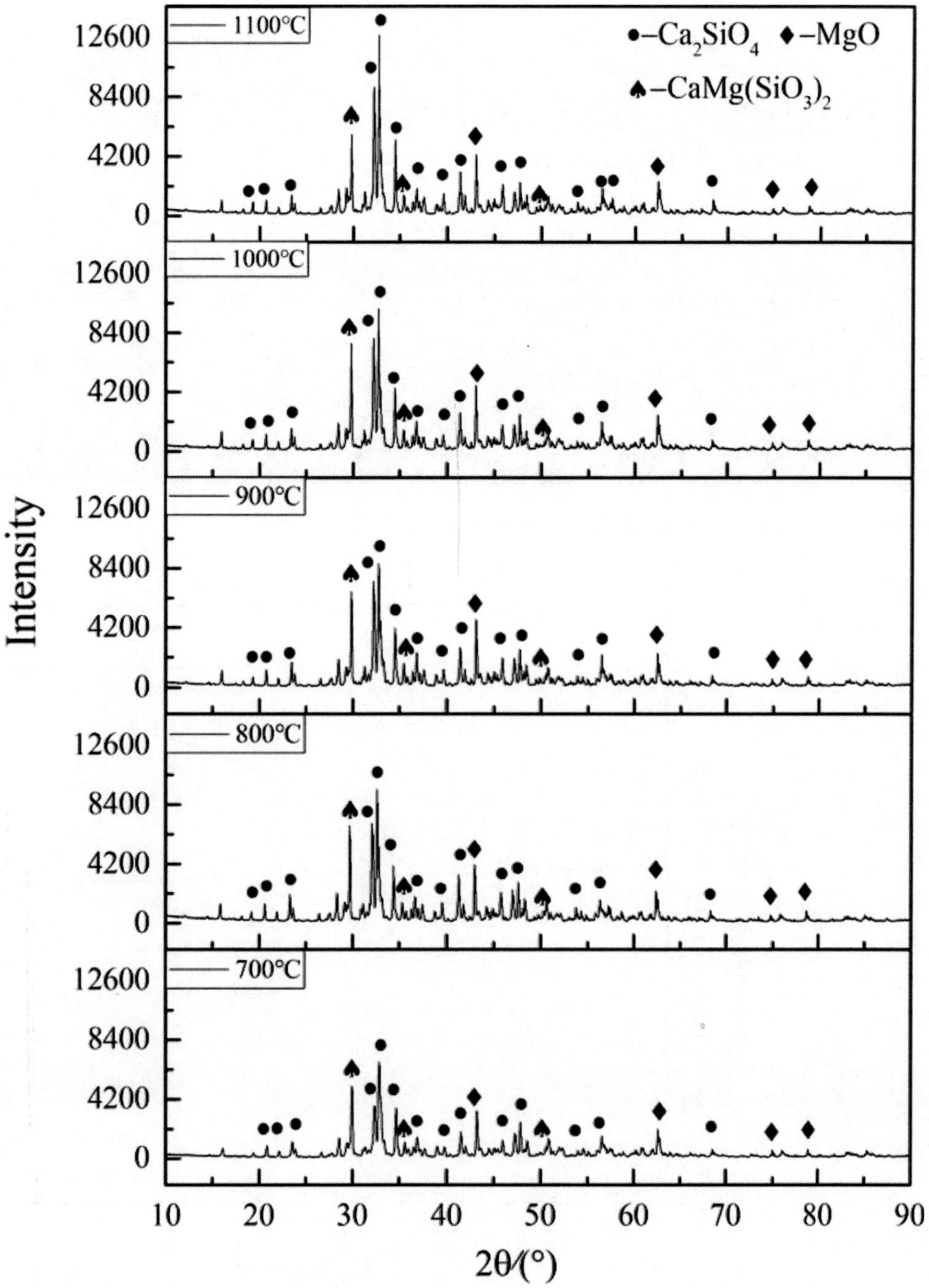

Fig. 7 XRD pattern of reduction slag at different calcination temperatures

8913 N and that of pellets after calcination was 1000 N at 25 MPa.

(2) The calcination temperature had a great influence on the reduction rate of magnesium. With the increase of calcination temperature, the reduction rate of magnesium decreased. This was due to the sintering of CaO and MgO at high temperature, which reduced the activity of CaO and MgO, and made MgO enter the slag in the form of $CaMg(SiO_3)_2$.

(3) When the molar ratio of calcium to magnesium was between 0.8 and 1.2, the reduction rate of magnesium increased with the increase of the molar ratio of calcium to magnesium. The reduction rate of magnesium was 87.95% when the molar ratio of calcium to magnesium was 1.2.

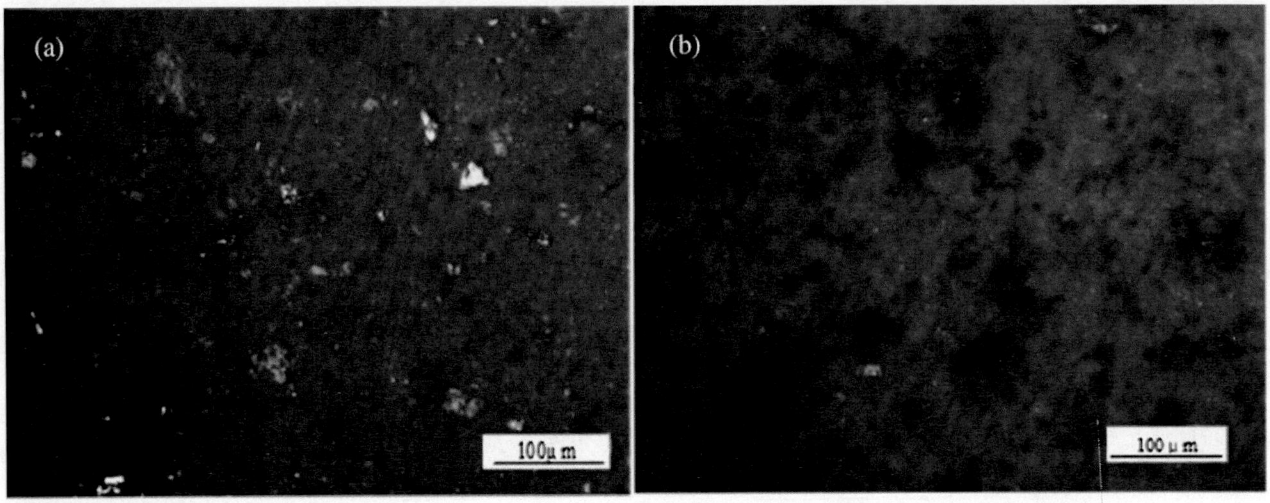

Fig. 8 Metallographic photos of pellets **a** before calcination; **b** after calcination

Fig. 9 Effect of Ca/Mg ratio on mass loss after calcination and reduction rate

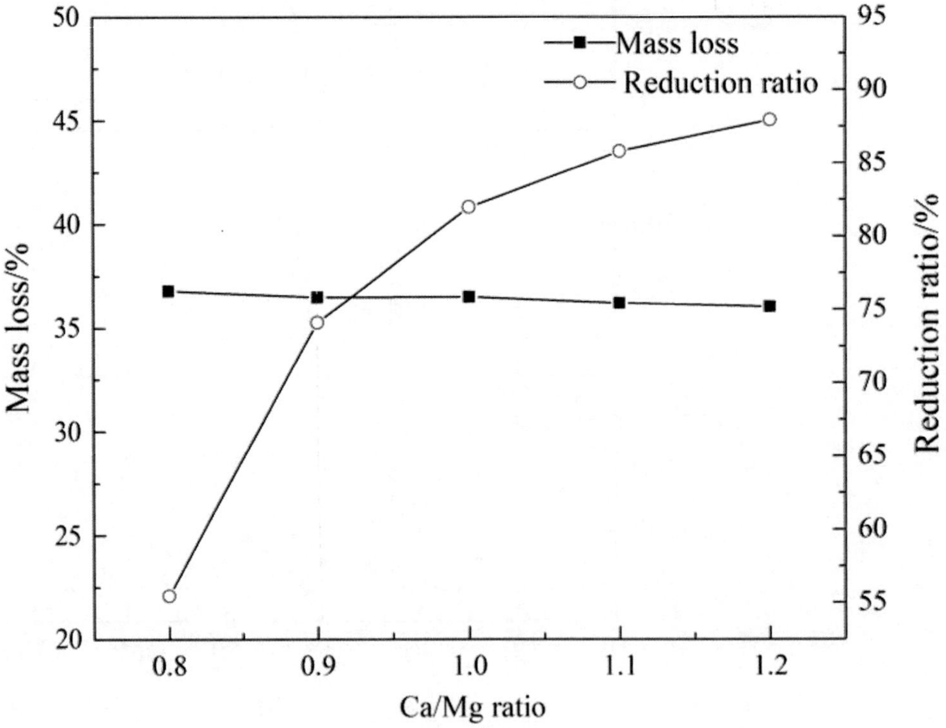

Acknowledgements This work was supported by National Natural Science Foundation of China under Grants (51504058; U1508217; 51404054; 51374064), the Fundamental Research Funds for the Central Universities of China (N162504003, N140204013), the Fund of Liaoning S&T Project (201601003, LZ2014021).

References

1. Mustafa Kemal Kulekci. Magnesium and its alloys applications in automotive industry [J]. The International Journal of Advanced Manufacturing Technology, 2008, 39: 851–865.
2. Yang Tian, Bao-qiang Xu, Bin Yang, Cheng-b Yang, Tao Qu, Da-chun Liu, Yong-nian Dai. Magnesium production by carbothermic reduction in vacuum [J]. Journal of Magnesium and Alloys. 2015, 3:149–154.
3. Fu DX, Zhang TA, Guan LK, Dou ZH, Wen M. Magnesium production by silicothermic reduction of dolime in pre-prepared dolomite pellets [J]. JOM 2016; 68: 3208–3213.
4. RY Xu. Magnesium production by silicothermic process [M]. 1st ed. Changsha: Central South University Press; 2003, pp. 28–47.
5. Wynnyckyj JR, Pidgeon LM. Equilibria in the silicothermic reduction of calcined dolomite [J]. Metall Trans B, 1971, 2(4): 979–986.

6. Wulandari W, Brooks GA, Rhamdhani MA, Monaghan BJ. Kinetic analysis of silicothermic process under flowing argon atmosphere [J]. Can Metall Q, 2014, 53: 17–25.
7. Daxue Fu, Zonghui Ji, Junhua Guo, Jibiao Han, Zhihe Dou, Yan Liu, Ting'an Zhang, Lukui Guan,. Diffusion and phase transformations during the reaction between ferrosilicon and CaO•MgO under vacuum [J]. Journal of Materials Research and Technology, 2020, 9(3):4379–4385.
8. Xin Wang. Brief introduction of comprehensive utilization of carbide slag [J]. China Chlor-Alkali, 2003, (8): 36–39.
9. Man Yi. Brief introduction of Resourcification Utilization Routes for Carbide Slag [J]. IOP Conference: Earth and Environmental Science, 2018, 208.
10. Huiqing Wang, Jihong Tong, Liping Shen. Resourcification Utilization Routes for Carbide Slag [J]. Chemical Production and Technology, 2007, 14(1): 47–51.
11. Guoqing Ma, Zhaoqian Li, Chonghua Pei. The Development of Comprehensive Utilization for Carbide Slag [J]. Journal of SWUST, 2005, 20(2): 49–52.
12. Dongfang Xie, Guoyuan Tian, Hui Liu, et al. Application of Carbide Slag Replacing Lime in the Treatment of Acidic Wastewater [J]. north environment, 2005, 30(1): 59–61.
13. Kai Liu, Bosheng Zhao, Yu Wu, Jianbin Zhang, et al. Bubbling synthesis and high-temperature CO_2 adsorption performance of CaO-based adsorbents from carbide slag [J]. Fuel, 2020, 269: 117481.
14. Yongjing Wang, Baofang Ye, Zengchun Hong, Yonghao Wang, Minghua Liu. Uniform calcite mircro/nanorods preparation from carbide slag using recyclable citrate extractant [J]. Journal of Cleaner Production, 2020, 253, 119930.
15. Bo Liang, Fenglan Han. Crystal transformation of calcium carbonate micro-particles prepared from carbide slags [J]. Inorganic Chemicals Industry, 2016, 48(10): 63–67.
16. Jianjun Cai, Feng Yan, Ming Luo, Shuzhong Wang, Highly stable CO_2 capture performance of binary doped carbide slag synthesized through liquid precipitation method [J]. Fuel, 2020, 280, 118575.
17. Zirui He, Yingjie Li, Wan Zhang, et al. Effect of re-carbonation on CO_2 capture by carbide slag and energy consumption in the calciner [J]. Energy Conversion and Management, 2017, 148:1468–1477.
18. XinWang, Yingjie Li, Wan Zhang, et al. Simultaneous SO_2 and NO removal by pellets made of carbide slag and coal char in a bubbling fluidized-bed reactor[J]. Process Safety and Environmental Protection, 22020, 134: 83–94.
19. Junhua Guo, Ting'an Zhang, Daxue Fu, et al. Research on properties of prefabricated pellets of silicothermic process after calcination in flowing argon atmosphere [J]. Magnesium Technology, 2020, 303–307.
20. Xiaoping Lin. Study on calcining active magnesium oxide with magnesite [J]. Science and Technology Innovation Herald, 2012, 27: 83–84.
21. Chong Wang, Yongbing Liu, Zhanyi Cao, et al. Structure transformation of calcining dolomite [J]. Transactions of Materials & Heat Treatment, 2013, 34(2): 23–26.

Part V
Poster Session

Optimization of Mechanical Properties in Magnesium Zinc Alloys

Christopher Hale, Zhigang Xu, HongLin Zhang, Sergey Yarmolenko, and Jagannathan Sankar

Abstract

Magnesium-based alloys are being used today in various lightweight applications. The mechanical properties of magnesium-based alloys can be enhanced for such applications through a combination of annealing temperature and holding time of the magnesium-based alloy. The magnesium-based alloy under investigation in this study is Mg-5%Zn. The results of this investigation show that optimizing annealing time and temperature can achieve homogenization and enhance the mechanical properties and formability of Mg-5%Zn as seen in hardness test results for the homogenized samples.

keywords

Magnesium alloys • Homogenization • Zinc • Mechanical properties • Hardness testing

Introduction

It is well known that magnesium-based alloys have significant potential in structural lightweight applications, particularly in automotive and aerospace applications. Magnesium-based alloys, however, have some limitations due to the Hexagonal Close Packed (HCP) crystal structure and a limited number of slip systems along which plastic yielding/deformation can occur [1, 2]. As a result, the wrought Mg alloys demonstrate poor formability at room temperature. The formability and mechanical properties of magnesium-based alloys can be greatly enhanced through grain refinement and deformation processing such as extrusion and hot rolling [3, 4].

It has been clearly shown that Mg alloy plates that have been processed and undergone plastic deformation have general problems, such as nonhomogeneous microstructure among other factors including internal stress and strain hardening that can influence the subsequent processing. One solution to this problem is proper annealing treatment that is a significant way to refine grains by static recrystallization and release residual stress and reduce the amount of work hardening. In their research, Chen and Yang et al. [5, 6] reported that a fine equiaxial grain microstructure and enhanced ductility can be obtained through controlling the annealing temperature and the holding time. Also, Yoshihara et al. [7–9] showed that appropriate annealing was even beneficial to eliminate the internal stress and decrease the work hardening and, as result, improve the formability of Mg alloys.

The aim of this paper is to study the effect of annealing temperature and holding time on the homogenized material and resulting microstructure and mechanical properties on magnesium-based alloys, particularly Mg5%Zn alloy. The effect of homogenization can be seen in evaluating the microstructure through optical microscopy (OM) and scanning electron microscopy (SEM). The effect of proper homogenization can be demonstrated in the mechanical properties such as hardness testing which exhibits a higher hardness with finer and more equiaxed grain size. The desired properties for the magnesium alloy material include maximizing the combination of strength and ductility and hence improved mechanical properties for lightweight applications.

Materials and Methods

The starting as-cast materials investigated in this study were Mg5%Zn with dimensions of 5 mm height, 3 mm length, and width of 0.5 mm. The heat treatment times were chosen evaluating the phase diagram of magnesium-zinc system.

C. Hale · Z. Xu (✉) · H. Zhang · S. Yarmolenko · J. Sankar
North Carolina A& T State University, Greensboro, NC 27411, USA
e-mail: Zhigang@ncat.edu

V. M. Miller et al. (eds.), *Magnesium Technology 2021*, The Minerals, Metals & Materials Series,
https://doi.org/10.1007/978-3-030-65528-0_25

The temperatures chosen were 400 and 375 °C for a varying amount of times from 0 up to 12 h. These temperatures are just below the eutectic temperature as seen in Fig. 1.

The samples were each cut from a larger ingot and then prepared for homogenization in a two-step process for two temperatures. The first set of samples were heat treated from 40 to 330 °C and held for 12 h and then the temperature was ramped up to 400 °C, and each sample individually pulled from a furnace at period of 0, 1, 2, 4, 6, 8, and 10 h (1 sample per time interval).

The second set of samples were heat treated from 40 to 320 °C and held for 12 h and then pulled from a furnace at a slightly lower temperature of 375 °C at a period of 0, 2, 4, 6, 8, and 12 h (1 sample per time interval). The heating diagram schematic can be seen in Fig. 2.

Once the sample sets were collected, they were prepared for metallography by standard mechanical grinding with SiC papers and final polishing with 0.05 mm Al_2O_3 suspension, then etched using 10% nitric acid and picric acid applied to highlight the grain boundaries. The samples were then cleaned with isopropanol and dried to prepare for Optical Microscopy.

The microstructural features of the 12 samples were then evaluated by optical microscopy (OM Zeiss AxioImager M2M) and some scanning electron microscopy (SEM, Hitachi SU8000). Grain size was measured by the mean linear intercept method.

Mechanical properties of the samples were evaluated by conducting hardness tests on the selected samples with a microhardness indenter (LECO M-400-H1). A total of 8 indentations were performed on each sample for calculation of average hardness. The hardness was calculated using the formula $HV = 0.1891F/d^2$, where F is the load in Newtons and d is the average of the diagonals d1 and d2. A load of 200 grams was used.

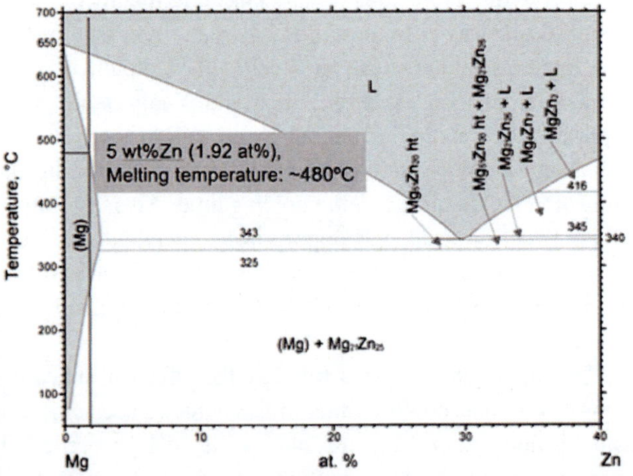

Fig. 1 Mg-Zn phase diagram used for determining heat treatment and homogenization temperature

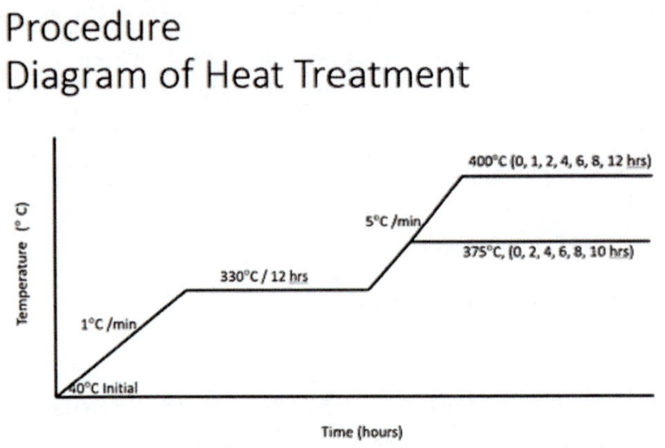

Fig. 2 Two-step heat treatment diagram for homogenization at 400 and 375 °C

Results and Discussion

Microstructure

Microstructure of the Initial Materials

Figures 3 and 4 show the micrographs of the as-cast Mg5% Zn alloy before homogenization. As seen in the figures, the second phase demonstrated by the bright dot features with OM and black ones with SEM was randomly discontinuously distributed in the Mg matrix.

Microstructure of the 400 and 375 °C Heat Treated Materials Before Homogenization

The images in Figs. 5 and 6 show that the microstructures at 400 °C for 0 h and 375 °C for 0 h are not fully

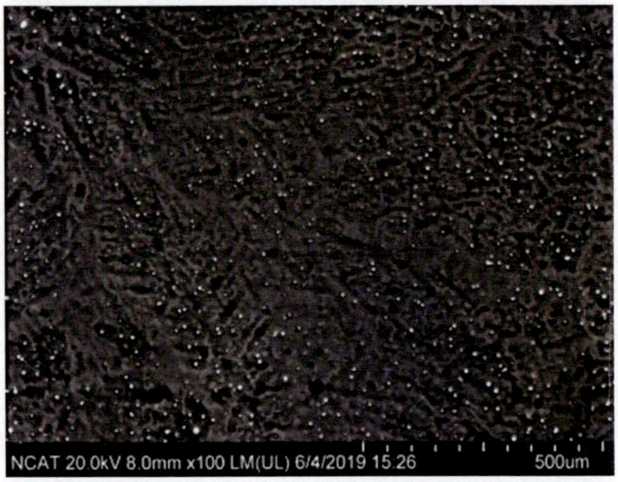

Fig. 3 SEM of the as-cast microstructure with secondary phase

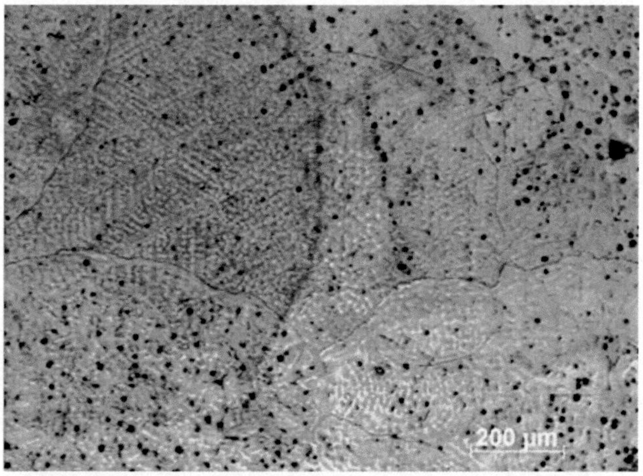

Fig. 4 OM of the as-cast microstructure with secondary phase

Fig. 5 OM of sample at 400 °C at 0 h demonstrating non-homogenous condition

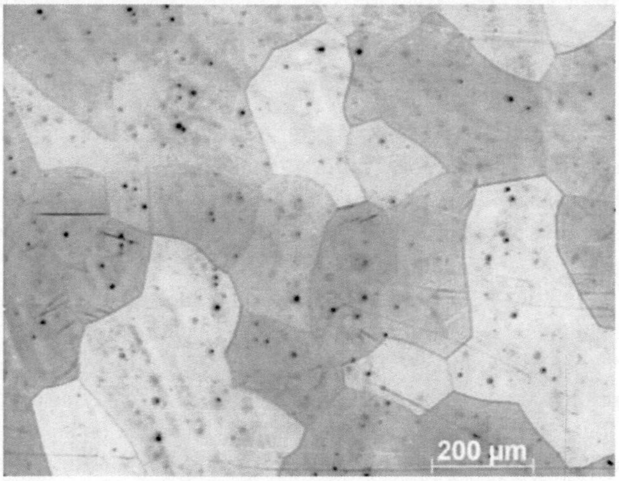

Fig. 6 OM of sample at 375 °C at 0 h demonstrating non-homogenous condition

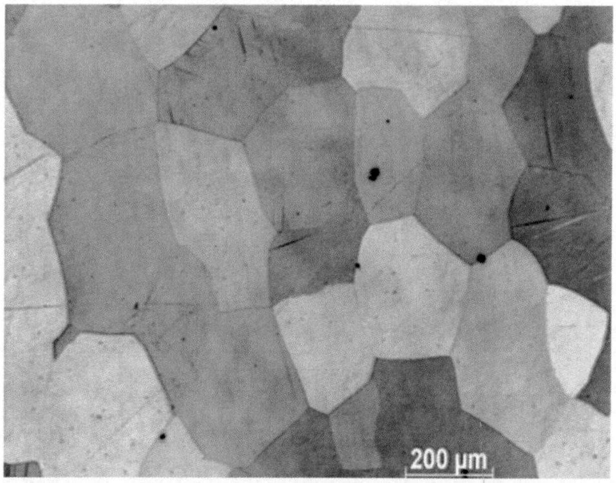

Fig. 7 OM of sample at 400 °C at 2 h demonstrating homogenous condition wherein the second phase shows homogeneity within primary phase of magnesium

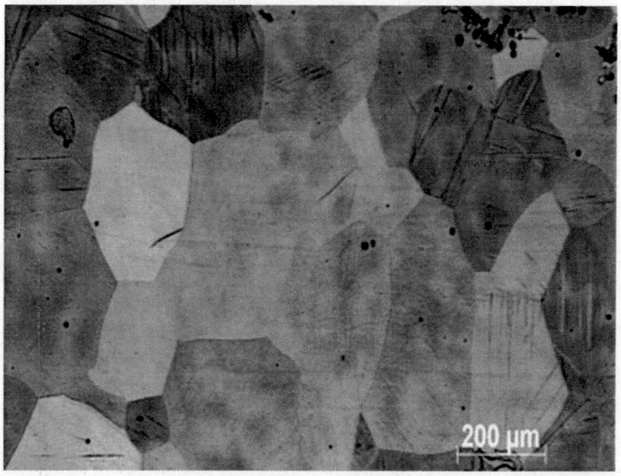

Fig. 8 OM of sample at 375 °C at 2 h demonstrating homogeneous condition wherein the second phase shows homogeneity within primary phase of magnesium

homogenized as indicated by the black dot features throughout the grains of the microstructure.

Microstructure of the 400 and 375 °C Heat Treated Materials After Homogenization

In contrast, the images in Figs. 7 and 8 show the homogenization of the microstructure at 400 and 375 °C, respectively, where the black dot features are no longer present and show a more uniform microstructure. The microstructure for the 2 h treatment for both 400 and 375 °C shows effective homogenization in which the black dot features essentially disappear. This indicates that the second phase (Mg5%Zn) has essentially diffused into the magnesium primary phase.

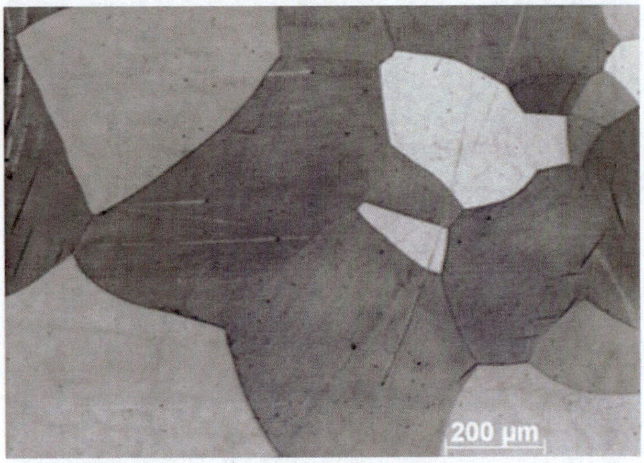

Fig. 9 OM of sample at 400 °C at 6 h demonstrating significant grain growth and less refined grain size

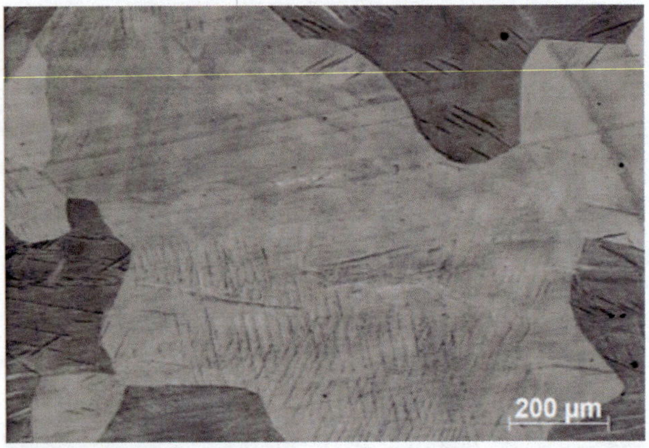

Fig. 10 OM of sample at 375 °C at 8 h demonstrating significant grain growth and less refined grain size

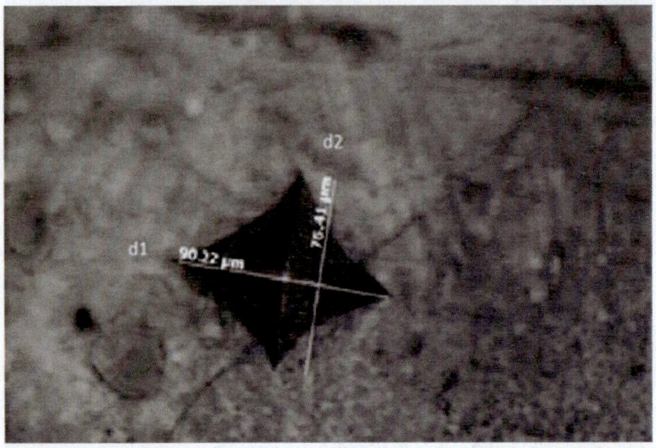

Fig. 11 Hardness indentation of Mg5%Zn at 375 °C and 2 h at 200x magnification

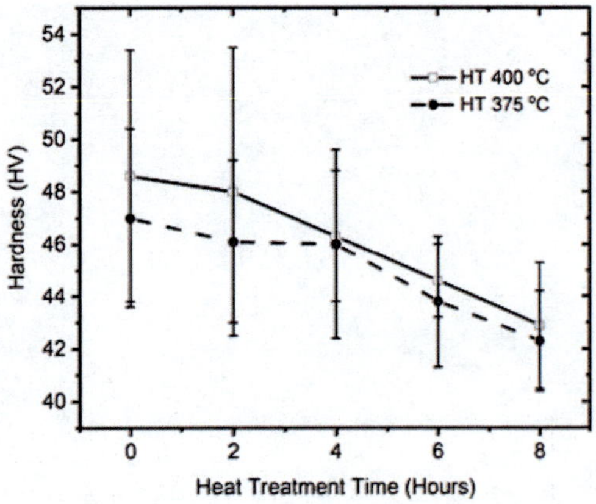

Fig. 12 Hardness values in HV for the 400 and 375 °C heat treatments

The microstructure for subsequent times greater than 4 h at both 400 and 375 °C were homogeneous but showed significant grain growth (and lower strength/hardness) as demonstrated in Fig. 9 for the 400 °C at 6 h and Fig. 10 for the 375 °C at 8 h.

Hardness Testing of the Heat Treated 400 and 375 °C Materials

The hardness of the as-cast material was determined to be 58.34 HV using an average of 8 hardness values for the as-cast material.

An example of the hardness indentation can be seen in Fig. 11 where the indent is made and then using the optical microscope the diagonals of the indent are measured and then converted into a corresponding hardness value.

The hardness of the Mg5%Zn at heat treatments of 400 and 375 °C can be seen in Fig. 12 The data suggests that there is obviously a drop as expected in the hardness at both 400 and 375 °C from the as-cast material (at an average of 58 HV) with values from 0 to 8 h for 400 and 375 °C as shown in the figure.

As stated previously, the Mg5%Zn at both heat treatments of 400 and 375 °C demonstrated homogenization and equiaxed grains (more uniform grains) around 2 h. Also, the two-hour window exhibited reasonably good hardness values at both temperatures before beginning to drop more for the 4, 6, and 8-hour periods.

The two-hour heat treatment for both the 400 and 375 °C exhibited the best combination of mechanical properties in terms of hardness and ductility (by achieving equiaxed refined grains). The heat treatment at 375 °C shows an advantage in obtaining higher hardness than that obtained at 400 °C. It is correlated to smaller grain size induced by the heat treatment at a lower temperature.

Conclusions

Magnesium-based alloys are being used today in various lightweight applications. The mechanical properties of magnesium-based alloys can be enhanced for such applications through a combination of annealing temperature and holding time of the magnesium-based alloy. The magnesium-based alloy under investigation in this study is Mg-5%Zn. The results of this investigation show that optimizing the combination of annealing time and temperature can achieve homogenization in an effective way and also enhances the mechanical properties and formability of Mg-5%Zn as seen in hardness testing results for effectively homogenized samples.

Further work is being done in this area with the Mg5% Zn samples being evaluated with SEM and CT (computerized tomography) to evaluate the homogeneity of the samples and reveal any voids that may be present. Other materials that will also be evaluated in comparison with less zinc are Mg3%Zn.

Acknowledgements This work was supported by the project of "Tailoring Mg-alloy Systems through Composition/Microstructure/ Severe Plastic Deformation for Army Extreme Dynamic Environmental Applications" which is funded by the Materials in Extreme Dynamic Environment (MEDE) cooperative agreement (W911NF-12-2-0022). NCA&T project number 280958.

References

1. Mordike, B.L., Stulíková, I. & Smola, B. Mechanisms of creep deformation in Mg-Sc-based alloys. Metall Mater Trans A 36, 1729–1736 (2005).
2. Kainer, K.U. Magnesium alloys and their applications, 1–46 (2000).
3. Homma, T. Fabrication of extraordinary high-strength magnesium alloy by hot extrusion, Scripta Materiala, 644–647 (2009).
4. Muralidhar, A. Effect of equal channel angular pressing on AZ31 wrought magnesium alloys, Journal of Magnesium and Alloys, 336–340 (2013).
5. Yang, P. Effect of heat treatment on microstructure and microhardness of rapidly solidified ZK60 magnesium ribbons, Transactions of Materials and Heat Treatment 95–99 (2010).
6. Chen, X. Effect of heat treatment on strain hardening of ZK60 Mg alloy, Materials and Design, 1526–1530 (2011).
7. Yoshihara, S. Formability enhancement in magnesium alloy deep drawing by local heating and cooling technique, Journal of Materials Processing Technology, 612–615 (2003).
8. Agnew, S.R. Plastic anisotropy and the role of non-basal slip in magnesium alloy AZ31B, International Journal of Plasticity, 1161–1193 (2005).
9. Berge, F. Influence of temperature and strain rate on flow stress behavior of twin-roll cast, rolled and heat-treated AZ31 magnesium alloys, Transactions of Nonferrous Metals Society of China, 1–13 (2015).

Quantitative Analysis of Impurity Elements in Pure Magnesium by Glow Discharge Mass Spectrometry (GDMS)

Jinyang Zhao, Jian Wu, Baoqiang Xu, QiMei Yang, and Bin Yang

Abstract

In this paper, the method of quantitative analysis by GDMS of Fe, Si, Cu, and other nine impure elements in pure magnesium was studied. The applicable method for the preparation of samples was confirmed. The isotopes, resolution, and the analysis conditions were optimized. In addition, RSF of these impure elements was obtained using standard samples in order to correct the standard RSF. The results showed that using wire cutting introduces fewer impurities. When the discharge current was 55 mA and the gas flow was 350 mL/min, matrix signal was stable and suitable. Quantitative analysis of samples with the RSF_{314} showed better precision and accuracy than using the standard RSF. Comparison with the results of ICP-AES, AAS, and ICP-MS, the testing results were closer to the standard value which were obtained by our studies. But our method was more convenient on sample preparation, faster analysis speed, and higher overall accuracy than other means.

Keywords

GDMS • Pure magnesium • RSF • Impurities

J. Zhao · J. Wu · B. Xu · Q. Yang
Faculty of Metallurgical and Energy Engineering, Kunming University of Science and Technology, Kunming, 650093, People's Republic of China

J. Zhao · J. Wu · B. Xu · Q. Yang · B. Yang (✉)
National Engineering Laboratory of Vacuum Metallurgy, Kunming University of Science and Technology, Kunming, 650093, People's Republic of China
e-mail: kgyb2005@126.com

B. Xu
The State Key Laboratory of Complex Nonferrous Metal Resources Clean Utilization, Kunming University of Science and Technology, Kunming, 650093, People's Republic of China

Introduction

Common detection methods for impurity elements in pure magnesium usually include ICP-AES and AAS [1–4]. These methods are inexpensive and have good precision [5]. However, due to the complexity of the sample dissolution process, impurity elements are easily introduced in the sample preparation process, which affects the accuracy of the analysis.

Glow Discharge Mass Spectrometry (GDMS) is a method that uses solid samples to directly inject samples and performs quantitative analysis based on the different isotope ratios of the elements to be tested [6–8]. It has the advantages of low detection limit and high detection efficiency. Since only the surface of the sample is sputtered during the test, it is easy to cause unstable or error matrix signals. This study takes pure magnesium as the research object, optimizes the instrument conditions and measurement conditions, and uses the certified reference materials to calibrate the internal relative sensitivity factor (RSF) value of the instrument to establish an accurate and rapid analytical method for the determination of trace impurity elements in high purity magnesium.

Experimental

Samples Preparation

Two pure magnesium samples were used in the experiment. One is the certified reference materials (CRMs) of pure Mg, they are G314, G315, and G316, which were provided by Southwest Aluminum (Chengdu, China).

The samples were processed into a small cylinder with 35 mm in diameter and thickness of 5 mm by lathe and wire cutting machine, one of them was ground to 2000 grid SiC-paper which was cut by a lathe. All samples were carefully cleaned with acetone, 20% dilute nitric acid,

V. M. Miller et al. (eds.), *Magnesium Technology 2021*, The Minerals, Metals & Materials Series,
https://doi.org/10.1007/978-3-030-65528-0_26

de-ionized water, and ethanol under ultrasonic conditions, sequentially. Then, washed more than 3 times using ultra-pure water and were dried by argon gas.

The samples used for inductively coupled plasma atomic emission spectrometry and Atomic Absorption Spectrophotometer are magnesium chips, which are processed by a milling machine and cleaned by ultrapure water under ultrasonic conditions.

Instrument

All the works of GDMS in this study were carried out with the Finnigan Element GD Glow Discharge Mass Spectrometer(Thermo Scientific, Germany), the main components of the instrument are shown in Fig. 1. In addition, OPTIMA 8000 inductively coupled plasma atomic emission spectrometry (PerkinElmer) and WFX-320 Atomic Absorption Spectrophotometer (Beijing Rayleigh Analytical Instrument Co., Ltd) were used to comparative analysis.

Procedure

Place the treated sample on the sample holder and put it into the GD system. Pass in high-purity argon gas and apply DC voltage under vacuum conditions. Determine the best test parameters and sample preparation method by analyzing the change of impurity signal intensity over time. Then, use the conditions and the principle of mass spectrometry to investigate the mass spectrum interference in the pure magnesium matrix and select appropriate impurity isotopes for the quantitative testing.

Methodology

Quantitative analysis by GDMS is based on the relationship between the ion beam signals of the measured element isotope and its content. It can be expressed as follows:

$$I_x = \frac{I_{xi}}{A_{xi}} \quad (1)$$

where I_x and I_{xi} is the ion beam signals of the element x and its isotope, A_{xi} is the isotope abundances of the element x.

C_x, the content of element x in sample can be calculated as:

$$C_x = \frac{I_x}{I_m} C_m \quad (2)$$

where I_m and C_m is the ion beam signals and the content of magnesium in the sample. Make

$$\frac{I_x}{I_m} = IBR_{x/m} \quad (3)$$

and consider the influence of the magnesium on different elements and the difference in sensitivity of different elements, a relative sensitivity factor (RSF) needs to be introduced in calculated, so the content of x in magnesium $Cx_{/m}$ can be expressed as

$$Cx_{/m} = RSFx_{/m} \bullet IBRx_{/m} \quad (4)$$

The values of $RSFx_{/m}$ can be obtained by analyzing the CRMs with Eq. (5).

$$RSFx_{/m} = \frac{I_{xi}}{A_{xi}C_x} \bigg/ \frac{I_{mi}}{A_{mi}C_m} \quad (5)$$

The repeatability and accuracy of the detection method are judged by the relative standard deviation (RSD) and the relative error (RE) from the CRMs. The calculation formula is as follows:

$$RSD = \frac{\sqrt{\frac{\sum_{i=1}^{n}(C_{xi}-\overline{C_x})^2}{n-1}}}{\overline{C_x}} \times 100\% \quad (6)$$

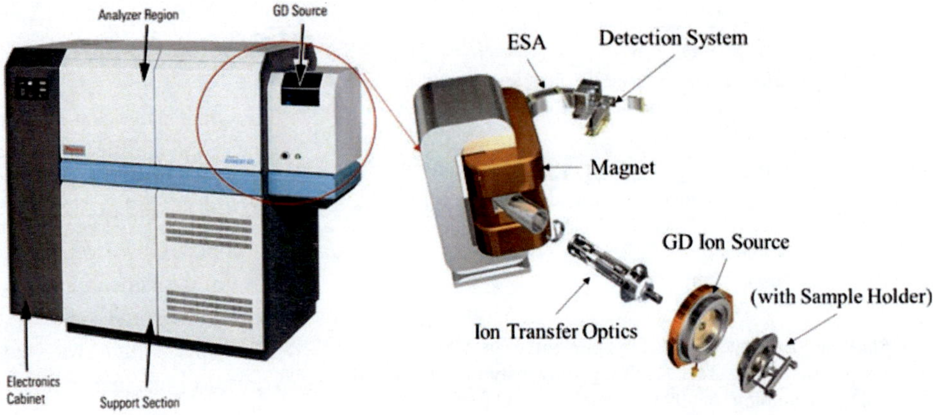

Fig. 1 The main structure of GDMS. (Color figure online)

$$RE = \frac{C_{xi} - \overline{C_x}}{\overline{C_x}} \times 100\% \quad (7)$$

where $\overline{C_x}$ is the average concentration of the same element in the same sample, n represents the continuous measurement of the same elements in the same sample five times.

Results and Discussion

Selection of Sample Preparation Method

GDMS uses solid direct sampling technology to sputter the surface material of the sample, and only obtain the surface information of the sample [9, 10]. Therefore, it is necessary to choose a suitable sample preparation method to avoid introducing contaminants during the sample preparation process, which can reduce the pre-sputtering time and improve the accuracy of the test results. Fe, Na, and Si are the most easily introduced pollutants during sample preparation. The content of these three elements in G314 samples processed by three different sample preparation methods varies with the pre-sputtering time as shown in Fig. 2.

It can be seen from the content of the three elements at the initial moment in Fig. 2 that the processing of raw materials by wire cutting introduces the least pollutants. Turning will introduce a large amount of iron into the sample. Grinding and polishing will eliminate some of the introduced iron, but it will bring into a large amount of silicon from the abrasive paper (SiC).

The content of all contaminants in the sample is unchanged after 6 min in Fig. 2a. That is, the pre-sputtering time is the shortest and the loss of the instrument is the least. Therefore, selecting the wire cutting method to process samples is the best sample preparation method. And pre-sputtering time could be selected at 8 min.

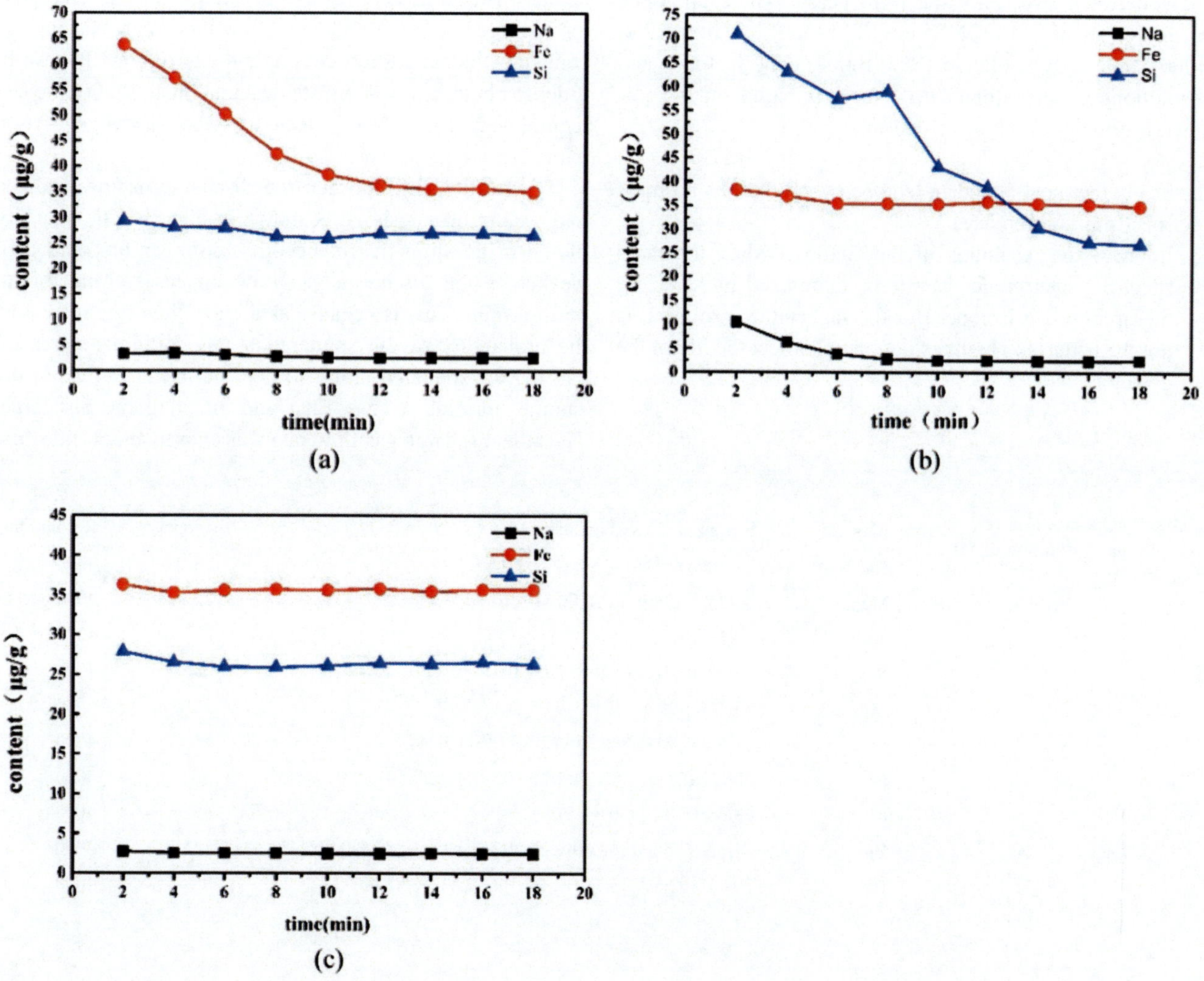

Fig. 2 Curve about the content of Na, Fe, and Si in samples with pre-sputtering time by different processing methods (**a** wire cutting; **b** turning; **c** turning, grinding, and polishing). (Color figure online)

Selections of Isotopes and Resolution for the Testing

GDMS qualitatively analyzes samples based on the principle that different elements have different mass charge ratios. In the GDMS system, Ar, O_2, H_2, N_2, and other elements will combine with Mg or other impurity elements in the sample to form ion interference, which has a similar mass charge ratio of the element to be measured.

The interference can be eliminated by choosing isotopes and resolution. The isotopes of common elements in pure magnesium are listed in Table 1. The resolution refers to the ability of the mass spectrometer to distinguish two adjacent mass spectra peaks,it can be expressed as:

$$R = \frac{m}{\Delta m} \quad (8)$$

where, R is the resolution, m is the average of the mass of the two mass spectrum peaks and Δm is the difference in mass between two mass spectrum peaks. There are three resolution modes in GDMS: low (R > 300), medium (R > 4000), and high resolution (R > 10000). The isotopes and resolutions selected for GDMS analysis based on the following principles are also listed in it.

1. Select the most abundant isotope, when there is no mass spectrum interference.
2. Improve the resolution of the analysis, when the mass spectrum interference cannot be eliminated by selecting the appropriate isotope. But the higher the resolution is not the better, too high resolution will affect the life of the detector.

Selections of Analysis Parameters for the Testing

When using GDMS to determines the content of impurity elements, a certain sensitivity is required which is usually characterized by the signal intensity (cps) of the matrix peak. Previous work found that an inaccurate measurement will be obtained when the signal strength of the magnesium matrix is below 10^{10}cps because there are no stable peak shape mass spectrum peaks formed in the instrument. The discharge current and gas flow rate are two important parameters that affect the matrix signal. Therefore, it is necessary to adjust these two parameters so that the matrix signal is greater than 10^{10}cps to meet the needs of obtaining accurate results. The change of matrix signal value with discharge current and discharge gas flow is shown in Fig. 3.

It can be seen from Fig. 3a that the magnesium matrix signals become stronger and stronger when the discharge current changes from 10 to 80 mA. The intensity of the magnesium can reach 1×10^{10} cps, when the current was greater than 50 mA. Excessive discharge current will cause the sputtered sample to adhere to the anode cap, causing the cone to be blocked, short circuit, and interrupting the test. In order to obtain a long-term stable signal that satisfies the test requirements, 55 mA is selected as the discharge current parameter.

The relationship between the discharge airflow and the magnesium matrix signal is shown in Fig. 3b. The intensity of the magnesium matrix becomes stronger first and then weaker as the discharge gas flow increases from 100 to 650 mL/min. That is because as the gas flow rate increases, the probability of the sputtered atoms colliding increases, causing the sputtered atoms to deposit on the surface of the sample, forming a cone plug and affecting the discharge. Thus the discharge gas flow rate was chosen at 350 mL/min.

Table 1 Optimization of isotopes and resolution of the instrument

Element	Isotope and its abundance	Isotope	Resolutions
Na	23(100.00%)	23	4000
Mg	24/25/26(78.99%/10.00%/11.01%)	24	4000
Al	27(100.00%)	27	4000
Si	28/29/30(92.23%/4.67%/3.10%)	28	4000
Mn	55(100.00%)	55	4000
Fe	54/56/57/58(5.80%/91.72%/2.2%/0.28%)	56	4000
Ni	58/60/61/62/64(68.27%/26.10%/1.13%/3.59%/0.91%)	58	4000
Cu	63/65(69.17%/30.83%)	63	4000
Zn	64/66/67/68/70(48.60%/27.90%/4.10%/18.80%/0.60%)	66	4000
Pb	204/206/207/208(1.40%/24.10%/22.10%/52.40%)	208	4000

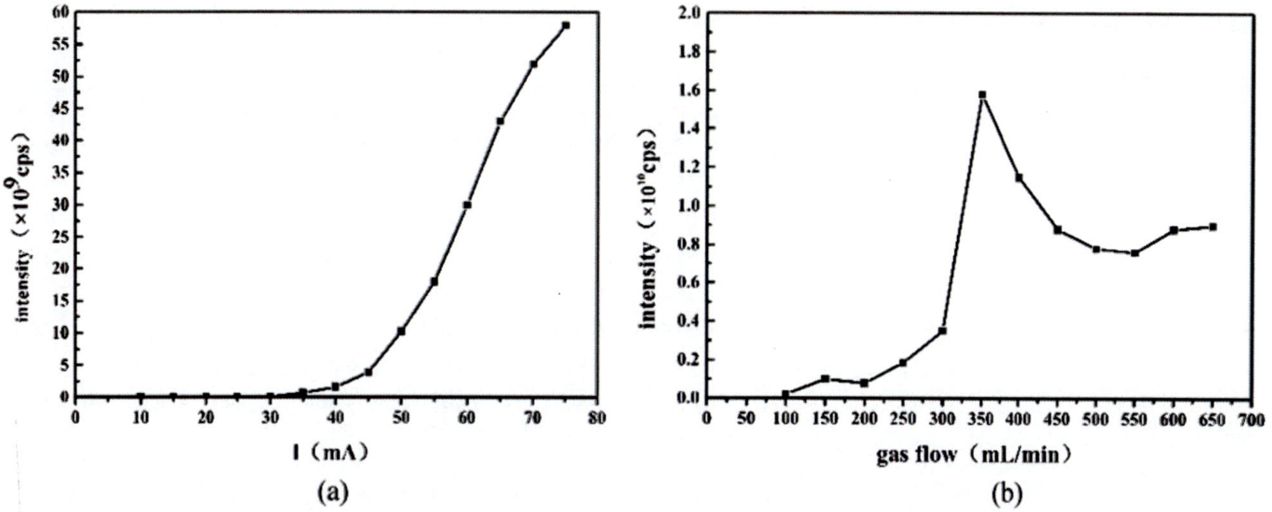

Fig. 3 Effect of analysis parameters on the intensity of magnesium (**a** discharge current versus intensity; **b** gas flow versus intensity)

Correct RSF Value and Quantitative Analysis of Pure Magnesium

The standard RSF values, which are calculated based on iron-based reference materials and the RSF corrected by CRMs are listed in Table 2.

It can be seen from Table 2 that there are significant differences between the standard RSF of each element and the corrected RSF, especially the difference of the Zn element. That means using CRMs to adjust the RSF for quantitative determination will have a greater correction effect on the Zn in the matrix of magnesium.

The measurement results before and after RSF value correction using CRMs are listed in Tables 3, 4 and 5. Comparing the relative standard deviations in the three tables, it can be observed that except for Ni, the RSD of other elements are all less than 10, indicating that the method of GDMS has high precision and reproducibility. The reason for the high RSD value of Ni element may be due to the small content of Ni in the sample and uneven distribution caused by segregation. The RE values in these three tables were indicated that the measured value of the sample corrected by CRMs is closer to the certified value in the CRMs. especially using the RSF_{314} to correct the results.

Comparison with Other Methods

The comparison of GDMS, AAS, ICP-AES, and ICP-MS analysis results are shown in Table 6. Since Si and Al in pure magnesium are elements that only can be excited at high temperatures, and it is easy to generate hard-to-volatile or hard-to-dissociate oxides in the flame atomizer, the atomization efficiency is low, so the AAS method cannot be used to determine these two elements. In addition, because the sample contains very little Ni, which is below the detection limit of ICP-AES, this method cannot be used to determine the element.

Table 2 The standard RSF and the RSF corrected by the G314 and G316 samples

Element	$RSF_{std.}$	RSF_{314}	RSF_{316}
Na	0.95	2.04	3.10
Al	1.27	1.36	1.40
Si	3.04	1.39	1.38
Mn	1.01	1.94	2.16
Fe	1	1.71	2.54
Ni	1.51	0.80	0.98
Cu	2.44	0.44	0.54
Zn	3.83	0.07	0.09
Pb	1.36	0.30	0.44

Table 3 The content of impurity elements in G315 standard sample (calculated by $RSF_{std.}$)

Element	C_{xi}(μg/g)					$\overline{C}_x$ (μg/g)	RSD (%)	RE (%)
	1	2	3	4	5			
^{23}Na	1.6917	1.5411	1.6009	1.5256	1.5851	1.5889	4.11	58.89
^{27}Al	46.6614	46.8398	47.5879	48.1671	46.8753	47.2263	1.34	23.83
^{28}Si	43.5599	41.4585	41.4560	42.7619	42.0726	42.2618	2.14	76.09
^{55}Mn	29.2163	32.0907	32.2113	33.0264	33.1248	31.9339	4.98	40.86
^{56}Fe	7.1933	7.8855	8.2844	8.2717	8.1504	7.9571	5.73	46.95
^{58}Ni	0.2429	0.1681	0.2483	0.2397	0.1339	0.2066	25.29	79.34
^{63}Cu	3.2512	3.2455	3.3790	3.0276	3.3320	3.2471	7.16	8.24
^{66}Zn	58.3145	58.3590	59.1492	58.0917	59.2467	58.2885	1.80	26.71
^{208}Pb	2.1393	2.7361	2.5363	2.6602	2.7729	2.5690	9.99	48.62

Table 4 The content of impurity elements in G315 standard sample (corrected by RSF_{314})

Element	C_{xi}(μg/g)					$\overline{C}_x$ (μg/g)	RSD (%)	RE (%)
	1	2	3	4	5			
^{23}Na	1.4245	1.2977	1.3481	1.2846	1.3347	1.3379	4.11	33.79
^{27}Al	62.2713	62.5094	63.5078	64.2808	62.5578	63.0254	1.34	1.94
^{28}Si	20.2050	19.2303	19.2292	19.8349	19.5155	19.6030	2.14	18.32
^{55}Mn	34.2461	37.6153	37.7568	38.7121	38.8280	37.4317	4.98	30.68
^{56}Fe	8.1183	8.8996	9.3498	9.3354	9.1987	8.9804	5.73	40.13
^{58}Ni	0.2154	0.1491	0.2202	0.2126	0.1188	0.1832	25.28	81.68
^{63}Cu	3.3331	3.3273	3.4642	3.1040	3.4160	3.3289	4.15	10.96
^{66}Zn	38.8097	37.6951	39.3649	38.6610	39.4303	38.7922	5.07	15.67
^{208}Pb	3.3552	4.2911	3.9777	4.1720	4.3489	4.0290	9.99	19.42

Table 5 The content of impurity elements in G315 standard sample (corrected by RSF_{316})

Element	C_{xi}(μg/g)					$\overline{C}_x$ (μg/g)	RSD (%)	RE (%)
	1	2	3	4	5			
^{23}Na	1.6548	1.5075	1.5660	1.4923	1.5505	1.5542	4.11	55.42
^{27}Al	57.2558	57.4748	58.3928	59.1035	57.5193	57.9492	1.34	6.53
^{28}Si	20.1789	19.2054	19.2043	19.8092	19.4902	19.5776	2.14	18.43
^{55}Mn	29.8984	32.8399	32.9634	33.7975	33.8987	32.6796	4.98	39.48
^{56}Fe	7.2142	7.9085	8.3086	8.2958	8.1743	7.9803	5.73	46.80
^{58}Ni	0.2082	0.1441	0.2128	0.2055	0.1148	0.1771	25.28	82.29
^{63}Cu	2.8810	2.8760	2.9943	2.6829	2.9526	2.8774	4.15	4.09
^{66}Zn	25.9414	28.1183	29.3638	28.8388	29.4126	28.3350	1.80	38.40
^{208}Pb	2.0928	2.6766	2.4811	2.6023	2.7127	2.5131	9.99	49.74

Table 6 Comparisons of the results between different methods

Element	ICP-AES		AAS		ICP-MS		GDMS	
	$\overline{C}_x$(μg/g)	RE (%)	$\overline{C}_x$(μg/g)	RE (%)	$\overline{C}_x$(μg/g)	RE (%)	$\overline{C}_x$(μg/g)	RE (%)
Na	53	5200	/	/	32.11	3111.00	1.3379	33.79
Al	64	3.23	–	/	63.56	2.52	63.0254	1.94
Si	6	75.00	–	/	18.62	22.42	19.6030	18.32
Mn	38	29.63	46	14.81	32.54	39.74	37.4317	30.68
Fe	6	60.00	33	120.00	23.68	57.87	8.9804	40.13
Ni	/	/	29	2800.00	1.9	90.00	0.1832	81.68
Cu	6	100.00	8	166.67	4.12	37.33	3.3289	10.96
Zn	51	10.87	73	58.70	36.33	21.02	38.7922	15.67
Pb	10	100.00	9	80.00	3.99	20.20	4.0290	19.42

Note "–" means the element cannot be detected using this method
"/" "." means that the element is not detected in this sample by using this method

Comparing the RE value, it can be found that GDMS has the highest accuracy for almost all impurity elements in this sample, especially to Na, Si, and Pb, which are easily introduced during sample preparation and dissolution.

Conclusions

The sample preparation method and the quantitative analysis method of GDMS for impurity elements in pure magnesium material have been established. Compared with other methods, the method proposed in this study has simple operation, fewer pollutants, high accuracy of results, and fast analysis speed, which could meet the requirement of rapid quantitative analysis of impurity elements in pure magnesium.

Acknowledgements This work was financially supported by the National Natural Science Foundation of China (grant No. 51734006).

References

1. Wang Y C, Tian Y, Qu T, et al. Purification of Magnesium by Vacuum Distillation and its Analysis[J]. Materials Science Forum, 2014, 788:52–57. https://doi.org/10.4028/www.scientific.net/MSF.788.52.
2. Dai Y, Tao M, Ma C. ICP-AES Determination of 9 Alloying Elements in Magnesium Alloys[J]. Physical Testing and Chemical Analysis Part B Chemical Analysis, 2004, 40(3): 155–158.
3. Ashy M A, Headridge J B, Sowerbutts A. Determination of trace and minor elements in alloys by atomic-absorption spectroscopy using an induction-heated graphite-well furnace as atom source—II[J]. Talanta, 1974, 21(6):649–652. https://doi.org/10.4028/www.scientific.net/MSF.788.52. https://doi.org/10.1016/0039-9140(74)80205-5,
4. Tan M E I, Mei-juan T A O, Guo-qiang Y A N. ICP-AES Determination of Silicon in Magnesium and Magnesium Alloys [J]. Physical Testing and Chemical Analysis (Part B: Chemical Analysis), 2009, 5.
5. Welz B. High-resolution continuum source AAS: the better way to perform atomic absorption spectrometry[J]. Analytical and Bioanalytical Chemistry, 2005, 381(1): 69–71. https://doi.org/10.1007/s00216-004-2891-8.
6. Bogaerts A, Neyts E, Gijbels R, et al. Gas discharge plasmas and their applications[J]. Spectrochimica Acta Part B Atomic Spectroscopy, 2002, 57(4):609–658. https://doi.org/10.1016/s0584-8547(01)00406-2.
7. Harrison W W, Hess K R, Marcus R K, et al. Glow Discharge Mass Spectrometry[J]. Analytical Chemistry, 1986, 58 (2):341A-356A. https://doi.org/10.1007/978-3-642-82724-2_17.
8. Chapman, Brian N. Glow discharge processes[M]. Wiley, 1980.
9. Beatriz Fernández, Costa J M, Pereiro R, et al. Inorganic mass spectrometry as a tool for characterization at the nanoscale[J]. Analytical & Bioanalytical Chemistry, 2010, 396(1): 15–29. https://doi.org/10.1007/s00216-009-2959-6.
10. Dong J, Qian R, Xiong W, et al. Determination of doping elements of synthetic crystals by direct current glow discharge mass spectrometry[J]. International Journal of Mass Spectrometry, 2014, 361:1–8. https://doi.org/10.1016/j.ijms.2014.01.018.

Author Index

V. M. Miller et al. (eds.), *Magnesium Technology 2021*, The Minerals, Metals & Materials Series,
https://doi.org/10.1007/978-3-030-65528-0

Subject Index

V. M. Miller et al. (eds.), *Magnesium Technology 2021*, The Minerals, Metals & Materials Series,
https://doi.org/10.1007/978-3-030-65528-0